T0345164

Emerging Low-Power Semiconductor Devices

This book gives insight into the emerging semiconductor devices from their applications in electronic circuits. It discusses the challenges in the field of engineering and applications of advanced low-power devices.

Emerging Low-Power Semiconductor Devices: Applications for Future Technology Nodes offers essential exposure to low-power devices and their applications in wireless, biosensing, and circuit domains. The book provides a detailed discussion on all aspects, including the current and future scenarios related to the low-power device. The book also presents basic knowledge about field-effect transistor (FET) devices and introduces emerging and novel FET devices. The chapters include a review of the usage of FET devices in various domains like biosensing, wireless and cryogenics applications. The chapters also explore device-circuit co-design issues in the digital and analog domains. The content is presented in an easy-to-follow manner that makes it ideal for individuals new to the subject.

This book is intended for scientists, researchers, and postgraduate students looking for an understanding of device physics, circuits, and systems.

Emerging Low-Power Semiconductor Devices

Emerging Low-Power Semiconductor Devices

Applications for Future Technology Nodes

Edited by

Shubham Tayal, Abhishek Kumar Upadhyay, Deepak Kumar, and Shiromani Balmukund Rahi

CRC Press is an imprint of the
Taylor & Francis Group, an informa business

First edition published 2023
by CRC Press
6000 Broken Sound Parkway NW, Suite 300, Boca Raton, FL 33487-2742

and by CRC Press
4 Park Square, Milton Park, Abingdon, Oxon, OX14 4RN

CRC Press is an imprint of Taylor & Francis Group, LLC

ISBN: 978-1-032-14729-1 (hbk)
ISBN: 978-1-032-14730-7 (pbk)
ISBN: 978-1-003-24077-8 (ebk)

DOI: 10.1201/9781003240778

Typeset in Times
by codeMantra

Contents

Editors

Dr. Shubham Tayal is an assistant professor in the Department of Electronics and Communication Engineering at SR University, Warangal, India. He has more than six years of academic/research experience teaching at UG and PG levels. He has received his PhD in Microelectronics & VLSI Design from the National Institute of Technology, Kurukshetra, MTech (VLSI Design) from YMCA University of Science and Technology, Faridabad, and BTech (electronics and communication engineering) from MDU, Rohtak. He has qualified GATE (2011, 2013, 2014) and UGC-NET (2017). He has published more than 25 research papers in various international journals and conferences of repute, and many papers are under review. He is on the editorial and reviewer panel of many SCI/SCOPUS-indexed international journals and conferences. Currently, he is the editor of four books from CRC Press (Taylor & Francis Group, USA). He is a member of various professional bodies like IEEE, IRED, etc. He was the recipient of Green ThinkerZ International Distinguished Young Researcher Award 2020. His research interests include simulation and modeling of multigate semiconductor devices, device-circuit co-design in digital/analog domain, machine learning, and IOT.

Dr. Abhishek Kumar Upadhyay obtained his PhD degree in electrical engineering from the Indian Institute of Technology (IIT), Indore, M.P., India, in 2019. After obtaining his PhD, he worked as a post-doctoral fellow in Model Group, Material to System Integration Laboratory (IMS Lab), University of Bordeaux, France. Currently, he is working as Scientific Staff in the Institute for Fundamentals of Electrical Engineering and Electronics, Technische Universität, D-01062, Dresden, Germany. He is actively involved in compact modeling, parameter extraction, 3D EM simulation, measurement, and characterization of RF devices.

Mr. Deepak Kumar works as an assistant professor (senior scale) in the Department of Electrical and Electronics Engineering at University of Petroleum and Energy Studies (UPES), Dehradun, India. He has six years of experience in teaching and research. He is pursuing his PhD from UPES, Dehradun, in the field of dye-sensitized solar cells (simulation and experimental study). He obtained his MTech degree from Indian Institute of Technology (IIT), Kanpur, India, in 2015 in the specialization of VLSI and microelectronics (electrical engineering) and received a BTech degree in Electronics Engineering from Harcourt Butler Technical University in 2011. He has published various research articles in reputed journals and conferences, and has also been awarded for the best paper presentation in the ICAMEES conference at UPES in 2018. In addition to academic responsibilities, he also serves the IEEE community as IEEE Student Branch (STB 10991) Counselor at UPES Dehradun and faculty coordinator (department level) for the SPOKEN TUTORIAL, IIT Bombay, MHRD Program. His research area includes VLSI, semiconductor devices (TFET, OSC, OTFT), and circuits and solar cell technology.

Dr. Shiromani Balmukund Rahi received his BSc (physics, chemistry, and mathematics) degree in 2002, MSc (electronics) degree from Deen Dyal Upadhyaya Gorakhpur University, Gorakhpur in 2005, GATE (2009), MTech (microelectronics) from Panjab University Chandigarh in 2011, and Doctorate of Philosophy in 2018 from Indian Institute of Technology, Kanpur, India. He also completed his master's project (MSc) in Central Electronics Engineering Research Institute (CEERI, 2005). He currently works at Mahamaya College of Agriculture Engineering and Technology, Akabarpur, Ambedkar Nagar, Uttar Pradesh, affiliated to Narendra Dev University of Agriculture and Technology, Uttar Pradesh, India. He has successfully published 15 research papers, 1 conference proceedings, and a book chapter. He has also addended and presented his research work in various international conferences and workshops. He has received various reviewer awards and recognizations such as IEEE Golden Reviewers (2018), AIP Advances (2020), and International Journal of Circuit Theory and Applications, Wiley (2020). He is also jointly working with Professor S.C. Misra (Indian Institute of Technology, Kanpur, India) for the development of IoTs for smart applications and Dr Naima Guenifi (LEA Electronics Department, University Mostefa Benboulaid of Algeria) for the development of ultra-low-power devices such as tunnel FETs and negative capacitance FETs.

Contributors

Abhishek Acharya
Department of ECE
SV National Institute of Technology
Surat, India

J. Ajayan
Department of ECE
SR University Warangal
Telangana, India

Naushad Alam
Department of Electronics Engineering
ZHCET, AMU
Aligarh, India

T. S. Arun Samuel
Department of Electronics and Communication Engineering,
National Engineering College
Kovilpatti, India

Asra Ansari
Department of Electronics Engineering
ZHCET, AMU
Aligarh, India

Babu Devasenapati, S.
SNS College of Technology
Coimbatore, India

S. Baishya
Department of Electrical Engineering
National Institute of Technology
Silchar, India

N. B. Balamurugan
Department of ECE
Thiagarajar College of Engineering
Madurai, India

K. Baruah
Department of Electrical Engineering
National Institute of Technology
Silchar, India

Saurabh Chaudhury
Department of Electrical Engineering
NIT Silchar
Assam, India

Shalini Chaudhary
Department of Electronics and Communication Engineering
Malaviya National Institute of Technology Jaipur
Jaipur, India

Rishu Chaujar
Department of Applied Physics
Delhi Technological University
Delhi, India

Yunho Choi
University of Texas at Austin
Austin, Texas

Basudha Dewan
Department of Electronics and Communication Engineering
Malaviya National Institute of Technology Jaipur
Jaipur, India

Neerja Dharmale
Department of Electrical Engineering
NIT Silchar
Assam, India

Daryoosh Dideban
University of Kashan
Kashan, Iran

Mekonnen Getnet
Applied Physics Department
Delhi Technological University
Delhi, India
and
Physics Department
Debre Tabor University
Debre Tabor, Ethiopia

Bijo Joseph
SRM Institute of Science and Technology
Chennai, India

Tripuresh Joshi
Govind Ballabh Pant Engineering College
Pauri, India

M. Karthigai Pandian
Department of Electrical Electronics and Communication Engineering
GITAM University
Bengaluru, India

A. Karthika
SNS College of Technology
Coimbatore, India

B. Kumar
Department of Applied Physics
Delhi Technological University
Delhi, India

B. Lakshmi
Centre for Nano Electronics & VLSI Design and School of Electronics Engineering
VIT Chennai
Chennai, India

G. Lakshmi Priya
School of Electronics Engineering (SENSE),
VIT University
Chennai, India

B. Lokesh
School of Electronics Engineering
VIT Chennai
Chennai, India

Anitha Mathew
IES College of Engineering
Chittilappilly, India

Menka
Department of Electronics and Communication Engineering
Malaviya National Institute of Technology Jaipur
Jaipur, India

D. Nirmal
Karunya Institute of Technology and Sciences
Coimbatore, India

Chandan Kumar Pandey
School of Electronics Engineering
VIT-AP University
Amaravati, India

Eswaran Parthasarathy
SRM Institute of Science and Technology
Chennai, India

S. Preethi
Department of ECE
Sri Krishna College of Technology
Coimbatore, India

Shiromani Balmukund Rahi
MCAET Ambedkar Nagar
Akbarpur, India

Chitrakant Sahu
Department of Electronics and Communication Engineering
Malaviya National Institute of Technology Jaipur
Jaipur, India

M. Saravanan
Sri Eshwar College of Engineering
Coimbatore, India

Nawaz Shafi
Department of Electronics and Communication Engineering
Malaviya National Institute of Technology Jaipur
Jaipur, India

Shashidhara, M.
Department of ECE
SV National Institute of Technology
Surat, India

B. Sivasankari
SNS College of Technology
Coimbatore, India

Young Suh Song
Department of Computer Science
Korea Military Academy
Seoul, South Korea

S. Sreejith
SNS College of Technology
Coimbatore, India

Shubham Tayal
Department of ECE
SR University
Warangal, India

Suman Lata Tripathi
School of Electronics and Electrical Engineering
Lovely Professional University
Phagwara, India

T. Venish Kumar
Department of ECE
Nadar Saraswathi College of Engineering & Technology
Theni, India

M. Venkatesh
School of Electronics and Communication Engineering
REVA University
Bengaluru, India

1 Role of TFET Devices and Their Performance Analysis for Wireless Communications

B. Lokesh and B. Lakshmi
VIT Chennai

CONTENTS

1.1 INTRODUCTION

Due to the scaling down of semiconductor devices in the nanometer region, conventional devices such as complementary metal oxide semiconductors (CMOSs) and Moore's law are approaching their limits fast [1]. CMOS-based transistors are no longer in frame in terms of efficiency and hence not suitable for applications requiring low supply voltages due to short channel effects, which in turn limit subthreshold swing (SS) and produce high I_{OFF}. In the coming decade, there can be an increasing demand for solid-state transistors with channel lengths as short as 1–5 nm for both analog and digital applications [2]. To satisfy Moore's law and achieve a high performance of transistors in nanometer regions, alternate devices such as the p-i-n structure, known as tunnel field effect transistors (TFETs), have been explored and proved to be efficient.

DOI: 10.1201/9781003240778-1

Band-to-band tunneling (BTBT) is the principle of mechanism for conduction unlike diffusion in MOS transistors [3]. The mechanism is due to the wave-like properties of electrons, which allows them to pass through barriers with little effort. Due to the mechanism of BTBT, the limitation of SS of 60 mV/decade and better resistance to variations in temperature can be achieved [4,5]. Though this improves the efficiency of the device, TFETs exhibit an ambipolar nature, which means that the device will be conducting for both positive and negative gate voltages. For an n-type TFET (n-TFET) when $V_{gs} < 0$, the tunnel junction starts to shift from the source to the drain side. This behavior can cause problems when used in applications such as complimentary logic circuit and hence limits the use of TFET in digital circuits. If this behavior can be suppressed, then TFETs can be used to obtain promising results in digital circuits. Many structures to suppress this behavior are explored in the literature already [6–8].

One such promising structure is an independent gate heterojunction TFET (IG HJ-DGTFET) [9]. The double gate provides control of gate leakage current and improves the magnitude of drain current, whereas the heterojunction structure improves tunneling by modifying the band energy levels in a way to reduce barrier energy for positive gate voltages, which improves I_{ON}, and increase barrier energy for negative gate voltages, thus reducing I_{OFF} [10,11]. This makes the devices compatible for applications such as a quadrature phase shift keying (QPSK) system, which demands low power. For TFETs, recent experiments have shown that on using lower-band-gap materials such as InAs or InSb, there is a huge increase in I_{ON} at lower voltages [12]. Also, to reduce the tunneling barrier and increase the probability for an electron to tunnel through, a combination of materials, GaSb for source and InAs for channel and drain, is used. Another reason for using these material combinations is that they allow lattice-matched growth, which helps in changing the band gap without altering the crystal structure [13].

In this chapter, to show the efficiency of IG HJ-DGTFET, the digital part of the QPSK system has been implemented and validated for low power consumption and less propagation delay. Also, the variations of rise and fall times of an inverter cell used in the QPSK system with respect to temperature have been provided to show temperature independency. The organization of this chapter is as follows: Section 1.2 provides a description of the device and its operation; Section 1.3 provides results and discussion of an implemented QPSK system using IG HJ-DGTFET, and finally Section 1.4 provides the conclusion.

1.2 DESCRIPTION AND SIMULATION OF THE DEVICE

The simulation of the device is carried out using Sentaurus technology computer-aided design (TCAD) simulator from Synopsys. Figure 1.1a shows the schematic of the device; Figure 1.1b shows the GaSb-InAs heterojunction TFET structure without doping and Figure 1.1c, with doping/meshing. The device consists of the source doped with GaSb and drain and channel doped with InAs, which is used to increase the I_{ON} considerably by increasing the energy of the valence band to that of the conduction band at the tunneling junction. This reduces the tunneling barrier and hence increases the probability of tunneling at the junction. The device dimensions, appropriate models for the device simulator, and the DC characteristics obtained are taken from our previous calibrated results [14]. The length of the device is 70 nm with

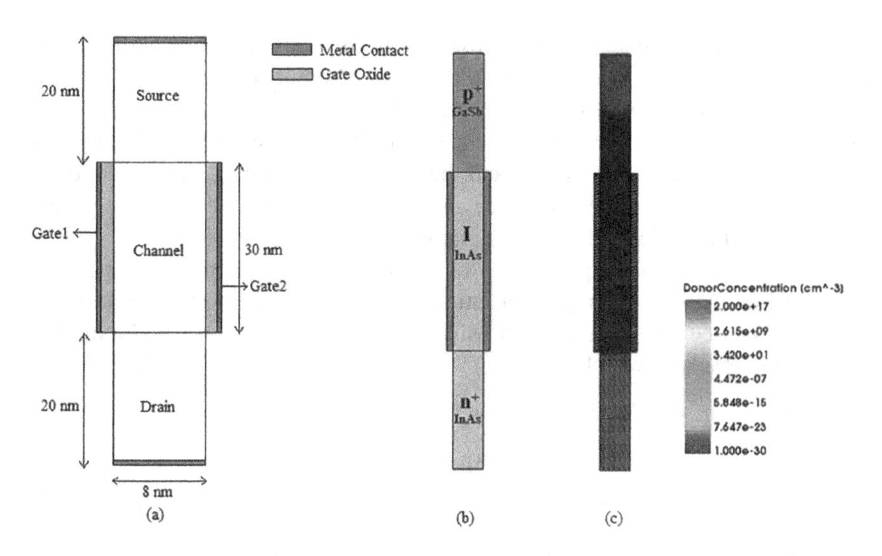

FIGURE 1.1 (a) Schematic of the device and (b) the simulated structure of IG HJ-DGTFET without doping and (c) with doping/meshing.

source and drain length of 20 nm each with a channel length of 30 nm and high-k dielectric of silicon nitride (Si_3N_4). The geometrical and doping parameters used in this simulation are listed in Table 1.1.

In the device physics, a hydrodynamic model, doping dependence mobility, effects of high and normal electric fields on the saturation of velocity, a Shockley–Read–Hall (SRH) recombination model, and mobility of electrons are used. Also, the Fermi–Dirac statistics is used instead of Boltzmann statistics because of the high doping concentrations used in source and drain. The effects of scattering are ignored as the doped n+ region remains active with the flow of charge carriers even with high doping concentrations. In the used model, it is assumed that only the voltage of gate-source decides band bending, which is necessary in the source region (for line tunneling) or

TABLE 1.1
Parameter Space of IG HJ-DGTFET

Geometrical/Doping Parameters	IG HJ-DGTFET (GaSb-InAs)
Gate length (L_g)	30 nm
Channel thickness (T_{ch})	8 nm
Front and back gate oxide thickness (T_{ox})	1 nm
Source dopant/doping concentration	Boron $-4 \times 10^{19}/cm^3$
Drain dopant/doping concentration	Arsenic $-2 \times 10^{17}/cm^3$
Channel doping concentration	Intrinsic $-1 \times 10^{15}/cm^3$
Source material	GaSb
Drain/channel material	InAs
Gate work-function	4.85 eV

in most of the source-channel region (for point tunneling). The effect of the voltage of drain is considered by assuming that the voltage of drain-source is what determines the location of the level of Fermi energy in the energy bands of drain. Specifically, the used model assumes that the carriers can tunnel into an energy level that is only at or above the level of Fermi energy and the electron Fermi level is assumed to be the same throughout the regions of source and drain from which electrons are generated and tunnel to the drain (considering a majority carrier density).

As it is known that the probability of transmission T_{WKB} is mainly dependent on the tunneling mass and bandgap, variations of these two parameters can improve the I_{ON} of the device and is calculated using the Wentzel–Kramers–Brillouin (WKB) expression, which is given by

$$T_{\text{WKB}} \approx \exp\frac{4\lambda\sqrt{2m^{*}E_g^3}}{3qh\left(E_g+\Delta\varnothing\right)} \tag{1.1}$$

where m^* is the electron's effective mass, E_g is the band gap energy, and q is the charge of electron. λ is the tunneling length, which defines the spatial extent of region of transition at the interface of source and channel, $\Delta\varnothing$ is the range of energy over which the tunneling can occur or the difference in energies of conduction band in the source and the valence band in the channel, and h is Planck's constant.

$$\lambda = \sqrt{\frac{\varepsilon_{Si}}{\varepsilon_{ox}}}\cdot\sqrt{t_{ox}t_{Si}} \tag{1.2}$$

where t_{ox} and t_{Si} are the thickness of oxide and silicon, respectively, and ε_{ox} and ε_{Si} are dielectric constants of oxide and silicon, respectively.

The tunneling barrier not only depends on the energy band gap but also on the carrier generation rate. As the gate voltage rises, the electric field across the channel also increases. The double-gate structure used in this work operates in independent gate (IG) mode, which means that both gates are simulated with different gate voltages ($V_{g1} \neq V_{g2}$). The transfer characteristics of IG HJ-DGTFET with a constant voltage of gate-1 at 1 V and a varying voltage of gate-2 from 0 to 1 V with a drain voltage of 1 V are shown in Figure 1.2. It can be observed that as the gate voltage rises, I_{ON} and I_{OFF} start to increase because of carrier generation dependency, which is given in equation (1.3).

$$G_{tun} = A\frac{|E^2|}{\sqrt{E_g}}\exp\left(-B\frac{E_g^2}{|E|}\right) \tag{1.3}$$

where G_{tun} is the carrier generation rate due to the interband tunneling at the junction, E is the electric field, E_g is band gap energy, and A and B are constants.

It can be said that by only increasing I_{ON} of TFETs, it is not sufficient to improve the performance metrics in the circuit level. In the circuit level, there are many performance parameters such as delay, power consumption and dissipation, and rise

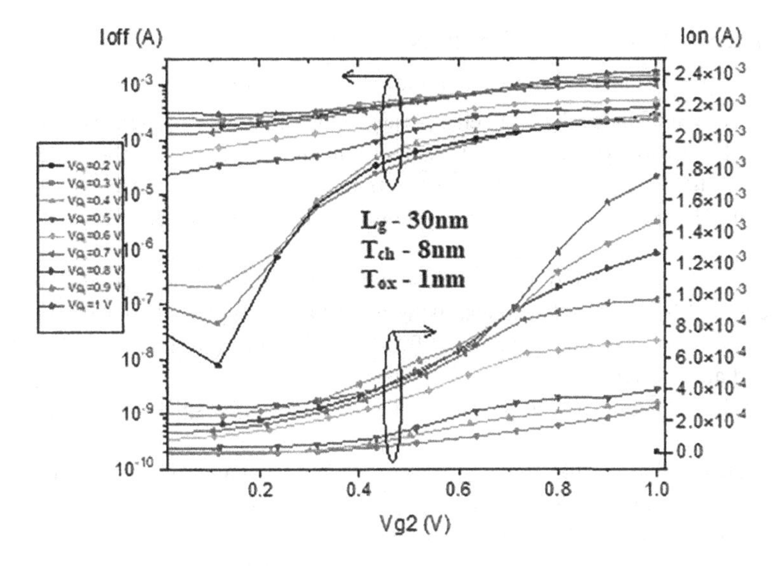

FIGURE 1.2 Transfer characteristics of IG HJ-DGTFET.

and fall times. These performance metrics are dependent on device parameters such as drain current in the linear regions and saturation of TFETs, gate and Miller capacitances, output conductance (g_d), and tunneling distance. The mentioned device parameters in turn affect SS. The expression for SS is shown in equation (1.4).

$$SS = \ln \frac{KT}{q} \left(1 + \frac{C_{dep}}{C_{ox}} \right) \tag{1.4}$$

From equation (1.4), it can be made clear that by improving SS, the performance of the device/circuit can be improved to an extent. One way to improve SS is by eliminating the temperature dependency of the device, which has various advantages shown in the results and discussion section. The circuit symbols of both n-type and p-type IG HJ-DGTFETs are shown in Figure 1.3.

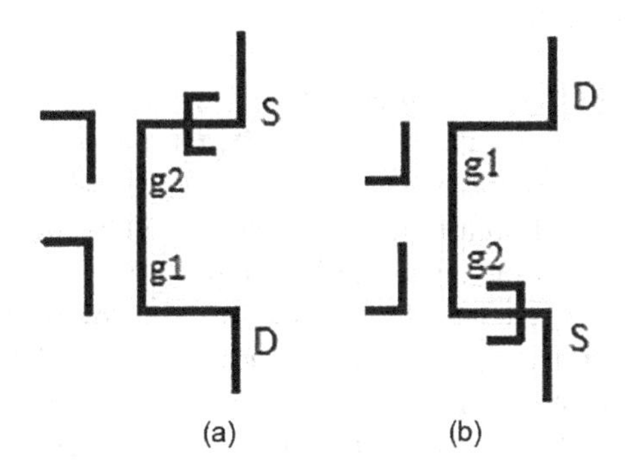

FIGURE 1.3 Symbol of IG HJ-DGTFET: (a) PTFET and (b) NTFET.

1.3 RESULTS AND DISCUSSION

This section provides the implementation and analysis of the QPSK system. The implementation includes the leaf cells, a pseudorandom binary sequence (PRBS) generator, and bit splitter circuits of the QPSK system and the analysis of the same circuits in terms of different performance parameters in both circuit and device levels.

The implementation of the QPSK system is done in three stages:

1. Realization and simulation of basic blocks/leaf cells, which are an inverter cell, NAND, and X-OR gates, using n-type and p-type IG HJ-DGTFETs.
2. Realization and simulation of D flip-flops (DFFs) using four NAND gates and an inverter cell.
3. Realization and simulation of PRBS generator and bit splitter using DFFs.

After each simulation, the values of power consumption and dissipation and propagation delay are extracted. For calculation of power consumption and dissipation, schematics of circuits are verified, and then the Analog Design Environment L (ADEL) window is accessed in which transient simulation is selected for analysis; the stimulus is then given to run the simulation, and from save-all options, all pwr signals are saved. Then, the results browser can be accessed to obtain the power consumption and dissipation.

For calculation of propagation delay, the calculator is accessed from the simulation window and from special functions, the function delay is selected in which the lower and upper threshold values are provided as 0.08 and 0.4 V, respectively, since the propagation delay is defined as 10% and 50% of supply voltage, which is 0.8 V. After applying all the data, an expression is displayed which is evaluated as propagation delay.

1.3.1 Realization of Leaf Cells Using IG HJ-DGTFET

The leaf cells of the QPSK system are NAND and inverter cells. In this section, these cells are realized using IG HJ-DGTFETs, which are the fundamental cells in the construction of D flip-flops. The constructed D flip-flop is used to realize PRBS and even/odd sequence generators.

1.3.1.1 Realization of NAND Logic Circuit

The circuit diagram of IG HJ-DGTFET based NAND is shown in Figure 1.4a. Two gates, g1 of both n- and p-type IG HJ-DGTFETs, are shorted together and given to V_{in1}. Similarly, gates g2 of both types of IG HJ-DGTFETs are shorted together and given to V_{in2}. When one of the inputs is given logic high or when both the inputs are given logic low, that is, V_{in1} to logic high and V_{in2} to logic low or V_{in1} and V_{in2} are given logic low, the transistor in the pull-up network is turned on and the output gets a logic high signal due to the shorted path between V_{DD} and the output. When both inputs are given logic high, both transistors in pull-up and pull-down networks are turned off and the output gets logic low. The timing diagram of the IG HJ-DGTFET-based NAND gate is shown in Figure 1.4b.

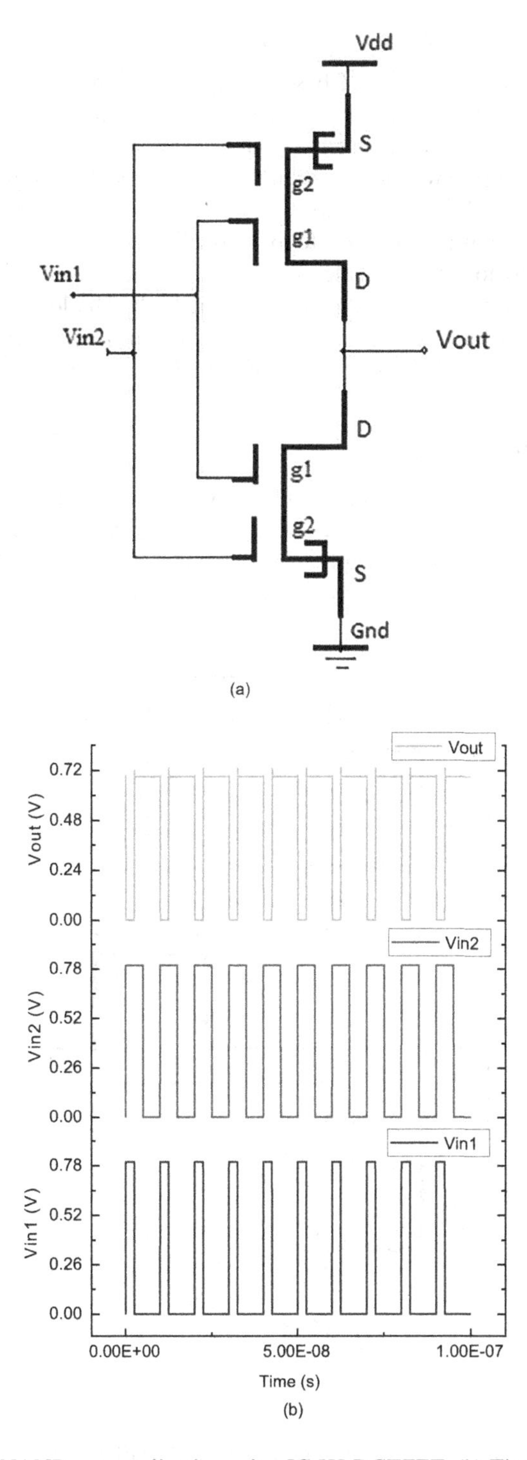

FIGURE 1.4 (a) NAND gate realization using IG HJ-DGTFET. (b) Timing diagram of IG HJ-DGTFET-based NAND gate.

1.3.1.2 Realization of X-OR Logic Circuit

The circuit of the IG HJ-DGTFET-based X-OR gate is shown in Figure 1.5a. This circuit requires only three transistors unlike conventional X-OR gate circuits [15]. This X-OR gate is based on the inverter and pass transistor logic (PTL). In PTL, the transistors act as simple switches between the nodes to pass the logic levels using the NTFET and PTFETs [16]. When inputs are given the same logic, either logic low/high, the output node is pulled down to logic low because of the transistor M3 being turned off owing to no potential difference between gate and source. When the inputs are different, the transistor M3 is pulled to logic high. The logic levels, logic high and logic low, for different combinations of input are shown in Table 1.2. The timing diagram of the IG HJ-DGTFET-based X-OR gate is shown in Figure 1.5b.

1.3.2 Realization of D Flip-Flop

In this section, D flip-flop realization using NAND and inverter logic gates is given and shown in Figure 1.6a. The input data is referred to as "D"; when the input D is given logic 1 or logic high, the flip-flop goes into a state called "SET", and when the input data is given as logic 0 or logic low, a state of "RESET" occurs in the flip-flop.

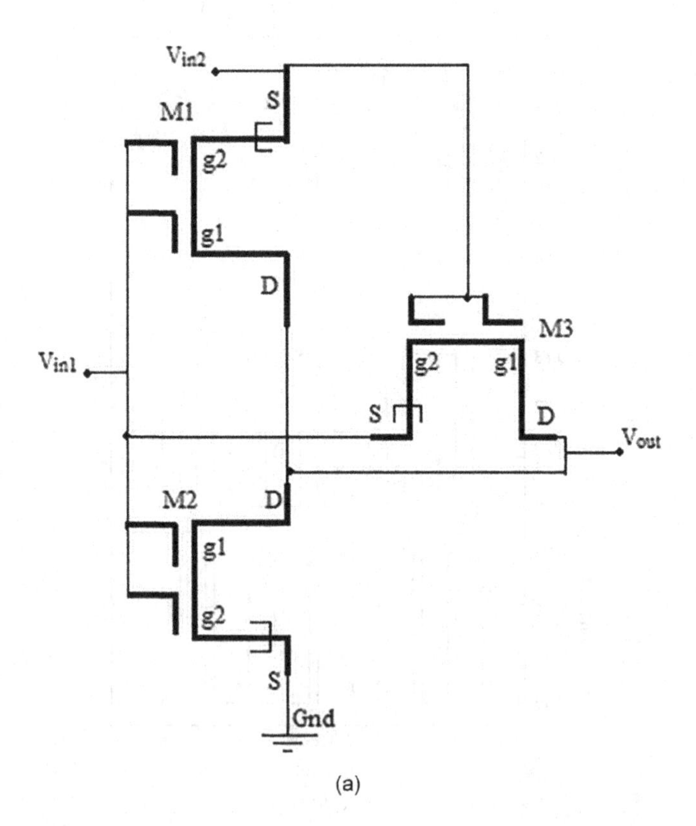

(a)

FIGURE 1.5 (a) XOR gate realization using IG HJ-DGTFET. (b) Timing diagram of IG HJ-DGTFET-based X-OR gate.

(Continued)

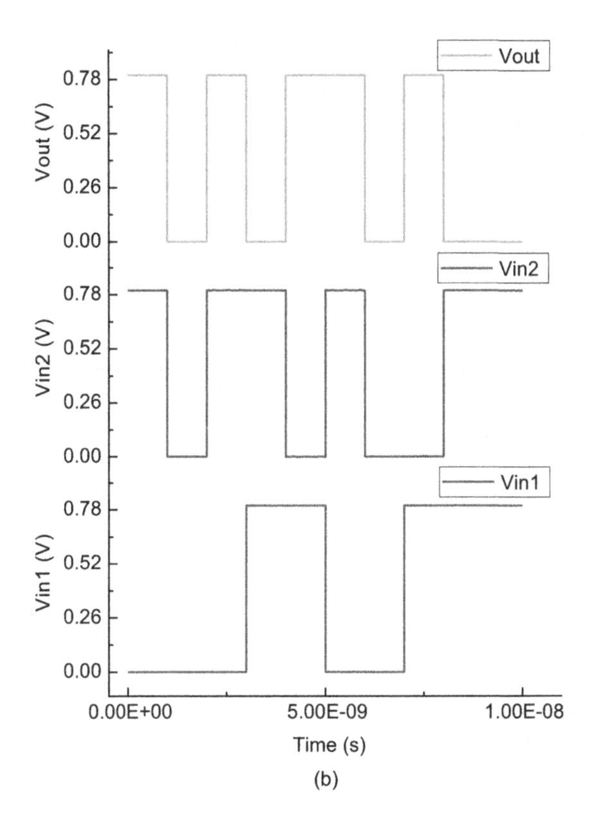

FIGURE 1.5 ***(CONTINUED)*** (a) XOR gate realization using IG HJ-DGTFET. (b) Timing diagram of IG HJ-DGTFET-based X-OR gate.

TABLE 1.2
Output Logic Levels of X-OR Gate for Different Input Combinations

Inputs		**Output**
A	**B**	**Y**
0	0	$\lvert V_{TP}\rvert$
0	1	V_{DD}
1	0	$\lvert V_{DD}-V_{TP}\rvert$
1	1	Gnd

But this kind of response is not desired for the QPSK system because the output of the flip-flop will always respond to the data input D on every pulse applied. To remove this undesired effect, an additional input called "CLK" is used to make the flip-flop go into "SET" state only when "CLK" transitions from logic high to logic low, making the flop positive-edge-triggered. The timing diagram of the D flip-flop logic circuit is shown in Figure 1.6b.

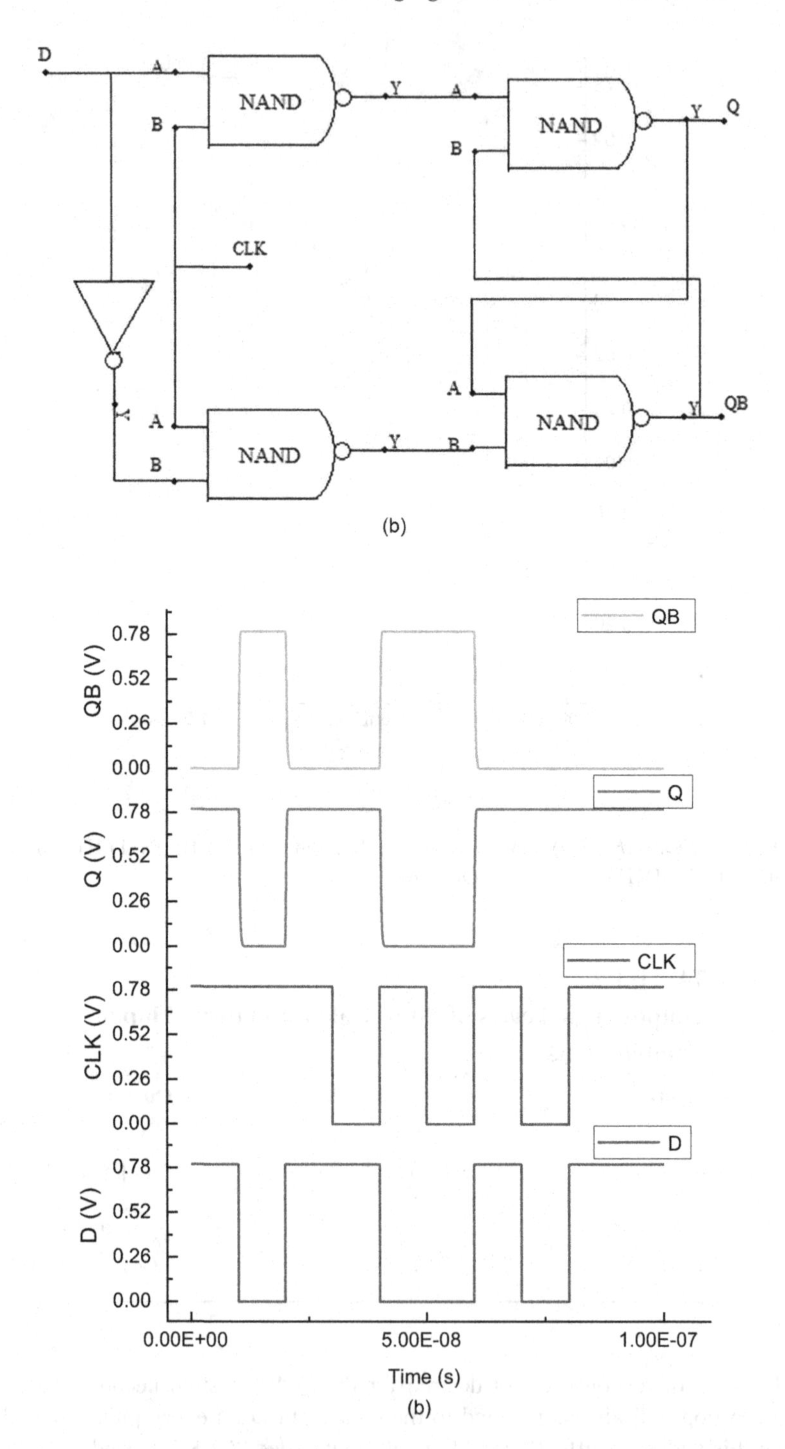

FIGURE 1.6 (a) D flip-flop realization using NAND and inverter logic gates. (b) Timing diagram of D flip-flop logic circuit.

1.3.3 Realization of the QPSK System

In this section, the realization of the digital part of the QPSK system is implemented. This can be achieved by connecting the following two systems in cascade:

1. PRBS generator.
2. Bit splitter or even/odd stream generator.

The output of the PRBS generator is given to bit splitter or the even/odd sequence generator [17,18]. PRBS is based on the concept of linear feedback shift register (LFSR) in which each bit is shifted from each D flip-flop. In the PRBS circuit, an X-OR gate is used between the last two flip-flops to give the difference output to the first D flip-flop. This creates a loop to generate random sequences. Once the random sequence is generated from the PRBS circuit, the PRBS output is given to the bit splitter circuit, which splits the sequence into odd and even stream of bits.

1.3.3.1 Realization of PRBS Generator

The PRBS circuit has been constructed using the leaf cells and is shown in Figure 1.7a. A PRBS bit stream can be generated using a linear feedback shift register (LFSR) [19,20]. The four flip-flops connected acts as a shift register. The LFSR is used to construct the PRBS sequence generator. D flip-flops three and four outputs are given to the XOR gate, and the output is given as feedback to the first D flip-flop input. Since there are four D flip-flops, there will be 2^4-1 bits. For example, if a 15-bit sequence is generated as 1 1 1 1 0 0 0 1 0 0 1 1 0 1 0, the output of the last flip-flop, 0 in this case, is sent to the even/odd sequence generator. Whenever the data bits change with time, the XOR gate is the one responsible for the change and is an important logic component in generating PRBS signals. The timing diagram of PRBS circuits is shown in Figure 1.7c.

1.3.3.2 Realization of the Even/Odd Stream Generator

The realization of the even/odd stream generator is shown in Figure 1.7b. This stage divides the input bit sequence from PRBS into even and odd bits. The PRBS output sent to the bit splitter splits the sequence as an even bit sequence 11000100 and an odd bit sequence 1 1 0 1 0 1 1 1. The first D flip-flop is used to divide the input clock into two, which is then sent to the second and third flip-flops which split the sequence. These divisions of clocks send Q to even-splitting flip-flop and QB to the odd-splitting flip-flop. The second flip-flop shifts the bits only in even positions and the third flip-flop shifts the bits only in odd positions. The timing diagram of the bit splitter circuit is shown in Figure 1.7c.

1.3.4 Performance Analysis of the QPSK System

This section deals with the performance analysis of the QPSK system. Both the device level and circuit-level parameters are extracted and tabulated.

1.3.4.1 Device Level Performance Metrics

Table 1.3 gives the device performance parameters of IG HJ-DGTFET.

In the device level, the mentioned performance metrics are affected by parameters such as drive current (I_{ON}) and SS, which are optimized to deliver the best performance. The optimization is done as follows. As shown in equation (1.2), λ is directly proportional to t_{ox} and affects SS; it is directly dependent on thermal voltage shown in equation (1.4). But despite the relation between SS, rise and fall times, and thermal voltage, the effect of oxide capacitance dominates as very thin oxide of 1nm is taken. Further effects of C_{ox} are discussed in the circuit-level performance metrics section.

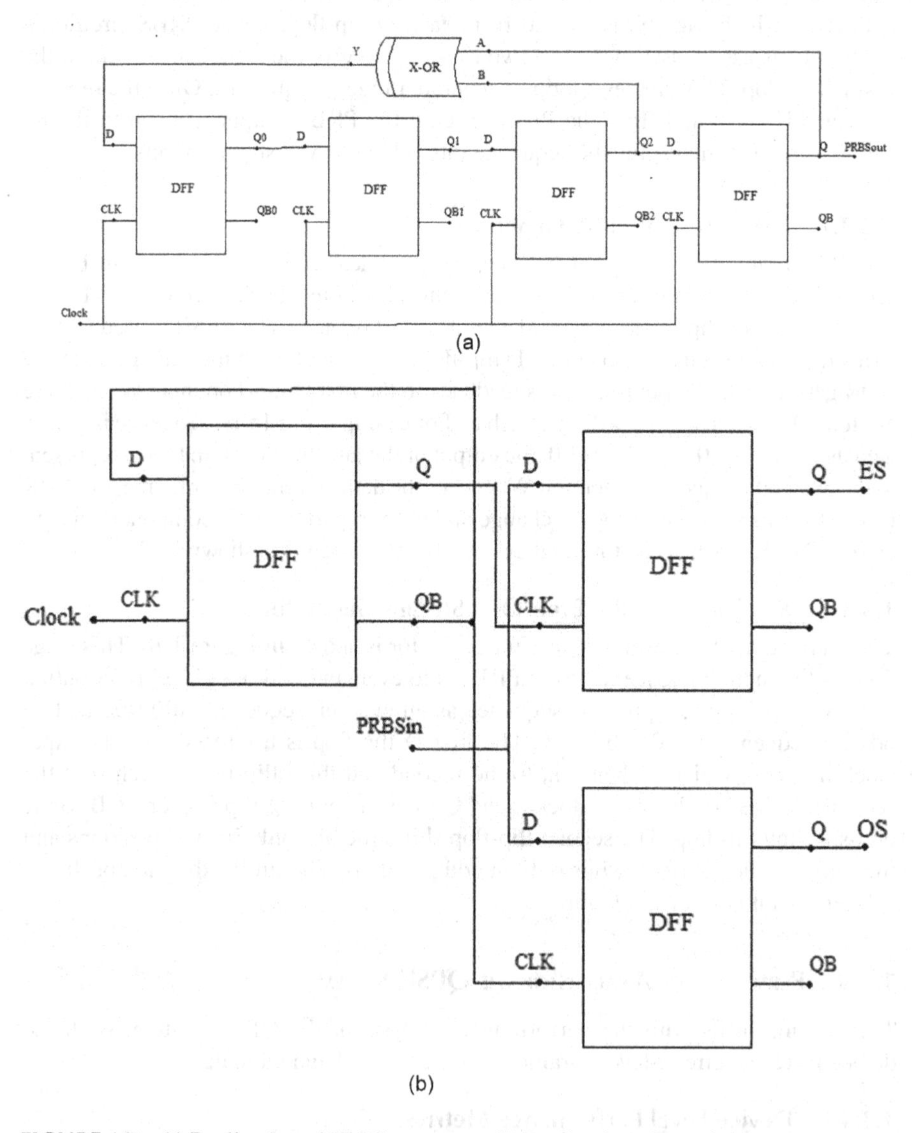

FIGURE 1.7 (a) Realization of PRBS generator. (b) Realization of even/odd stream generator. (c) Timing diagram of PRBS and bit splitter circuits.

(Continued)

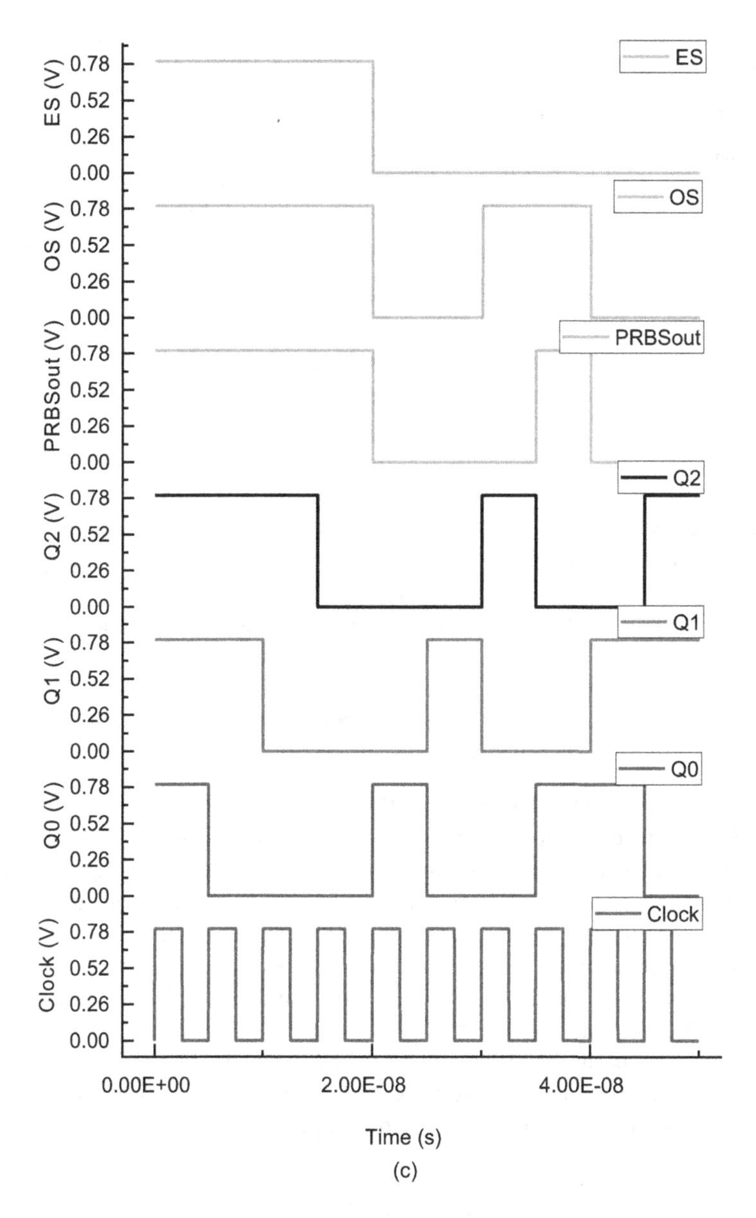

FIGURE 1.7 (*CONTINUED*) (a) Realization of PRBS generator. (b) Realization of even/odd stream generator. (c) Timing diagram of PRBS and bit splitter circuits.

The usage of a lower-band-gap material such as GaSb-InAs material affected the band gap, which directly points to SS. From Table 1.3, it can be seen that the extracted SS is very close to the ideal SS, which is 60 mV/decade, and this feature makes it to contribute to lesser delay and temperature independency. In addition to this, the other two factors, usage of the optimized oxide thickness and GaSb-InAs material, also contributed to less leakage current (I_{OFF}) in the device, which reduced the power dissipation in the device level. The output conductance (g_d) is optimized by obtaining

TABLE 1.3
Performance Metrics of IG HJ-DGTFET

Device Parameter	Value
I_{ON}	1.287 mA
I_{OFF}	20.08 nA
I_{ON}/I_{OFF}	0.6409×10^5
SS	58.8 mV/decade
C_{gd}	44.57 pF
g_d	0.162 μS
g_m	0.597 mS
f_t	979.87 GHz
GBP	213.5 GHz

a lower drain saturation voltage, V_{ds}, which is achieved by controlling the threshold voltage. Due to the higher transconductance (g_m), the unity gain cut-off frequency (f_t) was increased, which resulted in an optimal gain-bandwidth product (GBP) needed for the QPSK system. The GBP for the device is calculated using equation (1.5) [21].

$$\text{GBP} = \frac{g_m}{20\pi \times C_{gd}} \tag{1.5}$$

1.3.4.2 Circuit-Level Performance Metrics

The procedure to compute power and delay is explained in the Cadence manual [22,23]. The performance metrics considered here are propagation delay and power consumption and dissipation. The average power and the propagation delay are measured by performing the transient analysis of logic circuits. The parameter propagation delay is affected by the rise and fall times, and as previously discussed in device performance metrics, these rise and fall times are affected by C_{ox}, which can be seen from the rise time (t_{PLH}) relation with respect to C_{ox} shown in equation (1.6). The inverter used in the QPSK system produced rise and fall times of 0.09 ps. With respect to variations in temperature from 27°C to 54°C, these rise and fall times are varied by a maximum of 2.3 ps. The fall time equation is given as equation (1.7).

$$t_{PLH} = \frac{C_L}{\mu_p\, C_{ox} \frac{W}{L} (V_{in} - V_{TH})^2} \tag{1.6}$$

$$t_{PHL} = \frac{C_L}{\mu_n\, C_{ox} \frac{W}{L} (V_{in} - V_{TH})^2} \tag{1.7}$$

where t_{PLH} and t_{PHL} are rise and fall times, respectively. C_L is the load capacitance, μ is the mobility of the charge carriers, C_{ox} is oxide capacitance, W and L are width and length of the channel, respectively, and V_{in} and V_{TH} are input and threshold voltages, respectively. From equations (1.6) and (1.7), propagation delay is defined as the time measured between the 50% transition points of the input and output waveforms and is calculated using equation (1.8).

$$t_{pd} = \frac{t_{PHL} + t_{PLH}}{2} \quad (1.8)$$

In the parameter of power, two different power metrics are shown here: one is power consumption, which is the power drawn from the supply to obtain the desired output, and the other is power dissipation, which is unproductive and undesired, such as the power drawn from supply even when the transistor in a logic circuit is in the OFF state or simply the power consumed even when input is not given. The power consumption and dissipation are given in equations (1.9 and 1.10), respectively. Table 1.4 provides the overall performance metrics of the QPSK system.

$$P_C = \left(\Sigma\ (\text{leakage current}) \times V_{DD}\right) + C_L \times V_{DD}^2 \times f_i \quad (1.9)$$

$$P_D = C_{pd} \times V_{DD}^2 \times f_i \quad (1.10)$$

where P_c and P_D are power consumption and dissipation, respectively. C_{pd} is dissipation capacitance and is calculated using equation (1.11). V_{DD} is the supply voltage and f_i is the input frequency.

$$C_{pd} = \frac{I_{DS}}{V_{DD} \times f_i} - C_{L_{eff}} \quad (1.11)$$

$C_{L_{eff}}$ can be calculated by multiplying the number of switching bits (N_{SW}) and drain capacitance (C_d).

From Table 1.4, it can be inferred that the overall performance of the QPSK system can be said to be improved with the use of this IG HJ-DGTFET structure against the QPSK system, which is discussed in the literature [24]. In the circuit level, since the complexity of circuits mainly affects power consumption, it has been reduced as

TABLE 1.4
Performance Metrics of the QPSK System

Logic Gates	Propagation Delay	Power Consumption	Power Dissipation
NAND gate (IG HJ-DGTFET)	43.53 ps	38.02 pW	0.652 aW
X-OR gate (IG HJ-DGTFET)	13.31 ps	6.147 pW	21.78 pW
D flip-flop	346.1 ps	66.05 nW	7.731 pW
PRBS generator	200.09 ps	35.19 nW	2.600 nW
Odd/even stream generator	ES: 39.98 ps OS: 106.1 ps	205.3 nW	2.156 nW

a first step using IG mode in NAND gate, which ultimately reduced the complexity by half, and in X-OR gate, an inverter and PTL have been applied to reduce the complexity. In the X-OR gate, reusing signals and passing them using a PTFET are performed. This reduces the delay as the signals are only passed based on the difference between logic levels, and because of the reuse of signals, no external supply is needed, which means no active power is drawn and which further reduces power consumption. Therefore, as the performance metrics are enhanced considering both device and circuit-level parameters, it can be said that the overall performance of the QPSK system is totally enhanced.

1.4 CONCLUSIONS

In this chapter, an IG HJ-DGTFET has been designed in such a way that the device when used in the QPSK system produces an optimal subthreshold swing, high transconductance, and GBP required for the system at the cost of the I_{ON}/I_{OFF} ratio of the device. Also, the results show negligible rise and fall times of the inverter cell, which is an advantage for the system, if it is to be used in temperature-variant surroundings. SS being close to the ideal value, lesser I_{OFF} and the design of lower-complexity circuits combined produced a system with very low power consumption and less propagation delay. Circuit techniques such as IG mode and PTL helped in reducing power dissipation, which contributes to the increased lifetime of the entire QPSK system. With advantages such as temperature independency, high GBP, and low power consumption, the implemented QPSK system can be used in areas such as cellular phone systems, which demands low power, satellite transmission systems, or WIFI modems, which require temperature independency and high bandwidth. Also, the obtained results of less propagation delay enable higher frequency of operation for the QPSK system. Hence, it can be concluded that the implemented system consumes less power and produces low latency, which is the key feature required for any low-power RF wireless communications.

REFERENCES

1. S. Garg and S. Saurabh, "Implementing Logic Functions Using Independently-Controlled Gate in Double-Gate Tunnel FETs: Investigation and Analysis," in *IEEE Access*, vol. 7, pp. 117591–117599, 2019, doi: 10.1109/ACCESS.2019.2936610.
2. L. Barboni, M. Siniscalchi and B. Sensale-Rodriguez, "TFET-Based Circuit Design Using the Transconductance Generation Efficiency gm/Id Method," in *IEEE Journal of the Electron Devices Society*, vol. 3, no. 3, pp. 208–216, May 2015, doi: 10.1109/JEDS.2015.2412118.
3. S. Khandelwal, J. Duarte, A. S. Medury, V. Sriramkumar, N. Paydavosi, D. Lu, C. H. Lin, M. Dunga, S. Yao, T. Morshed, et al., *BSIM-CMG 109.0. 0 Multi-Gate MOSFET Compact Model Technical Manual.* Regents University California, Berkeley, 2015.
4. D. H. Morris, U. E. Avci, R. Rios and I. A. Young, "Design of Low Voltage Tunneling-FET Logic Circuits Considering Asymmetric Conduction Characteristics," in *IEEE Journal on Emerging and Selected Topics in Circuits and Systems*, vol. 4, no. 4, pp. 380–388, Dec. 2014, doi: 10.1109/JETCAS.2014.2361054.

5. W. G. Vandenberghe, A. S. Verhulst, G. Groeseneken, B. Soree and W. Magnus, "Analytical Model for a Tunnel Field-Effect Transistor," MELECON 2008- The 14th IEEE Mediterranean Electrotechnical Conference, 2008, pp. 923–928, doi: 10.1109/MELCON.2008.4618555.
6. C. Convertino, C.B. Zota, H. Schmid, D. Caimi, L. Czornomaz, A.M. Ionescu and K.E. Moselund, "A Hybrid III–V Tunnel FET and MOSFET Technology Platform Integrated on Silicon," in *Nature Electronics*, vol. 4, no. 2, pp. 162–170, 2021, doi: 10.1038/s41928-020-00531-3.
7. A. Sharma, A. A. Goud and K. Roy, "GaSb-InAs n-TFET with Doped Source Underlap Exhibiting Low Subthreshold Swing at Sub-10-nm Gate-Lengths," in *IEEE Electron Device Letters*, vol. 35, no. 12, pp. 1221–1223, Dec. 2014, doi: 10.1109/LED.2014.2365413.
8. H. Lu, T. Ytterdal and A. Seabaugh. Notre dame TFET model, Sep 2017.
9. A. Varghese, C. S. Praveen, A. P. Mani and A. Ravindran, InGaAs/GaAsSb HETEROJUNCTION TFET, 2015.
10. U. Dutta, M. K. Soni and M. Pattanaik, "Design and Analysis of Tunnel FET for Low Power High Performance Applications," in *International Journal of Modern Education and Computer Science*, vol. 10, no. 1, p. 65, 2018.
11. M.R. Salehi, E. Abiri, S. E. Hosseini and B. Dorostkar, "Analysis and Optimization of Tunnel FET with Band Gap Engineering," 2013 21st Iranian Conference on Electrical Engineering, 2013, pp. 1–4, doi: 10.1109/IranianCEE.2013.6599607.
12. K. Boucart and A. M. Ionescu, "Double-Gate Tunnel FET with High-κ Gate Dielectric," in *IEEE Transactions on Electron Devices*, vol. 54, no. 7, pp. 1725–1733, 2007. doi: 10.1109/TED.2007.899389
13. A. Madhukar and S. Das Sarma, "Intrinsic and Extrinsic States at Lattice Matched Interfaces between III-V Compound Semiconductors: The InAs/GaSb(110) System," in *Journal of Vacuum Science and Technology*, vol. 17, pp. 1120–1127, 1998, doi: 10.1116/1.570626
14. M. Pown and B. Lakshmi, "Performance Analysis of InAs- and GaSb-InAs-Based Independent Gate Tunnel Field Effect Transistor RF Mixers," in *Journal of Computational Electronics*, vol. 16, pp. 676–684, 2017, doi: 10.1007/s10825- 017-1005–8
15. K. Dhar, "Design of a Low Power, High Speed and Energy Efficient 3 Transistor XOR Gate in 45nm Technology using the Conception of MVT Methodology," 2014 International Conference on Control, Instrumentation, Communication and Computational Technologies, ICCICCT 2014, 2014, doi: 10.1109/ICCICCT.2014.6992931.
16. R. Zimmermann and W. Fichtner, "Low-Power Logic Styles: CMOS Versus Pass-Transistor Logic," in *IEEE Journal of Solid-State Circuits*, vol. 32, no. 7, pp. 1079–1090, July 1997, doi: 10.1109/4.597298.
17. R. Rameshkumar and J. N. Swaminathan, "A New Hardware Design and Implementation of QPSK Modulator in Normal and 4-QAM Mode for 5G Communication Networks," International Conference on Inventive Computation Technologies, 2020, doi: 10.1007/978-3-030–33846-6_29.
18. N. Haridas and M. N. Devi, "Efficient Linear Feedback Shift Register Design for Pseudo Exhaustive Test Generation in BIST," 2011 3rd International Conference on Electronics Computer Technology, 2011, pp. 350–354, doi: 10.1109/ICECTECH.2011.5941621.
19. M. JanakiRani and S. Malarkkan, "Design and Analysis of a Linear Feedback Shift Register with Reduced Leakage Power," in *International Journal of Computer Applications*, vol. 56, pp. 9–13, 2012, doi: 10.5120/8957–3159.
20. B. Singh, A. Khosla and S. Bindra, "Power Optimization of Linear Feedback Shift Register (LFSR) for Low Power BIST," 2009 IEEE International Advance Computing Conference, 2009, pp. 311–314, doi: 10.1109/IADCC.2009.4809027.

21. D. S. Yadav, D. Sharma, B. R. Raad and V. Bajaj, "Dual Workfunction Hetero Gate Dielectric Tunnel Field-Effect Transistor Performance Analysis," 2016 International Conference on Advanced Communication Control and Computing Technologies (ICACCCT), 2016, pp. 26–29, doi: 10.1109/ICACCCT.2016.7831593.
22. Cadence Tutorial2: Schematic Entry & Digital Simulation. Cadence Tutorial 1 (cu.edu.eg).
23. Power measurement with Cadence EDA: Power Measurement Guide (msu.edu).
24. Z. Ji, S. Zargham and A. Liscidini, "Low-Power QPSK Transmitter Based on an Injection-Locked Power Amplifier," ESSCIRC 2018- IEEE 44th European Solid State Circuits Conference (ESSCIRC), 2018, pp. 134–137, doi: 10.1109/ESSCIRC.2018.8494300.

2 Modeling and Simulation of Emerging Low-Power Devices

M. Venkatesh
REVA University

G. Lakshmi Priya
VIT University

S. Arun Samuel
National Engineering College

M. Karthigai Pandian
GITAM University

CONTENTS

DOI: 10.1201/9781003240778-2

2.1 INTRODUCTION

The human body is comprised of a huge number of biological cells, and an integrated circuit (IC) is comprised of a huge number of transistors. Transistors are a fundamental structure of present-day electronic devices. As far back as the appearance of complementary metal oxide semiconductor (CMOS) circuits, the components of the transistor have been ceaselessly downsized so as to pack more rationale on to a silicon wafer and furthermore to lessen power control utilization in circuits. In the recent scenario, with cell phones gaining prevalence, the quest for low-control gadgets with steep switching attributes has become significant. Profoundly scaled metal oxide semiconductor field effect transistors (MOSFETs) are rendered inadmissible for low-power-control applications because of their thermal point in their switching behavior. Consequently, the tunnel field effect transistor (TFET) is being investigated widely for ultralow-power applications. Tunnel FET has a steep switching trademark as it operates upon the principle of band-to-band tunneling (BTBT) effect. In the course of recent years, TFETs have been vigorously studied by researchers in the field of semiconductor gadgets over the world.

2.2 GORDON MOORE'S LAW SCALING

Scaling, as portrayed by Moore's perception, has multiplied the thickness with each generation rate with the related cost decrease of an integrated circuit. By the 1980s, complementary MOS (CMOS) was developed as the innovation of things to come, and by the 1990s, the CMOS overwhelmed all of the microelectronic applications and is dominant to date [1,2]. This is fundamentally a direct result of the low power utilization of CMOS circuits and their capacity to downsize to incredibly small scales as shown by Dennard of IBM. Figure 2.1 shows the component size and gate length as an element of time. One hundred nanometers is a principal innovation milestone. It is the division point between small-scale innovation and nanotechnology.

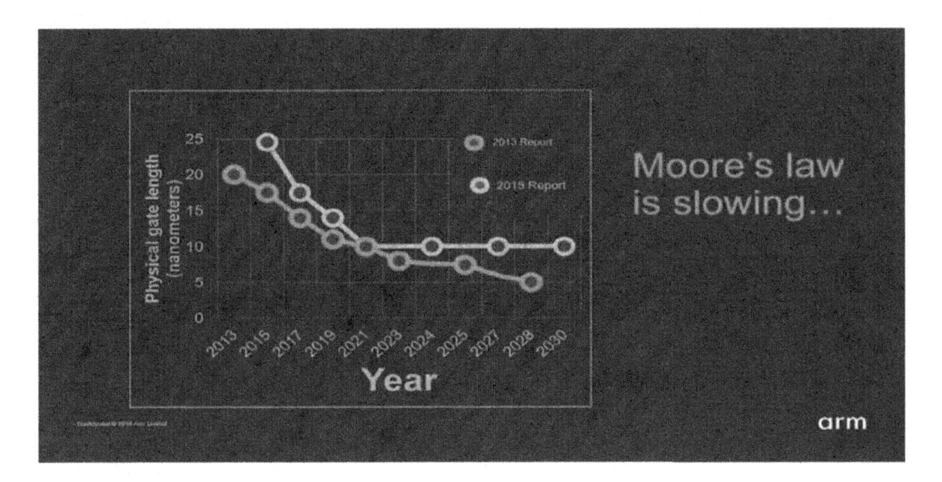

FIGURE 2.1 Moore's law.

The semiconductor manufacturing surpassed it in the twenty-first century from micro-ICs to nano-ICs. Moore's law has been the marker of progress in IC innovation throughout the five previous decades. According to Moore's law, the MOS innovation crosses a few innovation hubs during its downsizing venture.

2.3 BASICS OF TUNNEL FET

"Tunnel field effect transistor" (TFET) is a reverse-biased gated p-i-n diode. Because of the band-to-band tunneling of charge carriers between the source and the channel, the TFET's output current is determined by this BTBT process [3]. However, the carrier injection mechanism in MOSFETs is dependent on heat injection. Because of its decreased leakage current and fewer short-channel effects (SCEs), the TFET device is ideal for low-power applications. On the source side, the BTBT area is approximately 10 nm^2, and the device may be scaled up to 20 nm. These include reduced subthreshold swing (60 mV/decade), increased working speed, and a reduced threshold voltage roll-off ($V_{troll\text{-}off}$). In addition, the TFET does not have a punch-through effect because of its reverse-biased p-i-n architecture.

To replace MOSFET in low-power applications, TFET is being recognized as one of the most promising technologies. An N-type TFET schematic design is presented in Figure 2.2a. The p-i-n structure is reverse-biased, and a voltage is provided to the gate to turn the device on. There are p-type and n-type areas that are heavily doped at the source and the drain of the transistor. Intermediate channels consist of an intrinsic layer moderately doped [4]. Rather than using titanium dioxide (TiO_2) as a gate dielectric, SiO_2 is used. If a positive or negative voltage is connected to the gate terminal V_{GS}, the device functions as an N-type tunnel FET or a P-type tunnel FET. An N-TFET can be used with positive gate voltage, and a P-TFET can be used without it when a negative gate voltage is applied. As the gate voltage is increased, the energy barrier between the source and the intrinsic region is lowered, allowing for the n-type TFET to operate. A downward force is exerted on the intrinsic energy bands, as seen in Figure 2.2b.

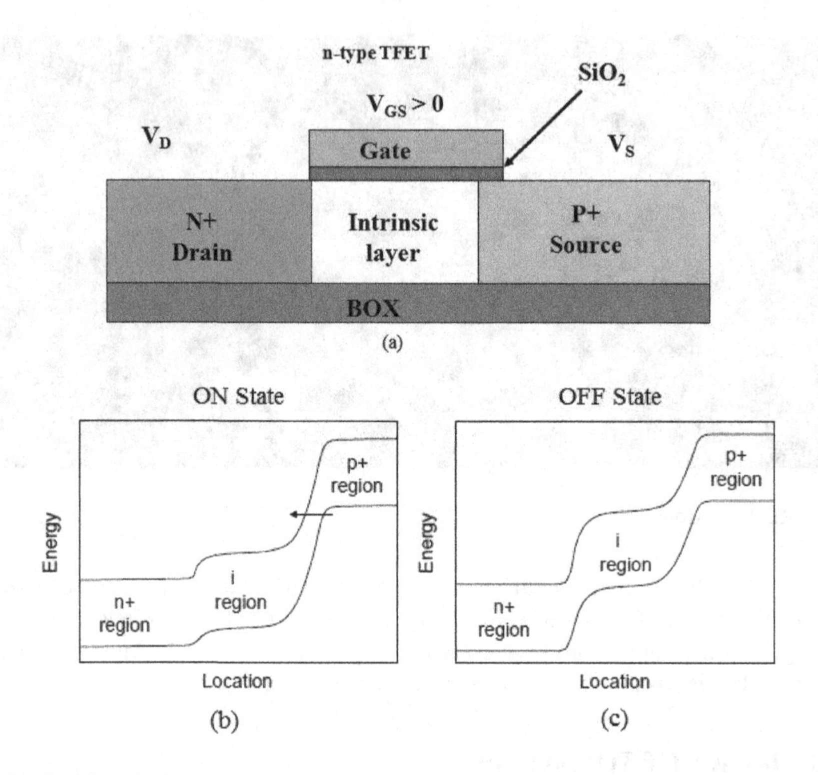

FIGURE 2.2 (a) N-type tunnel FET device structure. (b) ON state with a positive bias on the gate leading to N-TFET-type behavior. (c) OFF state where the only current is from p-i-n leakage.

Tunneling from one band to another is known as band-to-band (BTB) tunneling. This is followed by electron migration in the direction of an n^+-doped drain region. Electrons move to the n^+-doped drain area as a result of drift diffusion. Alternatively, the p-i-n diode leakage current (less than 1 fA/μm) flows between them when the source/drain voltage of the n-TFET is turned off (V_{GS}=0). Figure 2.2c illustrates the OFF state n-TFET energy barrier diagram (c).

The energy barrier between the drain and the intrinsic region is reduced when the negative gate voltage ($V_{GS}<0$) is increased in a p-type TFET as shown in Figure 2.3a. It is clear from Figure 2.3b that energy barrier bands in the intrinsic area under the gate have been pushed up. Band-to-band tunneling occurs between the intrinsic band and the n^+-region conduction band, generating the ON current. A p-tunnel FET is switched OFF ($V_{GS}=0$), as shown in Figure 2.3c, and the leakage current flows between the drain and source regions.

2.3.1 Components of Tunnel FET

The TFET device's operating concept relies on gate-controlled BTBT. As a result of line tunneling, BTBT occurs in a direction orthogonal to the gate, whereas the channel area is used for BTBT tunneling [5,6]. Figure 2.4 illustrates the two different

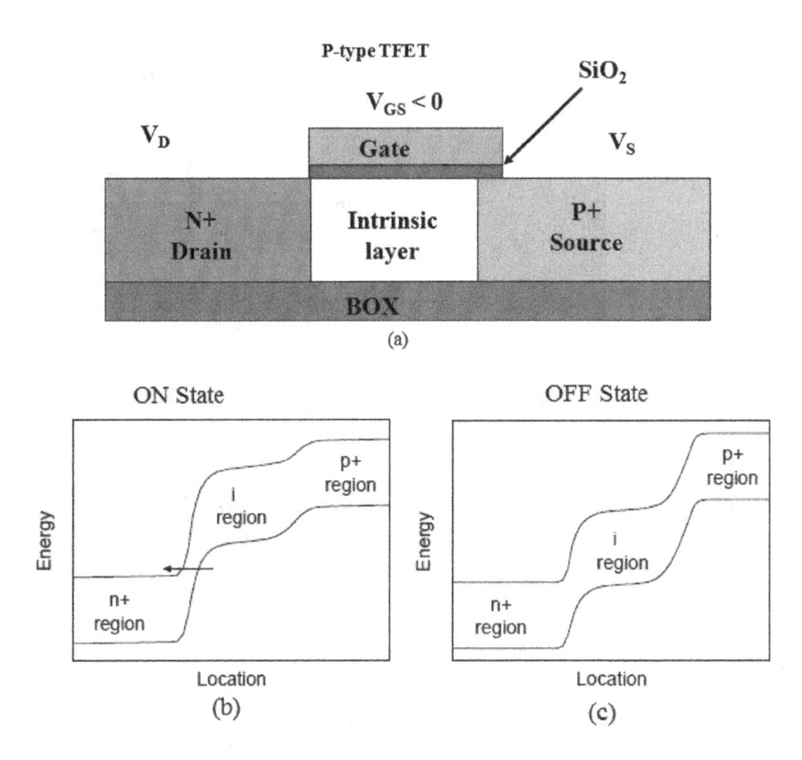

FIGURE 2.3 (a) P-type tunnel FET device structure. (b) ON state with a positive bias on the gate leading to P-TFET-type behavior. (c) OFF state where the only current is from p-i-n leakage.

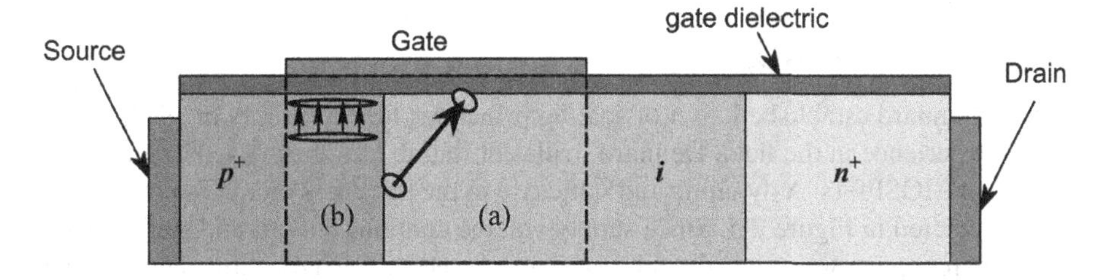

FIGURE 2.4 Illustration of (a) point tunneling contribution and (b) line tunneling contribution in a TFET with the gate over the source.

components of tunneling. Source–channel interface point tunneling is also known as lateral tunneling. Increasing the positive gate voltage results in a depletion zone that changes the channel potential profile. From the drain, a small amount of charge will flow into the channel. The source region is heavily doped and closely linked to the gate edge in order for BTB tunneling from the p+ source area to the channel region in order to improve the ON current. This is because dielectric thickness and band-gap width are important factors in determining the ON current in point tunneling. As a result of the source's excessive doping, a portion of the source's region overlaps

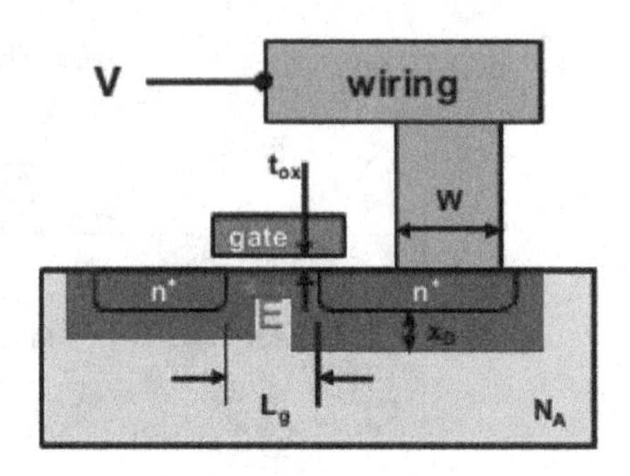

dimensions t_{ox}, L, W	1/a
doping	a
voltage	1/a
integration density	a^2
delay	1/a
power dissipation/Tr	$1/a^2$
Electric Field E	1

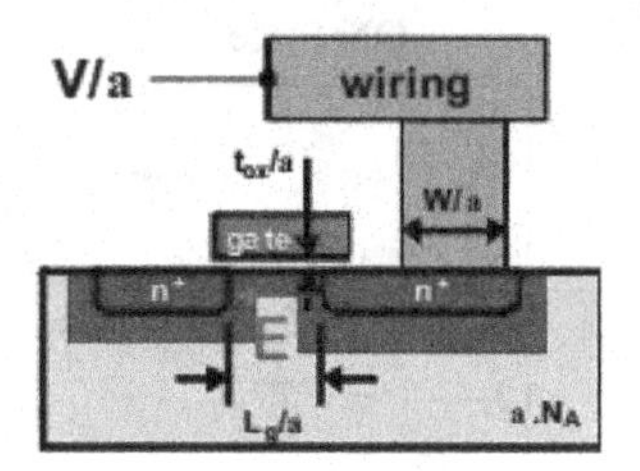

FIGURE 2.5 Constant field scaling of digital MOS circuits.

the gate region. Tunneling occurs in the p+ source region as a result. It starts in the valence band and tunnels to the gate dielectric, where it continues to tunnel [7]. As the gate voltage rises, the tunneling continues in the conduction band at the gate dielectric interface. For example, a longer gate overlap over the source allows for a larger line tunneling current, which in turn allows for a greater tunneling current.

2.3.2 Scaling of MOSFET

Robert Dennard established a set of scaling principles for MOSFETs in 1974, based on his experience in the field. Dennard's rules of thumb have been used to guide the scaling of MOSFETs. A dynamic and static rise in the thickness of a PC's power control is depicted in Figure 2.5. Since supply voltage continues to fall and static power usage increases, power control thickness is increased. It was Dennard's law that the supply voltage scaled according to the voltage necessary to produce a sufficiently high drive current [8,9]. Decreased voltages caused a linear drop in OFF state current (I_{off}) and threshold voltage.

2.4 SHORT-CHANNEL EFFECTS

To begin, FET dimensions are scaled by maintaining the device aspect ratio. A gadget's dynamic vertical element is defined as a ratio between the gateway length and the gadget's dynamic vertical element. In MOSFETs, the vertical measurement indicates the oxide thickness t_{ox}, the source and channel intersection $t_{channel}$, and the exhaustion profundities W_s and W_d at the source and channel intersection [10]. Small-channel conduct is characterized by a low aspect ratio.

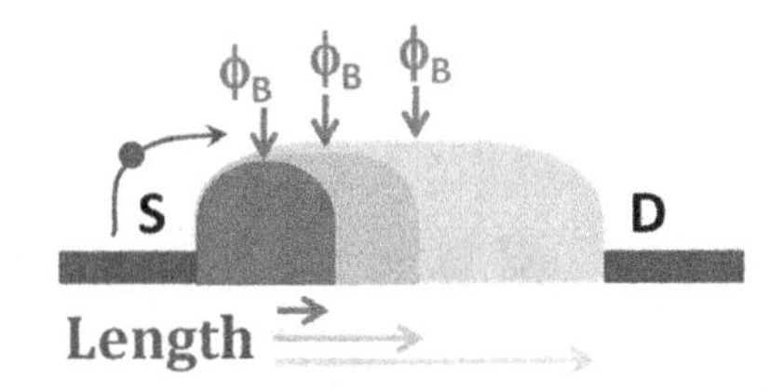

FIGURE 2.6 DIBL structure of MOS transistor.

2.4.1 Drain-Induced Barrier Lowering

As shown in Figure 2.6, threshold voltage is a measure of the quality of the barrier against the infusion of charge carriers from source to channel. $L<L_{min}$ short-channel systems with drain channel bias may be able to overcome this obstacle [11]. As the Vt of n-channel MOSFETs is lowered, the subthreshold current increases as V_{DS} is increased.

2.4.2 Channel-Length Modulation

When the drain channel–source bias of a FET advances near the drain channel saturation zone, an area of high electric field is framed close to the drain channel, and the electron speed in this locality is saturated (in long gadgets, we would rather have a pinch-off where N_S turns out to be slightly close to the drain area). In the saturation zone, the length L of the high-electric-field area increases toward the source with increasing V_{DS}, and the MOSFET continues to function as though the effective channel length has been shortened by L [12]. This is termed channel-length modulation (CLM), as seen in Figure 2.7.

2.4.3 Hot Carrier Effects

Impacts of hot carriers on FET measurements in deep submicrometer systems are a major problem. Stable voltage scaling, or decreasing the channel length while maintaining high-control supply levels, causes the electric field characteristics in the channel to increase, forcing the charge carriers to accelerate and warm up, as illustrated in Figure 2.8. For example, impact ionization leads to the development of

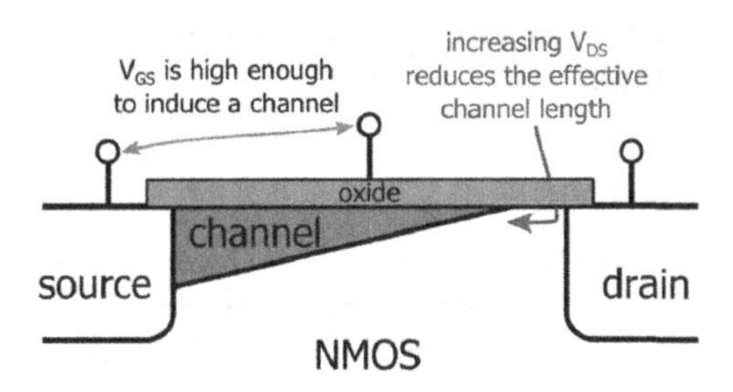

FIGURE 2.7 Illustration of channel-length modulation.

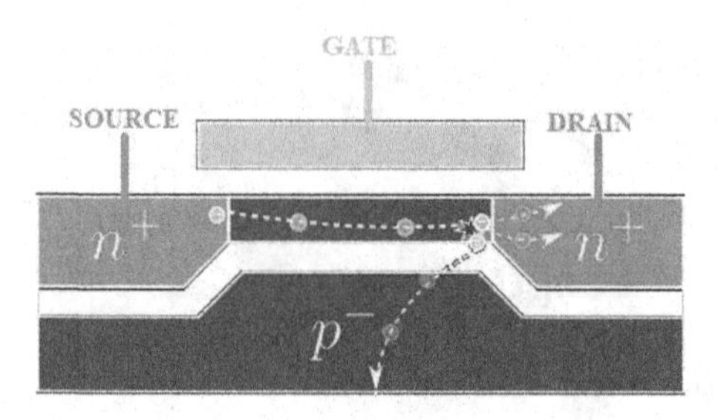

FIGURE 2.8 Illustration of impact ionization.

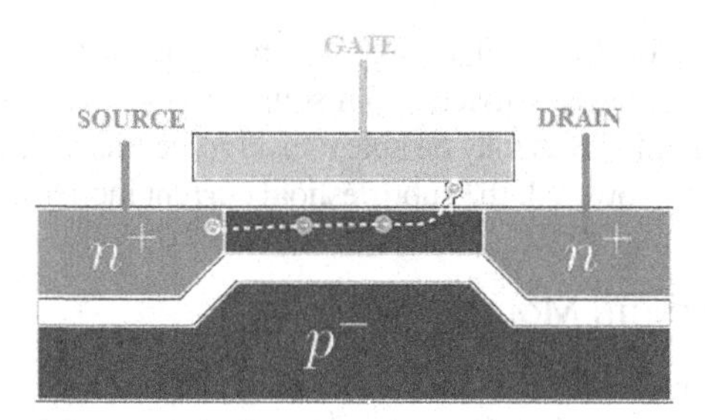

FIGURE 2.9 Illustration of hot electrons.

interface states and entering currents generated by hot-electron discharge over the interface boundary, whereas charge carrier tunneling results in oxide charges, and electron–hole recombination causes photocurrents, as shown in Figure 2.9.

2.4.4 Velocity Saturation

There is a tendency for charge carriers to slow down in high electric fields because of the mobility immersion that occurs when the field is high. To maintain the same electric fields in a shrunk transistor, a constant field scaling can be used [13]. Nevertheless, the scaling trend adopted in semiconductor manufacturing is not a field-wide one. In Figure 2.10, for example, the oxide thickness is scaled up and the supply voltages are scaled back. This results in a larger electric field in the nanoscale MOSFETs as a result of the increased current flow. As the electric field expands, the speed increases. This would have been a low-centrality issue if the immersion that occurs at channel voltages was higher [14,15].

2.4.5 Gate Oxide Leakage

As far as MOSFET devices are concerned, SiO_2 is a good insulator. However, when the gateway oxide's thickness is reduced to below 3 or 2 nm, the possibility of device tunneling increases, leading to an increased oxide overflow current in either case. Using a high-K dielectric can alleviate this problem by minimizing the direct tunneling excess of current [16].

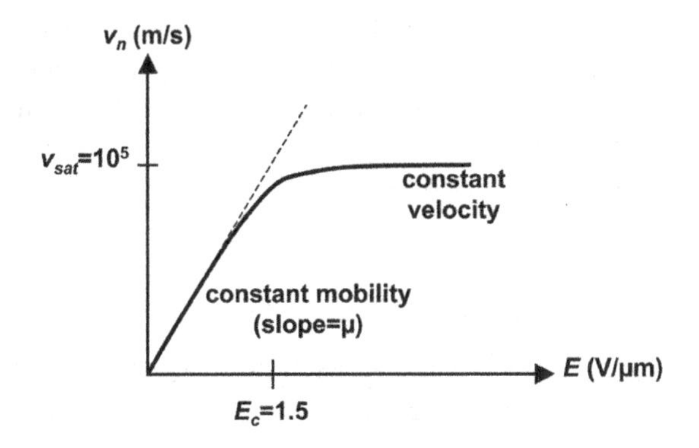

FIGURE 2.10 Illustration of velocity saturation effect.

2.5 NONCLASSICAL MOSFET STRUCTURES

Because of technological breakthroughs in process innovation, the basic structure of MOSFET devices has evolved throughout the years, although the fundamental structure has remained constant for some time. Bulk MOSFETs have altered in the drain channel and substrate design, leading to the depicted designs.

i. Lightly Doped Drain (LDD) structure: Using drain channel engineering, particle implantation is utilized to generate a gently doped channel. Thus, the electric field surrounding the channel is reduced in MOS devices and hot-electron-related quality issues are reduced.
ii. Substrate with a reversible doping pattern: There is a requirement for channel construction in this design in order to reduce the amount of mass punch through and leakage current.
iii. Structure of the ground plane with a thin coating of silicon: To grasp the properties of a retrograde substrate structure, chemical vapor deposition technique-based nuclear layer innovation of silicon on a mass substrate is substituted for embed innovation [17,18].
iv. Halo doping: The DIBL (drain-induced barrier lowering) influence can be reduced by creating a halo-doped pocket near the source and drain channels. A current expansion overflow might result as a result of band-to-band tunneling between the source and drain sides.

2.6 ENGINEERING TECHNIQUES

As microelectronics evolved into nanoelectronics, high leakage current and SCEs become a bottleneck to circuit designers. Colinge [19] has investigated a number of nonclassical devices, including double-gate MOSFETs, dual-material-gate MOSFETs, and surrounding-gate MOSFETs. But they face the thermionic limit of subthreshold slope limited to 60 mV/decade, which slowed down supply voltage scaling. Currently, the development of very large-scale integration (VLSI) technology is mostly directed toward the miniaturization of semiconductor devices. Scaling can be defined as the reduced feature size, which generally leads to better and faster performance and more gate per chip. With the help of scaling, more complex machines have been built with high speed. Nanostructures with high functionality, high device drive, and low power dissipation emerge as a result of device downscaling, which is critical for the advancement of electronics.

2.6.1 Multigate Architectures

The progressing semiconductor industry has greatly revolutionized human lives. Owing to various geometrical scaling issues in a semiconductor device, researchers are taking efforts to put forth new concepts of gate metal work function engineering. The ON state current is greatly improved by combining two or three distinct materials with various work functions in the gate metal. For several decades, planar bulk transistors have been an integral part of ICs, during which the size of these transistors have steadily decreased. Scaling the device dimensions has reached its physical limit and is on the verge of stagnation in various new technologies. The major limitations include undesirable SCEs, subthreshold leakage current, extreme heat dissipation, and power consumption of the device. To surmount all of these short-channel issues, the best solution can be to increase the electrostatic gate control over the channel. The gate has to be strong in establishing its control over the movement of electrons from source to drain. Therefore, double-, triple-, and surrounding/gate-all-around structures have been proposed to effectively suppress the subthreshold conduction and its corresponding leakage current. In general, having more than one gate can enhance the device characteristics, and such a configuration is called multigate device as shown in Figure 2.11.

The channel is entirely enclosed by several gates on many surfaces in a multigate device. There is an increase in the ON state current when many gates are used. With this structure, higher gain and transconductance ratio is achieved. Higher transconductance increases the electron generation efficiency and reduces the static power dissipation of the device. Improved mobility and reduced SCEs may be seen in Figure 2.12 using the double-gate (DG) MOSFET with top and bottom gates. For lightly doped devices, the subthreshold slope of DG MOSFET is less [20]. The front and back gates are biased separately providing more drive current. The gate electrode is highly successful at controlling the channel voltage; therefore, leakage current is kept to a minimum.

In triple-gate (TG) MOSFET, the channel is surrounded by the gate material on three sides as shown in Figure 2.13. As the number of gates increases, the electric field from the lateral and vertical side exerts more control on the top and bottom side of the channel. With this, the subthreshold swing of the transistor can be increased. Gate metal work function engineering is used in a variety of gate topologies to improve the overall performance of the device. The ON state current can be dramatically improved in gate metal engineering by combining two or three materials with varying work functions.

Figure 2.14 shows how the gate metals M1 and M2 are positioned with relation to the source and drain ends, respectively. The work function of the gate metals on

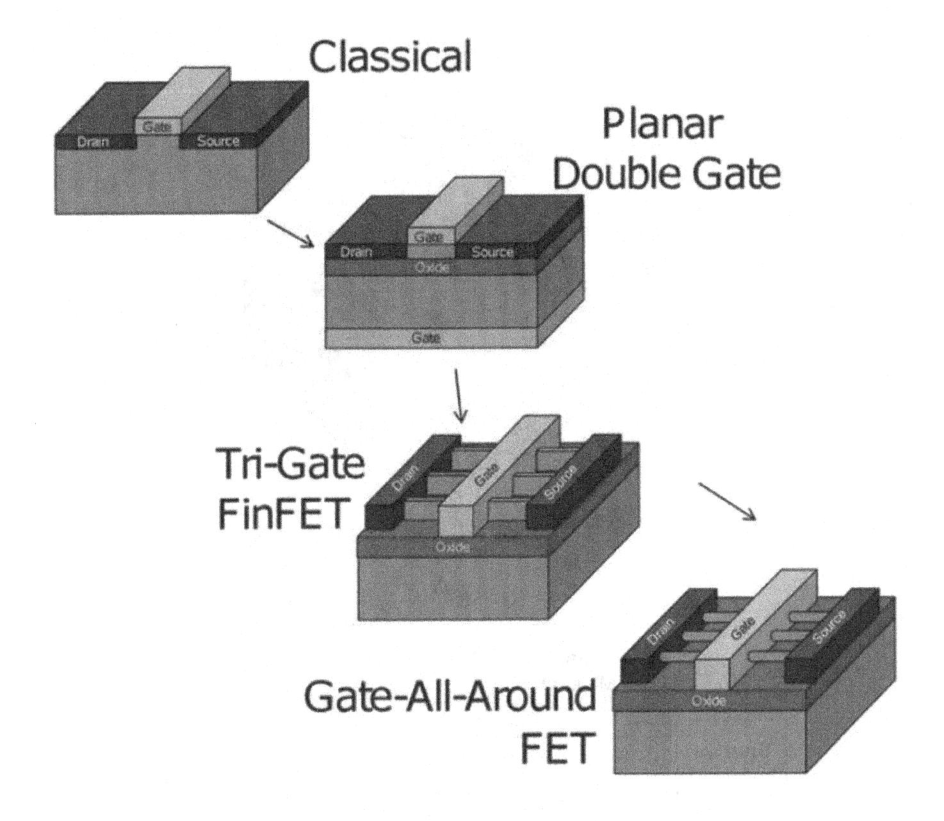

FIGURE 2.11 Various multigate configurations.

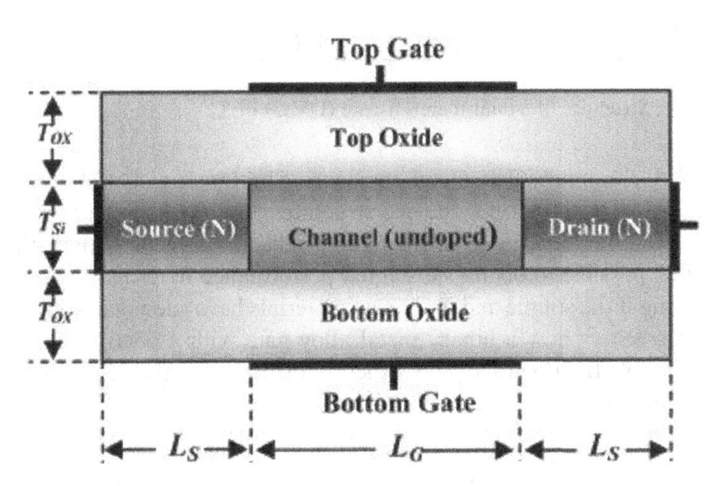

FIGURE 2.12 Schematic view of a double-gate MOS transistor.

either side of the source and drain is different. The drain end gate metal has a lower peak electric field near the drain because it has a smaller work function. As a result, many SCEs will be reduced.

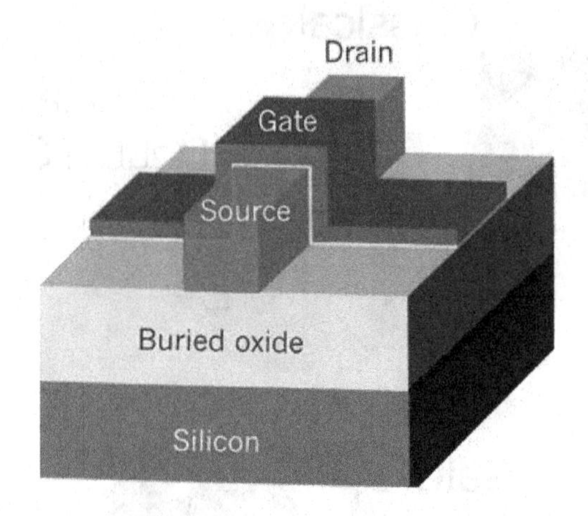

FIGURE 2.13 Schematic view of a trigate FET.

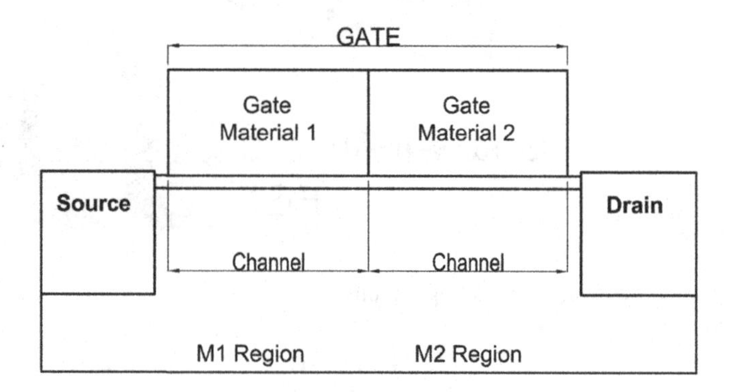

FIGURE 2.14 Structure of a dual-material-gate (DMG) FET.

Figure 2.15 shows how gate metal engineering and multigate architectures have been employed to construct new device structures. Both the top and bottom gates of the DMG DG TFET, as seen in the figure, include materials with distinct work functions. It is possible to obtain substantial performance in terms of minimal subthreshold swing if the source and drain gate materials have adequate work functions.

Figure 2.16 shows how a binary metal alloy gate with a continuous horizontal fluctuation of mole fraction in a MOS structure incorporates the new notion of work function engineering.

2.6.2 Gate Oxide Engineering

Gate oxide is the most significant element in designing short-channel transistors. The study of metal oxide semiconductor (MOS) breakdown and most of the short-channel issues begin with the gate oxide material. Gate oxide/gate dielectric region is the most fragile region as it has become extremely thin due to scaling. Downscaling

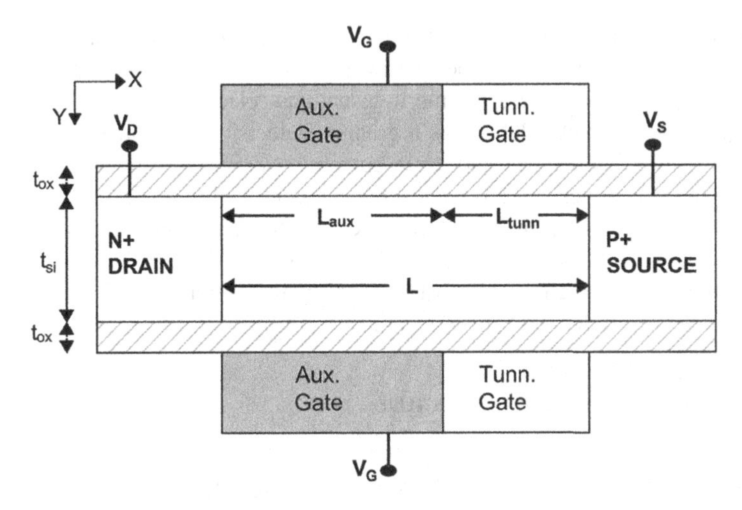

FIGURE 2.15 Schematic cross-sectional view of DM-DG TFET.

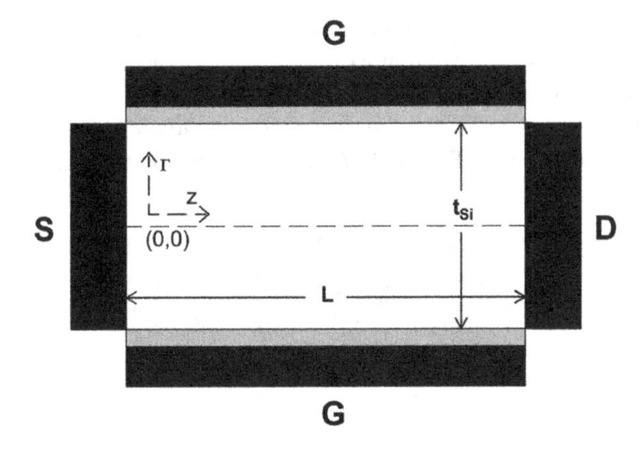

FIGURE 2.16 Schematic cross-sectional view of cylindrical MOSFET.

the thickness of gate oxide creates numerous problems such as hot carrier effect, subthreshold conduction, and threshold voltage roll-off. These issues make the device more susceptible to subthreshold leakage current and high dissipation of heat, and ultimately, this leads to the breakdown of the device. The present state of the art of the semiconductor industry is to utilize alternate gate dielectric materials and hetero-gate dielectric layers to enhance the device reliability.

2.6.3 High-K Gate Dielectrics

Many years ago, silicon dioxide was the preferred gate oxide material (SiO_2). Silicon dioxide has been steadily thickened to improve the driving current by shrinking transistors owing to gate oxide scaling. A high-K gate dielectric material can be used to increase gate capacitance by substituting silicon dioxide. As a result of this, certain dielectric materials have higher permittivity than silicon dioxide. Since high-K gate dielectric materials are used, the industry has "electric charge-holding capacity" [21]. Titanium oxide (TiO_2) has a greater dielectric constant, which increases the ratio of transconductance and drain current. An increase in carrier generation efficiency in the channel area has been achieved by replacing standard gate dielectric material (silicon dioxide, SiO_2) with high-K gate dielectric material. Smaller oxide thicknesses and high-K gate dielectric materials are used to remove the effect of the drain on the source side. Higher substrate doping and extremely efficient gate oxide materials are used to increase the current drive in FETs.

2.6.4 Stacked Gate Oxide Structure

The gate has increased electrostatic control over the channel when using high-K gate dielectric material. The only drawback in using high-K gate dielectric material is its imperfect expression due to the occurrence of interface traps, fixed bulk charges, and phase stability issues. Also, replacing SiO_2 with high- K gate dielectrics may result in difficulties in oxidizing other dielectric materials. Oxidation of silicon dioxide is direct and easier, as SiO_2 is a native oxide of silicon. But the oxidation of high-K materials involves a different oxidation process; hence, a novel approach of gate stack architecture is proposed. It is also known as a heterodielectric gate stack. Any high-K dielectric material may be generated by oxidation on top of the standard silicon dioxide, illustrated in Figure 2.17, as the foundation material. In the stacked

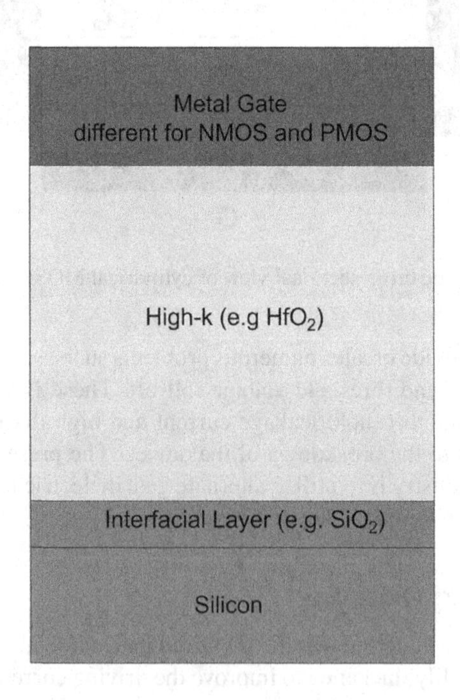

FIGURE 2.17 Gate stack architecture.

gate design, an ultrathin low-K interlayer is used in conjunction with a high-K layer to enhance device stability. Device properties may be affected because of the higher electrical focus on the low-K dielectric layer due to continuous displacement of the low-K dielectric layer by the fringing fields.

2.6.5 Channel Engineering

SCEs can also be reduced using the channel engineering technique. This includes halo doping and nonuniform channel profiles in a lateral direction. Using this approach, the threshold voltage's dependency on the channel length may be reduced. There are asymmetrical and symmetrical implants at the source or drain. There are two types of halo implants: vertical and slanted. A gate pattern is generally finalized before adding them. As a result, the implants provide a greater barrier between source drain junctions and the channel. The threshold voltage reduction in miniaturized transistors is the commonly known notification of the SCEs. This unfavorable roll-off effect is possibly the most frightening roadblock in the upcoming MOSFET design. On the other hand, for channel lengths below 50nm, DG MOSFETs still show evidence of considerable leakage currents. To overcome these SCEs, different channel engineering techniques have been deployed [22]. To reduce the threshold voltage roll-off effects, lateral channel engineering has been used to increase the doping concentration in the channel at source/drain junctions. Figure 2.18 shows the halo-implanted MOSFET channel area. The channel length is designated as L, and L is expanded as shown in equation (2.1)

$$L = L_{halo} + L_{channel} \tag{2.1}$$

where L_{halo} is the length of the halo-implanted and heavily doped channel region and $L_{channel}$ is the lightly doped channel region.

$$N_{halo} = N_{channel} + \Delta_N \tag{2.2}$$

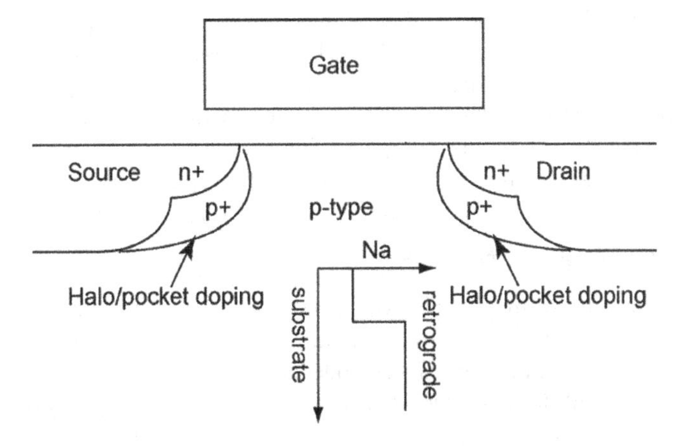

FIGURE 2.18 Halo-implanted MOSFET channel.

where N_{halo} is the doping concentration of the halo region, $N_{channel}$ is the doping concentration of the channel, and Δ_N is the difference in the channel doping measured in cm^2.

2.7 MODELING TECHNIQUES OF LOW-POWER DEVICES

Due to significant advances in semiconductor technology over the past 15 years, device modeling approaches have become increasingly popular. Instrumental in the design of verylarge-scale integrated (VLSI) devices and very high-speed IC devices for computers, device modeling is a critical component of design optimization. When it comes to new device architectures, modeling is a powerful tool. To develop transport equations, device modeling approaches can benefit from accurately defining the device shape, doping profile, carrier transport equations, and material properties of a device. You may use a number of analytical and numerical approaches to develop smaller semiconductor devices. Solving an equation (or a set thereof) that describes the behavior of real-world systems is required for a numerical or mathematical model. A large number of differential equations cannot be solved, even with starting and stopping conditions. These models require two- or three-dimensional solutions to solve the semiconductor transport equations. Especially when dealing with complex problems and real-world settings, device modeling is essential for success.

2.7.1 Analytical Modeling

In a mathematical model, the equations are derived directly from the device's physical attributes, which is called an analytical model. Many analytical models may be used to study the semiconductor device's transport mechanism. There is a charge sheet model based on surface potential analysis that has been around for quite some time. All working regions of a device may be accurately calculated using these models since they are continuous in all operating areas. Surface potential under the specified boundary conditions is calculated using complex equations including transcendental expression and multifrequencies in this model. Using a semiconductor device equation approximation, the findings of the second type of analytical model are investigated. To represent the device in its many working states, a variety of mathematical equations are required [23–25]. In addition to studying the behavior of first-order devices, these features are also utilized to analyze the impacts of higher-order devices. Sometimes, these models are referred to as semiempirical analytical models (SEAMs). Modeling the relationship between a physical process and a device's geometrical structure is a key advantage of this paradigm.

2.7.2 Quasi Transport Modeling

Carrier transport model may reach its ballistic limit, as semiconductor devices are scaled down. The widely examined quasi-ballistic transport model is used in GaAs N^+-N-N^+ structures. The quasi transport model describes the transport characteristics of electrons in a well-designed MOSFET. The current voltage (I_D–V_D) relations are derived based on Landauer's model for current through a ballistic conductor. Once, the channel length becomes comparable with the mean free path of the carriers, the analysis of nanoscale transistors becomes a quasi-ballistic regime [26]. The scattering events can also be explored with quasi transport model, which uses fewer fitting parameters to match the I_D–V_D characteristics. The model finds its practical applications in radio frequency (RF), microwave, and analog operations. For highly scaled FETs, the quasi transport model provides new insight to analyze the ballistic transport nature.

2.7.3 Charge-Based Modeling

Legitimate modeling of nanoscale MOSFET does require accurate and compact physics-based models. These models are useful for realizing VLSI circuit simulation. The modeling principles of nanoscale devices differ from those of conventional bulk MOSFETs as volume conduction is considered. At all operating regimes, the depletion charge and charge sheet approximation will not be included in the Poisson's equation for undoped DG and GAA FETs. Instead of charge sheet approximation, gradual channel approximation (GCA) can be used to obtain the expression for surface potential and channel current [27]. At present, many accurate models have been developed for undoped DG and GAA FETs using the above principles. These models show good agreement with numerical simulations. Here, the SCEs are neglected, and better electrostatic control of the channel is established. Recently, researchers have started to explore models by enveloping the SCEs in multigate MOSFETs.

2.7.4 Numerical Modeling

To design and optimize novel semiconductor devices, numerical device simulation is considered to be the most promising method. The added advantages in using numerical simulations are the calculation of the electrical behavior and inner-electronic values, fault detection in technological process, and the subsequent cost-saving of implementation. With rigorous miniaturization in the semiconductor device technology, we need to upgrade the conventional simulation methods [28]. Upgrading the current practices is necessary as several quantum mechanical effects appear in the nanoscale device structures. To have a better understanding of quantum well-based semiconductor devices, quantum hydrodynamic (QHD) models have been explored. For analyzing quantum mechanical-based devices, the initial wave function is used to obtain the macroscopic behavior of electrons. The hydrodynamic models describe continuous electron and hole distribution in a semiconductor device. The microscopic/macroscopic models based on the self-consistent solution of the Schrodinger and Poisson equations deliver a clear description of the electron distribution as the solution is highly time-consuming. But the quantum hydrodynamic model takes into account the quantum mechanical effects at adequate computation times.

2.7.5 Finite Element Modeling

The finite element method (FEM), used for decades, is a typical numerical method for addressing engineering and mathematics problems, which include heat transmission, fluid movement, and mass transport, to name a few. FEM results in a system of algebraic equations when it is formulated. A bigger system is subdivided into smaller and simpler pieces called finite elements. They are then integrated into a bigger system of equations using the finite components that have been previously created [29]. Variational techniques are used in FEM to estimate the solution by linking an error function with the solution. As part of FEM, a link between an error function and the solution is made using variational approaches to estimate the solution. You may estimate a boundary value problem using finite element analysis (FEA). It is sometimes referred to as a field problem.

2.7.6 Finite Difference Modeling

The finite difference method (FDM) is another numerical method for solving differential equations. As early as the 1950s, finite difference approaches began to gain popularity, providing engineers with a useful foundation for tackling difficult engineering issues of many types. In this approach, finite differences are used to estimate the derivatives of differential equations. One of the simplest and oldest ways to solve differential equations is to use finite difference approximations. Space or time intervals are used to calculate an approximation of the answer. As illustrated in Figure 2.19, a numerical solution based on finite differences provides us with values at discrete places known as grid points. Uniform or nonuniform grid spacing can be used for both the x and y axes of the grid. In addition, grid points can be spaced at different distances from one another. In order to make programming easier and produce more accurate results, the majority of the analysis assumes equal spacing in each direction for most of the calculations [30]. FDM's main premise is to replace the derivatives of governing equations with algebraic differential equations instead of derivatives.

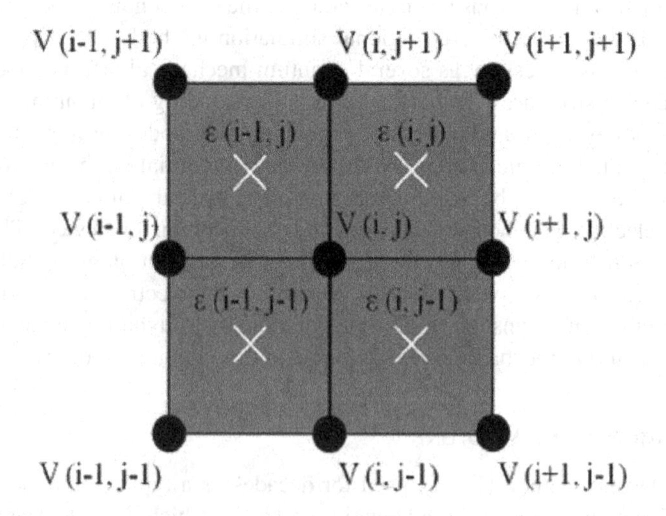

FIGURE 2.19 Discretized grid points.

2.7.7 Monte Carlo Modeling

The Monte Carlo technique, a class of computer methods, was created in the 1940s. Many complicated issues may be solved using this method. The Monte Carlo technique relies on a series of random samples. In 2006, Kurosawa presented approximate general solutions, and it is not a statistical instrument but a general approximation of answers. There is no analytical or numerical solution to the problem or it is too complicated to be executed;, therefore, they are commonly utilized.

As a general rule, Monte Carlo techniques follow the stages outlined below:

- determine the statistical characteristics of the potential inputs,
- have many possible inputs to consider,
- can use these sets to perform deterministic calculations, and
- perform a statistical analysis.

2.8 DOUBLE-HALO GATE-STACKED TRIPLE-MATERIAL DOUBLE L-GATE TFET: PERFORMANCE ANALYSIS

The diagram of a dual-halo stacked oxide triple-material double-gate tunnel FET is given in Figure 2.20, with L1 = 5 nm, L2 = 10 nm, L3 = 20 nm, L4 = 10 nm, and L5 = 5 nm. It is made up of a triple-gate material with unique work functions.

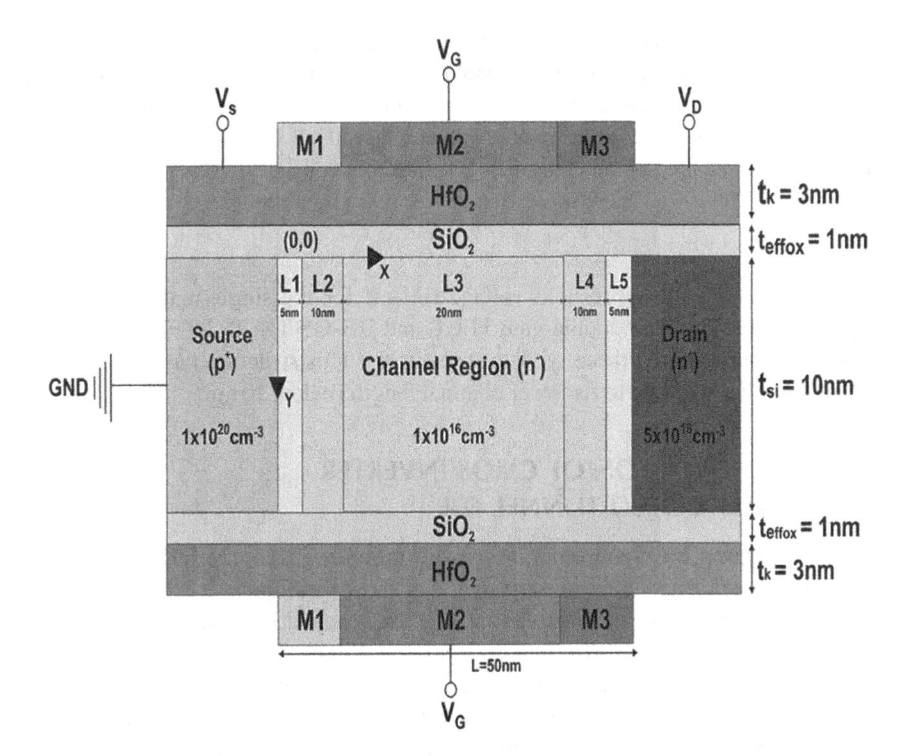

FIGURE 2.20 Cross-sectional view of DH-GS-TM-DG-TFET.

A dual-gate schematic with triple-gate material incorporates the innovative dual-halo doping symmetrical doping technique with the gate stacked to produce a novel device structure. The tunneling gate work function of titanium is 4.2 eV, whereas the control gate work function of the metal molybdenum is 4.6 eV, and the auxiliary gate function of the metal silver is 4.0. M2 must be greater than M1 and M3 so that the I_{ON}–I_{OFF} ratio can be as high as feasible.

2.8.1 Threshold Voltage Modeling

For example, a gate voltage applied in region 1 defines a threshold voltage when the energy barrier begins to saturate. Below the gate electrodes, copper, gold, and silver are used to create channels. For the second metal, the maximal work function is given by equations (2.3) and (2.4).

$$\varphi_3(L_1 + L_2 + L_3, 0) = \varphi_4(L_1 + L_2 + L_3, 0) \tag{2.3}$$

$$\varphi_{si}(x) = \frac{M_{ie}^{\tau x} + N_{ie}^{-\tau x} + \frac{\beta_i}{\tau^2} + \frac{r'_{oc}}{4}\left(V'_{GS} - V'_{fbi}\right)}{\left[1 + \frac{r'_{oc}}{4}\right]} \tag{2.4}$$

Metal 2 has the lowest channel potential. It follows thus that the amount of energy barrier is the greatest here. As a result, the threshold voltage action of the device is dependent on the channel region of the transistor. In other words, the flat band voltages in regions 2 and 3 are the same. According to mathematics, V_t roll-off is the difference between short-channel and a long-channel tunnel FET threshold voltages. In other words,

$$V'_{roll-off} = V'_{th} - V'_{thL} \quad (5)$$

Figure 2.21 displays the threshold voltage roll-off for the single-material double-gate TFET, triple-material double-gate TFET, and DH-GS-TM-DG-TFET, as well as comparisons between the three types of transistors. This structure has a lesser roll-off than the other two due to its lower channel length (below 10 nm).

2.9 IMPLEMENTATION OF CMOS INVERTER IN HALO-DOPED TUNNEL FET

Using an inverted CMOS amplifier, as shown in Figure 2.22a, the DH-GS-TM-DG TFET circuit performance is investigated. It is shown in Figure 2.22b as a voltage transfer characteristic of an inverter that uses halo-doped gate-stacked TFET and various dielectric materials. We can determine the gain by determining the slope of the voltage transfer characteristic. Table 2.1 shows the voltage gain of the CMOS inverter circuit for a variety of high-K dielectric materials. The gain of this circuit is 0.82 using HfO_2 as the gate oxide material, compared to 0.72 with SiO_2. A circuit with nanosized HfO_2 gate dielectric improves the performance of this inverter.

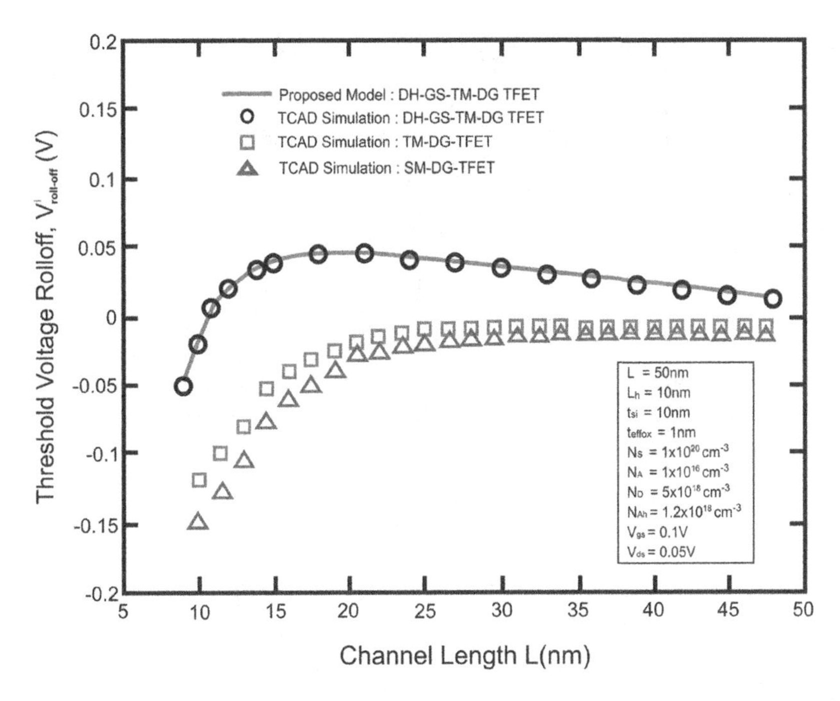

FIGURE 2.21 Deviation in threshold voltage roll-off with channel length for three different structures, with $V_{gs} = 0.1$ V and $V_{ds} = 0.05$ V.

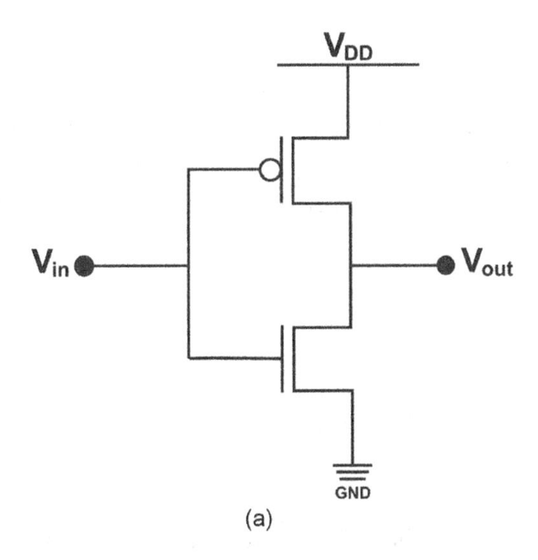

FIGURE 2.22 (a) CMOS inverter amplifier using DH-GS-TM-DG TFET. (b) Comparison of voltage transfer characteristics of the CMOS inverter with DH-GS-TM-DG TFET with different high-K dielectric materials.

(Continued)

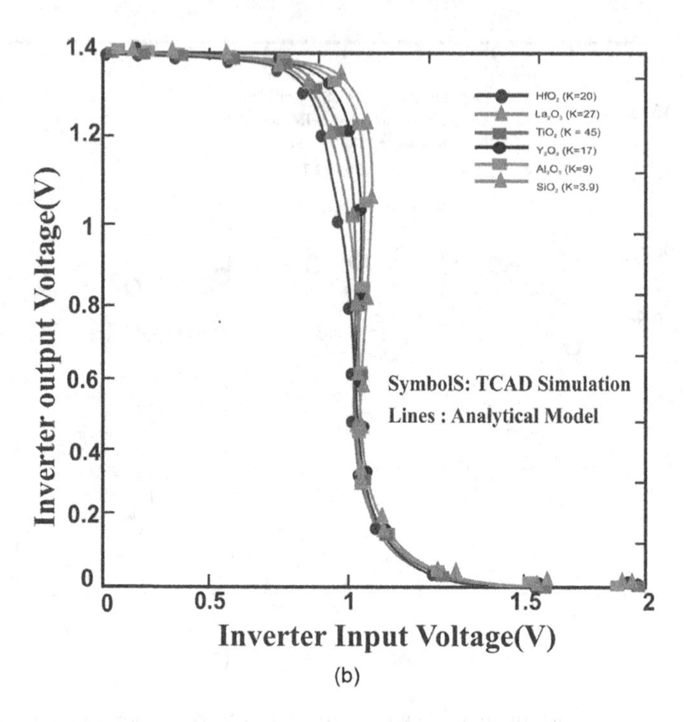

FIGURE 2.22 (*CONTINUED*) (a) CMOS inverter amplifier using DH-GS-TM-DG TFET. (b) Comparison of voltage transfer characteristics of the CMOS inverter with DH-GS-TM-DG TFET with different high-K dielectric materials.

TABLE 2.1
Dielectric Materials Utilized in CMOS DH-GS-TM-DG TFET Inverters Have Different Voltage Gains

Dielectric Material	Dielectric Value (k)	Voltage Gain V/v
SiO_2	3.9	0.72
Al_2O_3	9	0.78
Y_2O_3	17	0.80
HfO_2	22	0.82
$LaAlO_3$	25	0.83
TiO_2	40	0.86

2.10 PERFORMANCE INVESTIGATION OF DH-GS-TM-DG TFET

The device simulation incorporates mobility, bandgap narrowing, recombination, nonlocal BTBT, and Kane's tunneling models to illustrate the carrier transport mechanism in DH-GS-TM-DG TFET. Table 2.2 lists the device parameters that have been used for the proposed device.

TABLE 2.2
Parameters Used for DH-GS-TM-DG TFET Device Simulation

Parameters	DH-GS-TM-DG-TFET
Channel length, L	50 nm
Source doping, N_S (p-type)	1×10^{20} cm^{-3}
Drain doping, N_D (n-type)	5×10^{18} cm^{-3}
Halo doping concentration (N_{Ah})	1.2×10^{18} cm^{-3}
Channel thickness (t_{si})	10 nm
Metal M1 work function ($\phi M1$)	4.2 eV
Metal M2 work function($\phi M2$)	4.6 eV
Metal M3 work function ($\phi M3$)	4.0 eV
Effective oxide thickness (t_{effox})	1 nm
Exterior fringing field capacitance, C_{11}	0.63 fF
Direct overlap capacitance, C_{22}	3.23 fF
Internal fringing field capacitance, C_{33}	1.06 fF
Channel width of the capacitance, W_C	18.07 fF
Body effect coefficient, γ	0.58

2.10.1 Surface Potential Analysis

Figure 2.23 shows the change in surface potential from source to drain. Due to the device's gate engineering, there are two sets of rules in the surface potential. No matter how high the drain potential of metal 1 is, metal 2's surface potential is lower. This is what happens when the device's threshold voltage is modeled. At this point, activation of the channel is prevented by a high-energy barrier peak. As a result, the drain-induced barrier lowering is not significant.

2.10.2 Lateral Electric Field Analysis

Devices with different channel potentials have a different electric profile, which is due to the difference in drain to source channel voltage that creates the lateral electric field profile. Tunneling affects the vertex of the lateral electric field at the channel–source contact, as seen in Figure 2.24. The DH-GS-TM-DG TFET's ambipolar behavior is reduced because of the device's sluggish lateral electric field on the drain side.

2.10.3 Effective Oxide Thickness

For various gate-stacked dielectric material thicknesses, the threshold voltage deviation is depicted in Figure 2.25 using the effective oxide thickness (t_{effox}). There is also a correlation between dielectric materials in the gate stack architecture and their thickness variation.

2.10.4 Threshold Voltage Roll-Off

Threshold roll-off voltage for two nonidentical drain biases is shown in Figure 2.26. A bigger drain bias results in a shorter channel effect and a higher threshold voltage

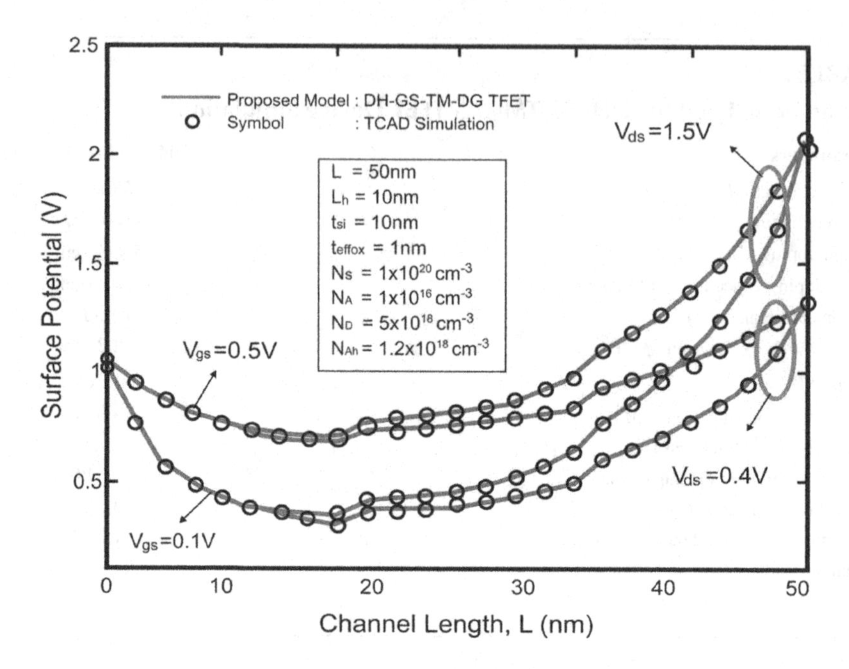

FIGURE 2.23 Surface potential profile of DH-GS-TM-DG TFET for different values of $V_{gs}=0.1$, 0.5 V and $V_{ds}=0.4$, 1.5 V.

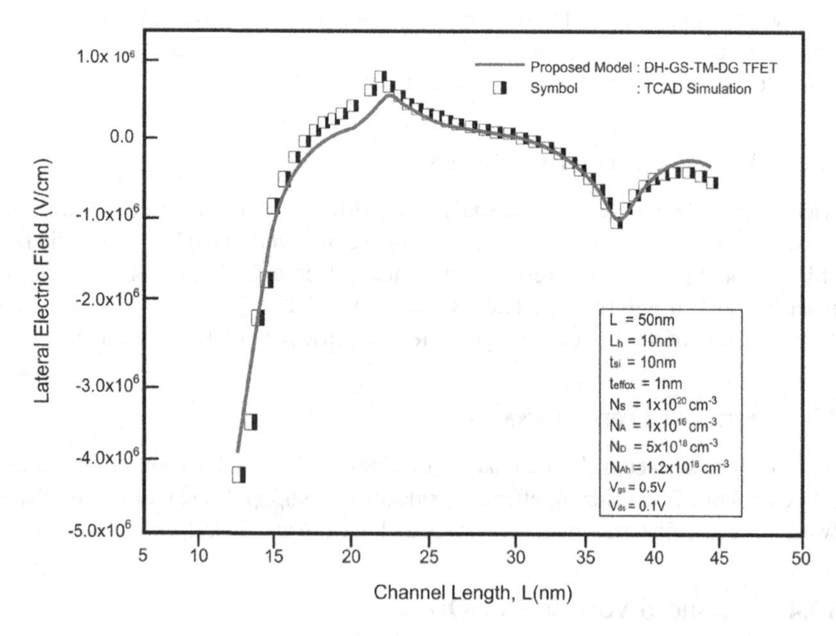

FIGURE 2.24 Lateral electric field profile of DH-GS-TM-DG TFET for $V_{gs}=0.1$ V and $V_{ds}=0.5$ V.

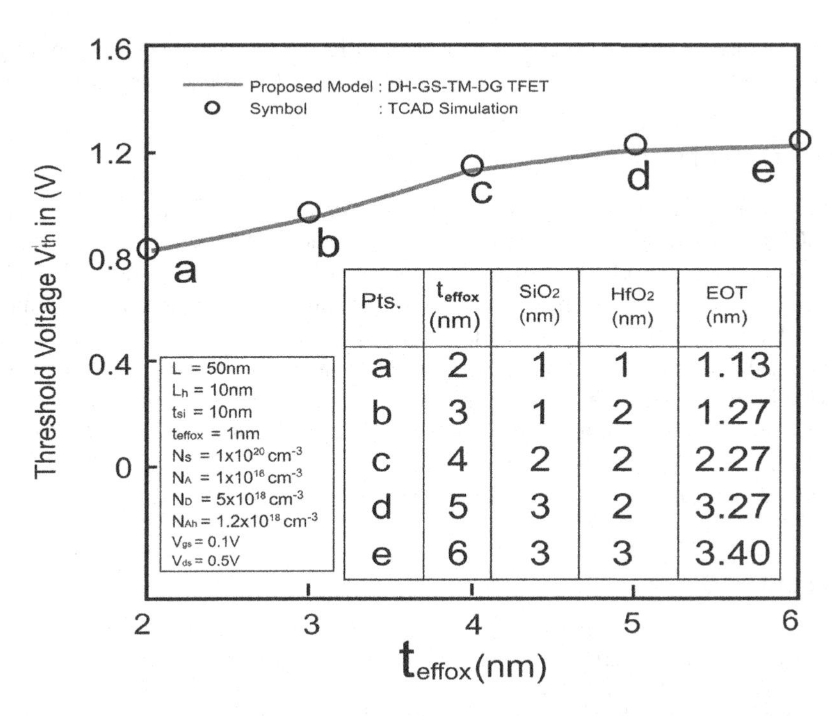

FIGURE 2.25 Variation of threshold voltage versus effective oxide thickness of DH-GS-TM-DG TFET.

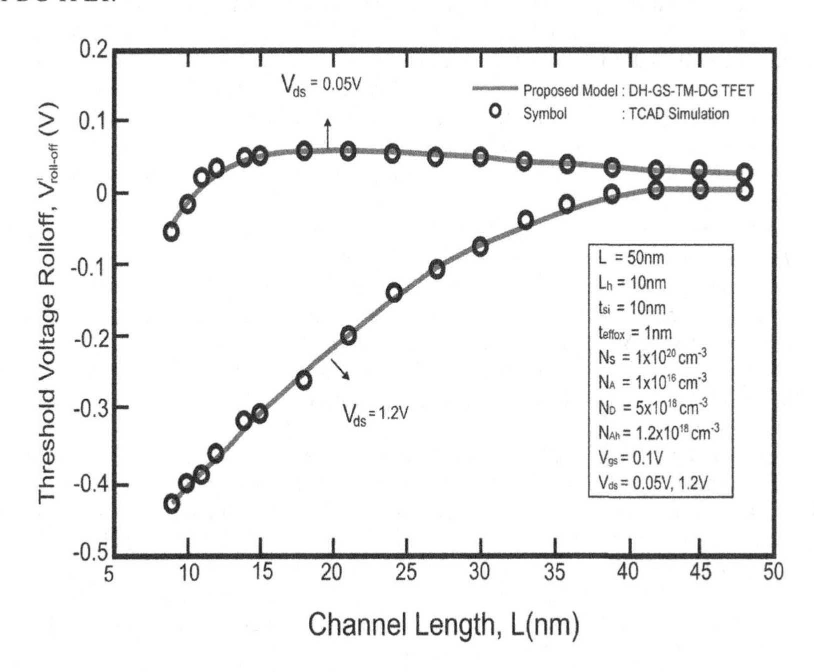

FIGURE 2.26 Plot of the threshold voltage roll-off versus channel length for DH-GS-TM-DG TFET for two different drain bias values, V_{ds} = 0.05 and 1.2 V.

roll-off voltage, as seen in the graph. As a result, DIBL (drain-induced barrier lowering) becomes more aggressive because of the significant drain bias, rather than the gate bias.

2.11 CONCLUSIONS

The tunnel FET has demonstrated to be an effective low-power device in suppressing the SCEs. Traditional FETs must be updated by adopting modern gate and channel engineering approaches in order to follow Moore's law and avoid SCEs. The dual-halo structure of the triple-material work function engineering is used to reduce the peak electric field at the drain end. Low-power tunnel FETs are being designed with high-K gate dielectric materials. Hafnium oxide has been substituted for silicon dioxide in order to improve the electrostatic control between the gate and channel. A substantial decrease in the electric field and threshold voltage of the DH-GS-TM-DG-TFET was observed. For ultra-ow-power and rapid-switching applications, tunnel field effect transistor has been shown to be the most promising technology.

REFERENCES

1. Yang, ES 1988. *Microelectronic Devices* (pp. 285–294). McGraw-Hill, New York.
2. Thompson, SE & Parthasarathy, S 2006. Moore's law: The future of Si microelectronics. *Materials Today*, 9(6): 20–25.
3. Skotnicki, T, Hutchby, JA, King, TJ, Wong, HS & Boeuf, F 2005. The end of CMOS scaling: Toward the introduction of new materials and structural changes to improve MOSFET performance. *IEEE Circuits and Devices Magazine*, 21(1): 16–26.
4. Suzuki, K, Tanaka, T, Tosaka, Y, Horie, H & Arimoto, Y 1993. Scaling theory for double gate SOI MOSFETs. *IEEE Transactions on Electron Devices*, 40(12): 2326–2329.
5. Kalra, S & Bhattacharyya, AB 2018. Scalable α-power law based MOSFET model for characterization of ultra-deep submicron digital integrated circuit design. *AEU - International Journal of Electronics and Communications*, 83: 180–187.
6. Moore, GE 1965. Cramming more components onto integrated circuits. *Electronics Magazine*, 3(8): 114–119.
7. Young, KK 1989. Short-channel effect in fully depleted SOI MOSFETs. *IEEE Transactions on Electron Devices*, 36(2): 399–402.
8. Venkatesh, M & Balamurugan, NB 2019. New subthreshold performance analysis of germanium based dual halo gate stacked triple material surrounding gate tunnel field effect transistor. *Superlattices and Microstructures*, 130: 485–498.
9. Veeraraghavan, S & Fossum, JG 1988. A physical short channel model for the thin film SOI MOSFET applicable to device and circuit CAD. *IEEE Transactions on Electron Devices*, 35: 1866–1875.
10. Lee, MJ & Choi, WY 2011a. Analytical model of a single-gate silicon-on-insulator (SOI) tunneling field-effect transistors (TFETs). *Solid-State Electronics*, 63(1): 110–114.
11. Venkatesh, M, Priya, GL & Balamurugan, NB 2021. Investigation of ambipolar conduction and RF stability performance in novel germanium source dual halo dual dielectric triple material surrounding gate TFET. *Silicon*, 13: 911–918.
12. Preethi, S, Venkatesh, M, KarthigaiPandian, M & Priya, GL 2021. Analytical modeling and simulation of gate-all-around junctionless Mosfet for biosensing applications. *Silicon*, 13(10): 3755–3764.
13. Venkatesh, M & Balamurugan NB 2020. Influence of threshold voltage performance analysis on dual halo gate stacked triple material dual gate TFET for ultra low power applications. *Silicon*, 13(1), 275–287.

14. Wallace, RM & Wilk, GD 2003. High-K dielectric materials for microelectronics. *Critical Reviews in Solid State and Materials Sciences*, 28(4): 55.
15. Saurabh, S & Kumar, MJ 2011, Novel attributes of a dual material gate nanoscale tunnel field-effect transistor. *IEEE Transaction on Electron Devices*, 58(2): 404–410.
16. Suveetha, P & Balamurugan, NB 2014, A 2D subthreshold current model for single halo triple material surrounding gate MOSFETs. *Microelectronics Journal*, 45(6): 574–577.
17. Suveetha, P, Balamurugan, NB, Chakaravarthi, GCV, Ramesh, RP, & Kumar BRS 2014, A 2D analytical modeling of single halo triple material surrounding gate MOSFET. *Journal of Electrical Engineering and Technology*, 9(4): 1355–1359.
18. Vanitha, P, Samuel, TSA & Nirmal, D 2019. A new 2D mathematical modeling of surrounding gate triple material tunnel FET using halo engineering for enhanced drain current. *AEU-International Journal of Electronics and Communications*, 99: 34–39.
19. Wu, J, Min, J & Taur, Y 2015. Short-channel effects in tunnel FETs. *IEEE Transactions on Electron Devices*, 62(9): 19–24.
20. Zhang, Q, Zhao, W & Seabaugh, A 2006. Low-subthreshold-swing tunnel transistors. *IEEE Electron Device Letters*, 27(4): 297–300.
21. Arun Samuel, TS, Balamurugan, NB, Bhuvaneswari, S, Sharmila, D & Padmapriya, K 2013. Analytical modelling and simulation of single-gate SOI TFET for low-power applications. *International Journal of Electronics*, 101: 779–788.
22. Narimani, KS, Glass, P, Bernardy, N, Von Den Driesch, QT, Zhao, S, & Mant, L 2018. Silicon tunnel FET with average subthreshold slope of 55 mV/dec at low drain currents. *Solid-State Electronics*, 143: 62–68.
23. Nirschl T, Wang PF, Hansch W, & Schmitt-Landsiedel D 2004. The tunnelling field effect transistors (TFET): The temperature dependence, the simulation model, and its application. In *IEEE International Symposium on Circuits and Systems (IEEE Cat. No.04CH37512)*, Vancouver. BC: III–713.
24. Kumar, D 2019. Performance evaluation of double gate tunnel FET based chain of inverters and 6-T SRAM cell. *Engineering Research Express*, 1(2): 025055.
25. Kumar D, Rahi SB, Kuchhal P 2021. Investigation of analog parameters and miller capacitance affecting the circuit performance of double gate tunnel field effect transistors. In *Intelligent Communication, Control and Devices* (pp. 335–349). Springer, Singapore.
26. Venkatesh M, Suguna M & Balamurugan NB 2019. Subthreshold performance analysis of germanium source dual halo dual dielectric triple material surrounding gate tunnel field effect transistor for ultra low power applications. *Journal of Electronic Materials*, 48(10), 6724–6734, https://doi.org/10.1007/s11664-019-07492–0.
27. Venkatesh, M, Suguna, M & Balamurugan NB 2020. Influence of germanium source dual halo dual dielectric triple material surrounding gate tunnel FET for improved analog/RF performance. *Silicon*, 12(12), 2869–2877, http://link.springer.com/article/10.1007/s12633-020-00385-6.
28. Priya, GL, Venkatesh, M., Balamurugan, NB & Samuel TS 2021. Triple metal surrounding gate junctionless tunnel FET based 6T SRAM design for low leakage memory system. *Silicon* 13, 1691–1702, https://doi.org/10.1007/s12633-021-01075–7.
29. Kumar, TV, Venkatesh, M, Muthupandian, B & Priya GL 2021. Charge density based small signal modeling for InSb/AlInSb asymmetric double gate silicon substrate HEMT for high frequency applications. *Silicon*. https://doi.org/10.1007/s12633-021-01383-y.
30. Samuel, TSA, Venkatesh, M, Pandian, MK & Vimala P 2021. Investigation of ON current and subthreshold swing of an InSb/Si heterojunction stacked oxide double-gate TFET with graphene nanoribbon. *Journal of Electronic Materials*. 50(12), 7037–7043, https://doi.org/10.1007/s11664-021-09244-5.

3 Tunnel Field-Effect Transistor

An Energy-Efficient Semiconductor Device

Chandan Kumar Pandey
VIT-AP University

Saurabh Chaudhury and Neerja Dharmale
NIT Silchar

Young Suh Song
Korea Military Academy

CONTENTS

3.1 EVOLUTION OF MOSFET

In 1928, Lilienfeld [1] filed a patent on a device named as "device for controlling current", which was later called as metal-oxide semiconductor field-effect transistor

DOI: 10.1201/9781003240778-3

(MOSFET). During early days, fabrication of MOSFET appeared to be very tough because of the difficulty in obtaining a decent interface between the semiconductor and the oxide layer. Finally, the first MOSFET based on the Si/SiO_2 interface was successfully fabricated in 1959 by Kahng and Atalla at Bell Telephone Laboratory [2]. After many years of continuous advancement and research, now a single chip is capable of containing more than 19 billion of these devices. When we look at the journey of MOSFETs, we find continued device scaling as the driving force behind its success story. Scaling of MOSFETs now follows an exponential trend although for the last five decades it was mainly governed by Moore's Law [3]. It is amazing that despite facing many technical challenges, the trend of device miniaturization for exponential growth in the device density has continued for such a long period, and at times, when this trend appeared to be discontinued, other innovative techniques were introduced to overcome the obstacles. However, this continued device miniaturization starts facing serious threats when the dimensions of the device are scaled down to the sub-10nm regime, which is mainly due to the incapability of MOSFETs to switch between OFF- and ON-states without causing excessive power dissipation.

3.1.1 Moore's Law

In 1965, Gordon E. Moore published a paper entitled "Cramming More Components onto Integrated Circuits" in Electronics magazine [3]. Based on his observation, he made a prediction on the process of miniaturization of integrated circuits and wrote in the article

> The complexity for minimum component costs has increased at a rate of roughly a factor of two per year. Certainly over the short term this rate can be expected to continue, if not to increase. Over the longer term, the rate of increase is a bit more uncertain, although there is no reason to believe it will not remain constant for at least 10 years.

In 1975, he revised his earlier prediction (as shown in Figure 3.1) after redrawing the plot of component densities from 1975 onward and stated that the device density will double every 18 months. His revised plot was later called Moore's Law [3].

3.1.2 Methods of Scaling

Scaling of MOSFETs may be defined as the process of reducing the device dimensions and interconnecting wires in such a manner that the functionality of integrated circuits (ICs) does not change. Dennard et al. [4] first proposed and validated a set of certain rules termed as Dennard's Scaling Law for complementary metal-oxide semiconductor (CMOS) device scaling. In fact, the proposed scaling law is popularly known as constant field scaling, which states that supply voltage must be scaled down by same factor with which device dimensions are scaled in order to maintain a constant field in the scaled device. However, there exists a major issue with constant field scaling, which enforces a change of the power supply voltage by a different scaling factor, thus raising a conflict with the protocols set for the requirement of chip interface. Therefore, to overcome the problem faced by constant field scaling, a different scaling mechanism known as constant voltage scaling was proposed, which suggests

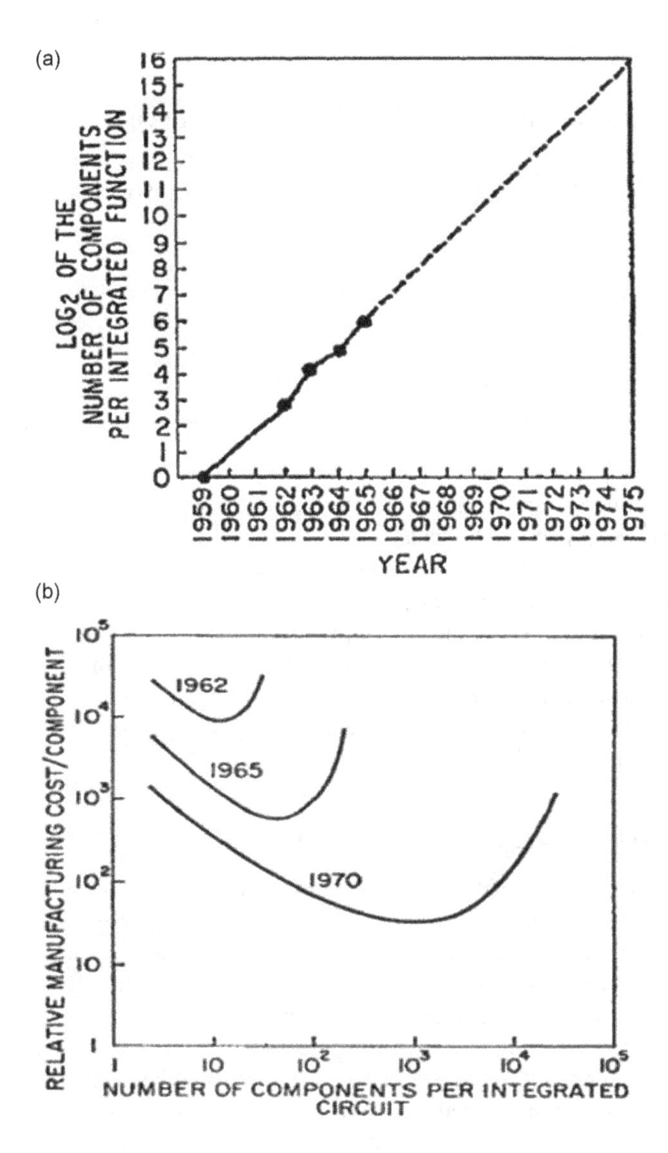

FIGURE 3.1 (a) Number of components per integrated function predicted along the years (the first phase of Moore's Law) and (b) cost per component versus number of components per circuit [3].

a scale-down of the device dimensions by keeping the voltage levels intact [5]. But the scaling of the oxide layer thickness along with an unchanged supply voltage causes a higher electric field in the channel, which eventually degrades the mobility of charge carriers. Table 3.1 lists and compares the set of rules for these two methods of scaling showing scaled parameter with their impact on MOSFET-based circuit parameters. Finally, a more generalized scaling method was introduced in Ref. [5], which suggest

TABLE 3.1
Comparison between Primary and Derived Scaling Principles

	Parameters	Constant Field Scaling	Constant Voltage Scaling	Generalized Scaling
Primary Scaling	t_{ox}, L, W	1/k	1/k	1/k
	N_A, N_D	k	K^2	αk
	V_{DD}	1/k	1	α/k
Derived Scaling	Electric field	1	k	α
	Capacitance	1/k	1/k	1/k
	Drain current	1/k	k	α^2/k

a different scaling factor for device dimensions and power supply. It was suggested that the power supply should be scaled down relatively with a smaller factor than the device dimensions to overcome the demerits faced by the other two scaling methods stated earlier. Initially, the scaling of device dimensions and power supply was mainly carried out to improve the device performances without a massive change in the structure of the device. Thereafter, many innovations were incorporated in the structure of conventional MOSFET, which allowed for it to further scale down the dimensions with improved electrostatic control of gate over the channel and with reduced short-channel effects (SCEs).

3.1.3 Challenges in CMOS-Device Scaling

As the technology scales down to sub-10 nm, quantum-mechanical effects along with SCEs, such as drain-induced barrier lowering (DIBL), gate-induced drain leakage (GIDL) , gate-leakage current due to vertical tunneling, direct tunneling of charge carriers from source to drain, threshold voltage roll-off, mobility degradation, etc., become more prominent, which adversely affects the device performance, and thus cannot be interpreted using theory of long-channel devices. For instance, when the drain voltage is high and the gate voltage is at a low potential during the OFF-state, a high electric field is induced at the drain terminal due to large V_{gd} ($\approx V_{dd}$), which further causes GIDL due to tunneling of electrons from drain to substrate. Similarly, when the channel length is scaled to a shorter length, then charge carriers directly tunnel from source to drain through the channel despite the presence of a potential barrier. As a result of these SCEs, continued device scaling for further technological enhancement becomes difficult with CMOS technology.

Apart from the abovementioned nonideal effects faced by MOSFETs during channel length scaling, there is an important fundamental physical limit faced by MOSFET, which restricts the scaling of power supply beyond a certain value. Subthreshold swing (SS), which is a measure of necessary gate voltage required to increase/decrease the drain current by a factor of 10 [5–7], mainly decides the minimum possible value of power supply to sustain the switching ability of a transistor between ON- and OFF-states. The current switching mechanism in a MOSFET is based on the thermal broadening of the Fermi distribution tail in the valance band of the source region to the conduction band of the channel region causing an injection of charge carriers over the potential barrier at the source–channel interface.

Based on the Landauer approach [8], the 1D model of drain current in MOSFETs can be expressed as:

$$I_D = \frac{2e}{h}\int_{-\infty}^{+\infty} dET(E)(f_{se}(E) - f\,\partial_{dn}(E)) = \frac{2e}{h}\int_{\Phi_f^0}^{+\infty} dE(f_{se}(E) - f_{dn}(E)) \quad (3.1)$$

where f_{se}, f_{dn}, e, and h are the distribution of Fermi–Dirac function in source and drain regions, respectively, electron charge, and Plank's constant, respectively, while Φ_f^0 represents the potential barrier faced by electrons. T(E) is the tunneling probability of charge carriers, which may be zero if $E < \Phi_f^0$ and one if $E > \Phi_f^0$. Since, the value of Φ_f^0 is much more than the chemical potentials μ_{dn} and μ_{se} in drain and source regions, respectively, f_{se} and f_{dn} can be approximated by Boltzmann distribution. Due to applied V_{ds} and with $f_{se} \gg f_{dn}$ in the subthreshold regime, f_{dn} can be neglected from equation (3.1). Using these approximations, equation (3.1) can be simplified to:

$$I_D = \frac{2e}{h}\int_{\Phi_f^0}^{+\infty} dE \exp\left(-\frac{E - \mu_S}{k_B T}\right) = \frac{2e}{h} k_B T \exp\left(-\frac{\Phi_f^0 - \mu_s}{k_B T}\right) \quad (3.2)$$

As a function of gate voltage, $\partial\Phi_f^0$ as in Ref. [9] can be expressed as follows:

$$\partial\Phi_f^0 = e\left(\frac{C_{ox}}{C_{ox} + C_{dep}}\right)\partial V_g \quad (3.3)$$

where C_{ox} and C_{dep} are oxide and depletion capacitances, respectively. From the I_D–V_G characteristics of MOSFETs, SS can be derived using equations (1.2) and (1.3) as in Refs. [7,10]; thus,

$$SS = \left[\frac{\partial \log I_D}{\partial V_g}\right]^{-1} = 1n(10)\frac{k_B T}{e}(1 + \frac{C_{dep}}{C_{ox}}) \quad (3.4)$$

At 300K, the minimum value of SS considering $C_{ox} > C_{dep}$ is approximately equal to:

$$P_D = C_{pd} \times V_{DD}^2 \times f_i$$

This value is certainly a physical lower limit of SS for the MOSFET devices, which has a severe impact on the switching characteristics of the MOSFET. A minimum of 5 orders of magnitude for the current switching ratio I_{on}/I_{off} is required for modern VLSI devices. Therefore, the gate voltage must be increased by at least 300 mV to turn on the device even if an ideal MOSFET is considered exhibiting a minimum slope of 60 mV/decade. Since thermal voltage ($^{k}\underline{B}_{e}{}^{T}$) cannot be scaled and owing to the C_{dep}/C_{ox} ratio, there is no impact of capacitance scaling. Thus, SS does is not scaled during the scaling of device dimension and power supply. As the SS exerts a limit on the minimum voltage required to switch ON/OFF the device, therefore power supply cannot be scaled beyond a certain value, which eventually causes more power dissipation in the device. A solution to reduce the supply voltage further was suggested by introducing the gate-all-around structure for the MOSFETs, which could not be sustained for continuous downscaling [11]. Therefore, no techniques other than steep-SS devices remain to be investigated for achieving further power supply voltage scaling. Since most of the devices exhibiting steep SS have not been found to show a satisfactory performance as compared to MOSFETs at nominal supply voltages, researchers need to come up with a novel device structure to match with the performance of existing CMOS technology.

3.1.4 Power Dissipation in Integrated Circuits

There are two types of power that are dissipated by the ICs. One type is static power that is dissipated by the transistors during the OFF-state is:

$$P_{static} = I_{off}.V_{dd} \tag{3.6}$$

which depends on the leakage current of the device and power supply V_{DD}. The second dissipation is dynamic power, which is consumed during the charging and discharging of C_{load} when the state of the transistors is changed:

$$P_{dynamic} = C_{load}.V_{dd}{}^{2}.f \tag{3.7}$$

The OFF-state leakage current I_{off} can be related with V_{dd} and SS as:

$$I_{off} \approx I_{on}.10^{-\frac{V_{dd}}{SS}} \tag{3.8}$$

From this equation, it is observed that when V_{dd} is decreased for a constant value of I_{on} and SS, there is a rapid increase in the value of I_{off}. Therefore, novel devices showing steeper SS are needed to achieve low OFF-state leakage current when V_{dd} is scaled down.

3.2 EMERGENCE OF NEW DEVICES

On looking back, we see that the semiconductor industry has progressed through vacuum tubes to theCMOS technology of today replacing existing technology by its successor to make the devices/circuits power-efficient [12,13]. Similarly, due to the inability of scaling power supply and device dimensions further in MOSFETs, a number of emerging devices have been investigated as a prospective successor to MOSFET and consequently to improve the existing CMOS technology. In its 2015 edition, the ITRS has already presented a road map for the emerging devices having potential to substitute the silicon technology used in conventional MOSFET devices for technological advancements. As shown in Figure 3.2 [14], the classification of devices is carried out on the basis of conventional or novel materials and/or structure and the way by which information is interpreted either as an electron or noncharge entity. A number of novel devices have been proposed of which only the devices that

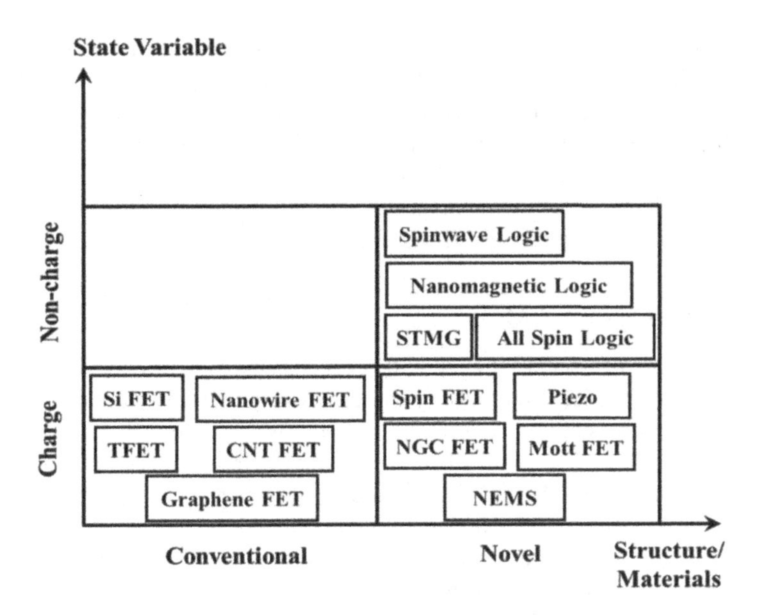

FIGURE 3.2 Emerging novel devices mentioned in the ITRS 2.0 2015 edition [14].

are most promising and relevant to the thesis are discussed in this section. Among all these devices, some are considered as the extension of the current CMOS technology while others are clearly possibilities to think beyond CMOS. Since thesis work is based on a charge-controlled device, therefore the main focus is confined to this type of devices only.

3.2.1 Devices for Extension of CMOS

Nearly 20 years ago, carbon nanotube field-effect transistors (CNT FETs) were shown to be a promising devices in the sub-10 nm regime due to their energy bandgap ranging from 0.6 to 0.8 eV and excellent transport properties of electrons and holes, which was mainly attributed to its ultrathin body of approximately 1 nm diameter. When all other devices were found to have many shortcomings when their dimensions were scaled down to the sub-10 nm regime, CNT FETs [15,16] were observed to offer solutions in most of the concerned areas.

Nanowire transistors [17–22] were found to provide the flexibility of using an extensive variety of materials such as silicon, germanium, group III-V (GaN, InN, AlN, InP, GaAs, etc.) and II-VI (CdS, ZnS, CdSe, ZnSe) compound semiconductors, and semiconducting oxides (ZnO, In_2O_3, TiO_2) as a channel in the form of a nanowire. The diameter of such semiconducting nanowires replacing the planner channel of a conventional MOSFET has been scaled to 0.5 nm. These nanowires have been found to exhibit ballistic conduction due to quantum confinement.

By employing the concept of band-to-band tunneling (BTBT), tunnel field-effect transistors (TFETs) have been demonstrated to have potential to overcome the limitation of SS to 60 mV/decade at room temperature faced by MOSFETs [22–27].

TFETs in their simplest form are just a reverse-biased gated p-i-n diode, which allow us to scale the power supply further to reduce the power dissipation. TFETs were found to show a smaller turn-on voltage as compared to MOSFETs, which may be realized by modulating the tunneling barrier at the source–channel interface with gate voltage and by modulating the electric field at the tunneling interface to regulate the barrier width. However, achieving a steep tunneling barrier at the input interface needs an extremely abrupt doping profile for TFETs.

3.2.2 Devices beyond CMOS

Negative gate-capacitance FETs [28–30] are considered to provide values of SS lower than 60 mV/decade and amplify the gate voltage similar to a step-up transformer if the thickness of the ferroelectric insulator is approximately chosen in place of oxide in MOSFETs. The main advantage of this device is its geometrical similarity with conventional MOSFETs. The ferroelectric material that is mainly used as gate dielectric is hafnium oxide doped with Al, Zr, or Si.

Spin FETs [12,31,32] are yet another class of charged-based nonconventional semiconductor devices that can further be classified into two groups: Spin FETs [32] and Spin-MOSFETs. For making the source and drain to act as a spin injector and detector, respectively, a ferromagnetic material is used for both the regions in spin-TFETs and spin-MOSFETs. Despite the similarities in the structure of both the spin transistors, they work on different operating principles. In Spin FETs, the gate is used to regulate the spin direction of charge carriers with the help of Rashba spin–orbit interaction, whereas in spin-MOSFETs, the gate works exactly the same as in conventional MOS devices to switch on/off the current in the channel. Both the spin transistors have been demonstrated storing nonvolatile information with the help of magnetization configuration, which is a great advantage using these devices.

3.3 TUNNEL FIELD-EFFECT TRANSISTORS (TFETs)

As mentioned in the previous section, a TFET is basically a gated p-i-n diode in which source and drain regions are doped with an opposite type of dopants. The cross-sectional view of a TFET is shown in Figure 3.3. The biasing conditions for p- and n-type TFETs are tabulated in Table 3.2. In contrast to MOSFETs, the transport mechanism of charge carriers in a TFET device depends on BTBT through barriers at the source–channel interface instead of thermal emission of charge carriers over the barriers. The operating principle of a TFET is mainly governed by the modulation of electrostatic potential and field at the source–channel interface for

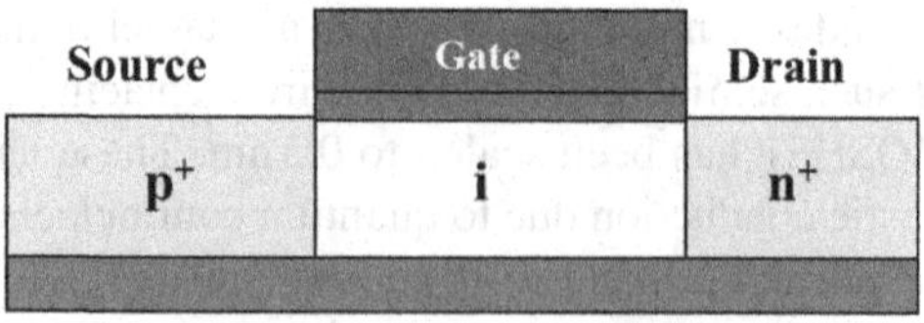

FIGURE 3.3 Cross-sectional view of a n-type TFET.

TABLE 3.2
Modes of Operation in TFET and Convention of Source and Drain Regions and Biasing Conditions

Type of TFET	Mode of Operation	Source	Drain	Bias
p-type	ON	N^+	P^+	$V_{gs} < 0; V_{ds} < 0$
	OFF			$V_{gs} > 0; V_{ds} < 0$
n-type	ON	P^+	N^+	$V_{gs} > 0; V_{ds} > 0$
	OFF			$V_{gs} < 0; V_{ds} > 0$

disabling or enabling the BTBT process while the operating principle of MOSFETs relies on modulation of charge carriers in the channel. Again, the OFF-state current in MOSFETs is mostly due to the remaining charge carriers in the channel, whereas the subthreshold leakage current in TFETs is due to tunneling at the channel–drain interface during the OFF-state. This is why TFETs unlike MOSFETs do not suffer from the limitation of SS of 60 mV/decade at 300 K. Due to steeper SS than 60 mV/decade, TFETs allow for reduction in power supply voltage while keeping the OFF-state current at a low value. Maintaining a lesser supply voltage results in a small active power consumption while a low OFF-state current leads to a low static power dissipation in TFETs. Eventually, this property of TFET makes it a more suitable transistor for low-power applications than conventional MOSFETs. The energy band diagram of Figure 3.4 shows the edges of conduction and valance band in p-channel TFET. When a negative voltage is applied to the gate terminal, then the band in the channel region is lifted up, which leads to an overlap of conduction band energy of the source with valance band energy of the channel region. As shown in the band diagram, a tunneling window is confined at the source–channel interface, which acts as a band-pass filter for the charge carriers trying to tunnel from the source to the channel region. The energy bandgap of the channel region filters out the high-energy tail of

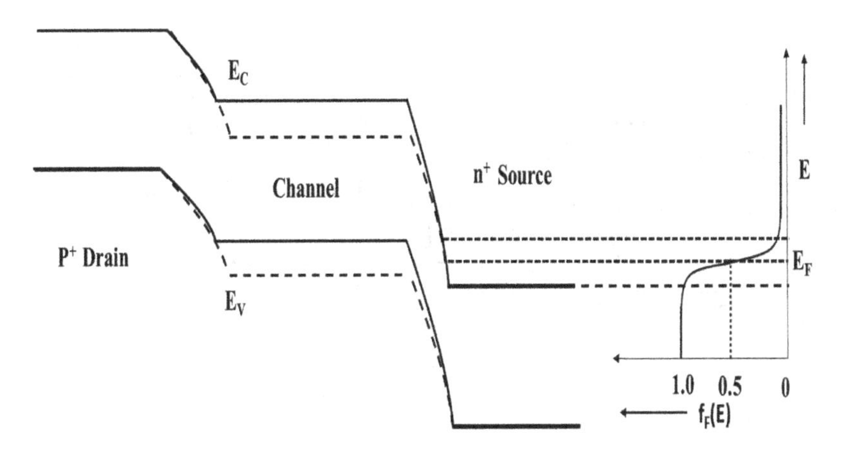

FIGURE 3.4 Energy band diagram of p-type TFET showing the band-pass filter property of TFET during the ON-state with the help of the Fermi–Dirac distribution function.

the Fermi–Dirac distribution function while the energy bandgap of the source region prevents the charge carriers with energies in the low-energy tail of the Fermi–Dirac distribution function. This particular arrangement of energy band edges of TFETs allows for only the charge carriers having energy around the Fermi-energy level to tunnel through the barrier at the source–channel interface [33]. As the change in the Fermi–Dirac distribution function around the Fermi-energy level is high as compared to the exponential tails, most of the charge carriers contributing conduction current must have energy around the Fermi-energy level to a steeper SS in TFETs.

3.3.1 Tunneling Probability

In contrast to classical mechanics, particles, such as electrons and holes, are treated as a wave function in quantum mechanics, which penetrate through the potential barrier rather than terminating on a finite potential barrier as considered in classical mechanics. In TFETs, the energy barrier formed at the source–channel interface is approximated by a triangular shape and the expression for tunneling probability is formulated by applying Wentzel–Kramers–Brillouin (WKB) approximation. The ON-state current in TFETs mainly depends on the value of this tunneling probability, which defines the probability that a charge carrier can tunnel through the triangular potential barrier of a finite width formed at the source–channel interface. The energy band edges of a p-TFET at the source–channel interface during ON- and OFF-states are shown in Figure 3.5. It can be clearly seen from the band diagram that there is no band overlapping between the conduction band of the source and the valance band of the channel during the OFF-state while a tunneling path is created at the source–channel interface during the ON-state due to a band-to-band overlapping of $\Delta\Phi$. Tunneling probability based on WKB approximation [34] can be expressed as:

$$T_{WKB} = exp\left(-2\int_{x1}^{x2} |k(x)|\, dx\right) \tag{3.9}$$

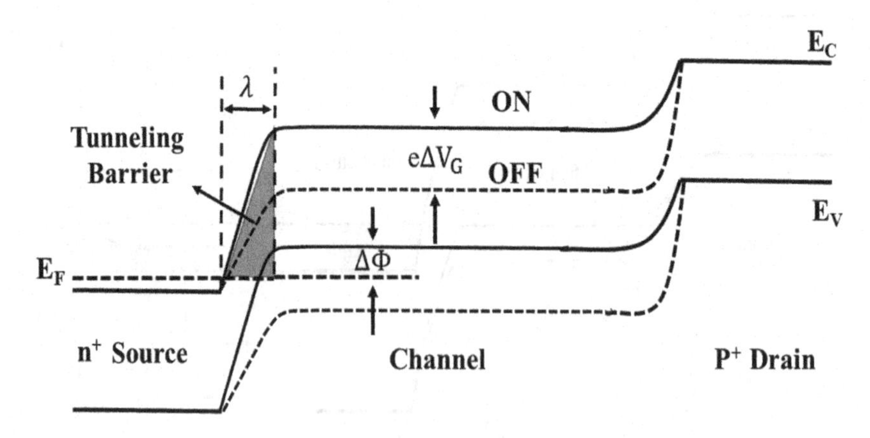

FIGURE 3.5 Energy band diagram of p-type TFET showing the formation of a triangular tunneling barrier when switching from the OFF- to the ON-state.

where x_1 and x_2 represent the two ends of the tunneling path while $k(x)$ is used as a wave vector for a charge carrier. By using the expression of $k(x)$ for an electron with energy E in the conduction band as $k(x)=\sqrt{2m^*_n/\hbar^2(E-E_c)}=\sqrt{2m^*_n/\hbar^2(-e\varepsilon x)}$ and considering the values of x_1 and x_2 to be 0 and $E_g/e\mathcal{E}$, respectively, where $\mathcal{E}$ represents the electric field at the source–channel interface, the tunneling probability can now be expressed as:

$$T_{WKB}=\exp\left(-\frac{4\sqrt{2m_n^*E_g}^{3/2}}{3he\varepsilon}\right) \tag{3.10}$$

The barrier width at the tunneling interface is calculated by adding the screening length in channel λ_{ch} and source λ_{dop} [35,36]. After substituting the expression for ε as $(E_g+\Delta\Phi)/e(\lambda_{ch}+\lambda_{dop})$, equation (3.10) becomes

$$T_{WKB}=\exp\left(-\frac{4(\lambda_{ch}+\lambda_{dop})\sqrt{2m_n^*E_g}^{3/2}}{3h(E_g+\Delta\Phi)}\right) \tag{3.11}$$

It can be observed from this expression that the tunneling probability can be increased by reducing the screening length in both source and channel regions. While λ_{dop} can be reduced using a steep doping profile at the source region, λ_{ch} depends on the electrostatic control of gate on the channel. Again, the value of total screen length can be expressed as $\lambda=\sqrt{(\epsilon_s/\epsilon_{ox})t_s t_{ox}}$ [35]. Thus, the expression of T_{WKB} suggests that using semiconductor materials with low bandgap, such as germanium, and compound semiconductors, such as $Si_{1-x}Ge$ (whose bandgap can be modulated by varying mole fraction x), at the input tunneling interface, the tunneling probability of charge carriers can be enhanced as also the tunneling current in TFETs.

3.3.2 Band-to-Band Generation and Conduction Current in TFETs

Since the tunneling probability depends on the screening length in the channel region λ_{ch}, controlled by gate potential, band-to-band generation of charge carriers is found to be high at the tunneling junction closer to the channel–gate dielectric interface. To mathematically represent this band-to-band generation in TFETs, the existing BTBT models use the expression as given by Kane [37] which is:

$$G(E)=A\frac{|\varepsilon|^2}{\sqrt{E_g}}\exp\left(-\frac{BE_g^{3/2}}{|\varepsilon|}\right) \tag{3.12}$$

where A and B are popularly known as Kane's BTBT model parameters and are dependent on material properties and ε is the net electric field at the input tunneling interface. To increase the BTBT generation rate in TFETs, the net electric filed in the tunneling region needs to be enhanced. Based on the Landauer approach [35], the drain current in TFETs can be expressed as:

$$I_D=\int_0^{\Delta\Phi} dED(E)v(E)T(E)[f_S(E)-f_{ch}(E)] \tag{3.13}$$

where $D(E)$, $v(E)$, and $T(E)$ are density of states, group velocity, and tunneling probability, respectively. $f_s(E)$ and $f_{ch}(E)$ are the Fermi–Dirac distribution function in the source and channel regions, respectively.

The most generalized method to find drain current in TFETs is to integrate the BTBT generation rate of charge carriers given as equation (3.12) over the volume of the device [37] and can be expressed as

$$I_D = q \int G(E)dV \tag{3.14}$$

3.3.3 Behavior in the Subthreshold Region

To analyze the behavior of TFETs in the subthreshold regime, the Landauer expression as in equation (3.13) along with the tunneling probability as given in equation (3.11) can be used to derive the expression for SS. By neglecting the charge carriers injected from the drain and considering the band interval in which tunneling occurs as $\Delta\Phi = E_{V,C} - E_{C,S}$ (assuming $E_F = E_{S,C}$), drain current, I_D can be represented as:

$$I_D = \frac{2qT(E)}{h}\int_0^{\Delta\Phi} f_s(E)dE = \frac{2qT(E)}{h} F_S(\Delta\Phi) \tag{3.15}$$

With the assumption of $\partial V_G/\partial\Delta\Phi = 1/e$, and the Taylor series expansion of integral function $F_S(\Delta\Phi)$ for small value of $\Delta\Phi$, the SS can be expressed as:

$$\mathrm{SS} = \left[\frac{\partial \log I_D}{\partial V_G}\right]^{-1} \approx In(10)\frac{\Delta\Phi}{e} \tag{3.16}$$

It can be observed from the expression that SS is independent of temperature unlike MOSFETs and varies linearly with gate voltage as $\Delta\Phi$ increases with increasing gate voltage in contrast to MOSFETs in which SS does not depend on gate voltage. In fact, there are two mechanisms in tunnel FETs that result in SS steeper than 60 mV/decade. In the first mechanism, the voltage at the source–channel junction is controlled by the gate bias when tunnel FETs are biased in the Zener breakdown. With an increase in the gate bias, the electric field at the interface of source and channel increases due to increasing interface voltage, and eventually tunneling current exponentially increases.

This phenomenon is referred to as the "voltage-controlled tunneling" [38]. In the second process, the source–channel junction behaves as a low-pass filter as the valance band edge of the source (for n-type channel) filters out the tail of nonequilibrium electron distribution having high energy while being injected from source to channel. This low-pass filter behavior causes an effective cooling to the Fermi function, and this process is known as "cold-carrier injection" [39]. In the devices with longer channel lengths ($L_{CH} \geq 10\,\mathrm{nm}$), the current transport is dominated by the "cold-carrier injection", whereas the second process "voltage-controlled tunneling" dominates the transport in shorter-channel-length devices [39]. When the gate bias is low in log channel devices, "cold-carrier injection" determines the current transport in tunnel FETs as equation (3.13). As the gate voltage increases, the current transport mechanism is shifted to the "voltage-controlled tunneling" due to an increase in $\Delta\Phi$ in the post-subthreshold regime. In tunnel FETs with short channel lengths, the electrons of the source region find a thin and small tunneling barrier during the OFF-state and, eventually, empty states of the channel are filled up due to direct tunneling to the channel–drain interface. The direct tunneling of charge carriers from source to drain reduces the probability of carriers crossing from $E_{C,\,CH}$ to $E_{V,\,S}$, in the low-gate-voltage regime, while switching the device from the OFF- to the ON-state. This eventually causes the overpassing of the low-pass filtering act of source Fermi function, and as a result of this, the device is forced to exhibit "voltage-controlled tunneling" [33]. Then, the subthreshold is mathematically calculated by the expression given as:

$$SS \approx \frac{In(10)}{|e|} \frac{3qh(E_g + \Delta\Phi)^2}{4\Delta\sqrt{2m^*}\,(E_g^{3/2})} \tag{3.17}$$

In the case of the "voltage-controlled tunneling", T(E) (given as equation (3.8)), is initially small but increases rapidly with increasing gate voltage, and SS is actually decided by a change in the value of T(E). In contrast to "voltage-controlled tunneling", the value of T(E) is close to unity and independent of the gate voltage when "cold-carrier injection" dominates. SS now becomes independent of T(E), which slightly changes with gate voltage and only depends linearly on $\Delta\Phi$ as shown by equation (3.13). This is why the process of "cold-carrier injection" provides us the possibilities for a larger range of SS steeper than 60 mV/decade.

3.4 AMBIPOLAR CONDUCTION: A MAJOR DRAWBACK OF TFETs

Since BTBT of charge carriers is responsible for current transport, which is mostly confined to a narrow tunneling area at the source–channel interface, TFETs do not suffer from SCEs during the downscaling of the device dimensions, thus rendering it more worthy for future requirements of the device with compact on-chip device integration and low standby power dissipation. Apart from all of these merits, however, TFETs have few critical roadblocks that include low ON-state current, inferior high-frequency (HF) performances, and ambipolar conduction when compared to conventional MOSFETs. Low ON-state current along with high parasitic capacitances is responsible for the degradation of switching speed while ambipolar conduction is found to limit the application of tunnel FETs in inverter-based digital logic circuits. In the last decade, most of the research efforts have been focused on improving the ON-state current to match with that of the conventional MOSFETs. Some of these techniques, which are mainly based on the optimization of charge carrier injection by narrowing the tunneling width at the source–channel interface, include double-gate structure, strained source, lateral heterostructure, high-*k* dielectric of gate, stacked source, using delta pocket adjacent to source, etc. The action of TFET devices is mainly governed by the electrostatic phenomenon at the source–channel interface but due to the symmetric structure, BTBT is also observed at the channel–drain interface for negative (positive) gate biasing if n-TFET (p-TFET) is considered. As a consequence of this, the TFET remains switched-on both during ON- and OFF-states, that is, when the applied gate voltage is positive (for source-side BTBT) and

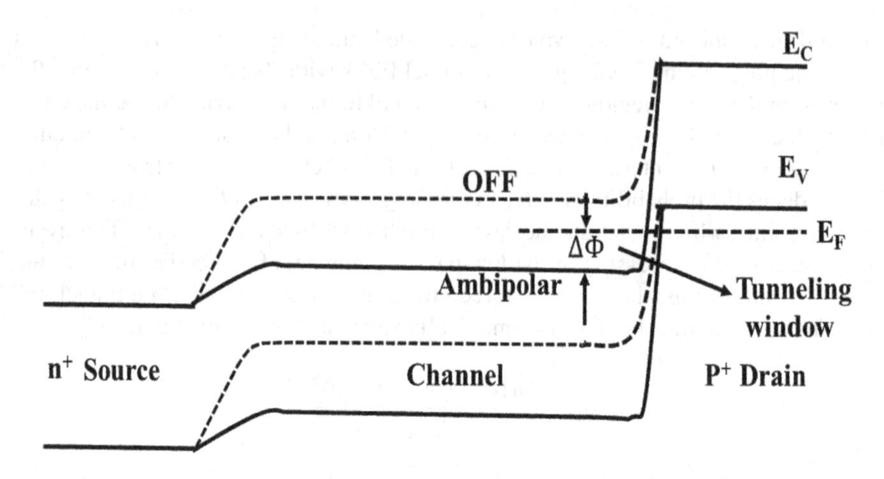

FIGURE 3.6 Energy band diagram of p-type TFET showing the formation of a tunneling window when the gate is biased with a positive value.

negative (for drain-side BTBT), keeping the same V_{DS} same (positive) if n-TFET is considered. This behavior of TFET is known as ambipolarity, which causes more static power dissipation during the OFF-state. The ambipolar conduction can even cause circuit failure due to high standby power dissipation in CMOS-based digital circuits such as inverters. Figure 3.6 shows the band diagram of p-TFET at a positive applied gate voltage. Further, it can be observed that there is a band overlapping between the conduction band of the channel and the valance band of the drain region causing tunneling of charge carriers during the OFF-state as well. However, the current conduction at positive gate voltage is mainly due to the tunneling of electrons (OFF-state), and when the gate voltage is negative, drain current is dominated by the holes (ON-state). In general, the BTBT at the channel–drain interface can be suppressed by reducing the doping level for the drain region as compared to source. However, this ambipolarity is not completely eliminated; therefore, we need some special techniques to mitigate it. In this chapter, two such techniques have been discussed, which are found to be very effective in reducing the ambipolar current up to a large extent along with an improvement in the RF performances as well.

3.4.1 Asymmetric Gate-Drain Overlap

The symmetric gate-drain overlap (SGDO) [40] can eliminate the ambipolarity in double-gate TFETs (DG-TFETs) but at the cost of increased parasitic gate-drain capacitance. In addition to degradation in HF performances, this technique needs a minimum overlap length of 30 nm for a visible reduction in ambipolarity, which eventually limits the device scaling. To overcome the demerits of SGDO, a promising technique of asymmetric gate-drain overlap (ASGDO), which not only eliminates ambipolarity but also enhances analog/HF performances, can be applied to the DG-TFET [41]. This ASGDO DG-TFET structure has its back gate overlapped with the drain region, as shown in Figure 3.7. ASGDO DG-TFET avoids the trade-off

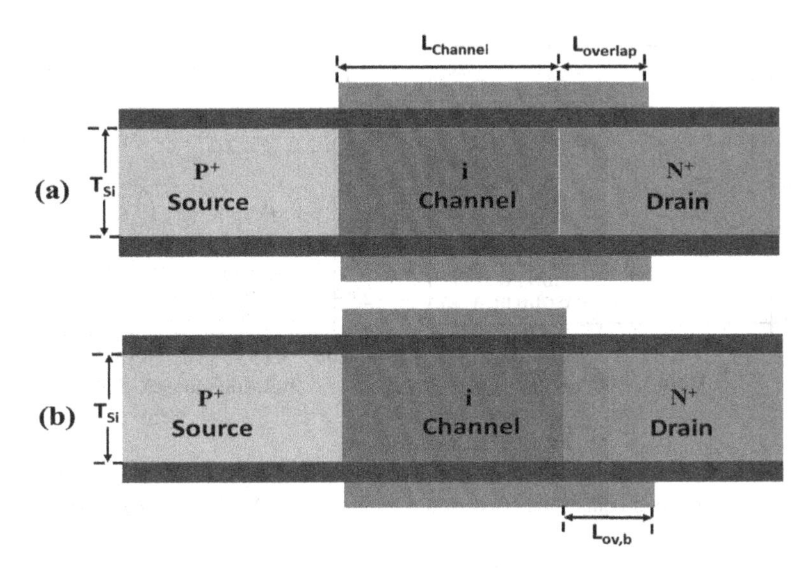

FIGURE 3.7 Cross-sectional view of (a) symmetric and (b) asymmetric GDO DG-TFET [41].

between ambipolarity and HF performances by taking the merit of gate-drain overlap in terms of reduction in ambipolarity and suppressing its demerit with reduced gate-drain parasitic capacitance. In ASGDO DG-TFET, only the back gate is overlapped with the drain region, which shows a remarkable reduction in ambipolar conduction for a smaller thickness of the channel when compared with SGDO DG-TFET as proposed in Ref. [40].

Figure 3.8 shows the I_{ds}–V_{gs} plot of the conventional, SGDO, and ASGDO DG-TFET. It is clearly visible that ASGDO TFET provides a huge reduction in ambipolar conduction compared to the other two devices. For the optimum value of L_{ov} (i.e., 20 nm), the ambipolar current (I_{amb}) in ASGDO DG-TFET is found to be 2.47×10^{-15} A/µm, which is approximately 3 orders lower than that in SGDO DG-TFET at $V_{gs} = -1$ V.

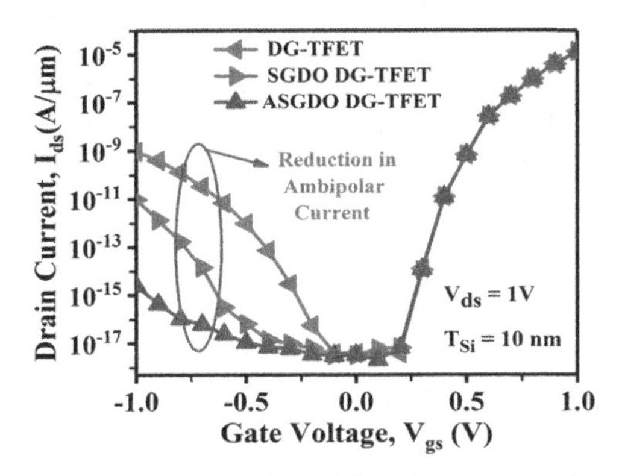

FIGURE 3.8 Comparison of transfer characteristics between conventional, SGDO, and ASGDO DG-TFET [41].

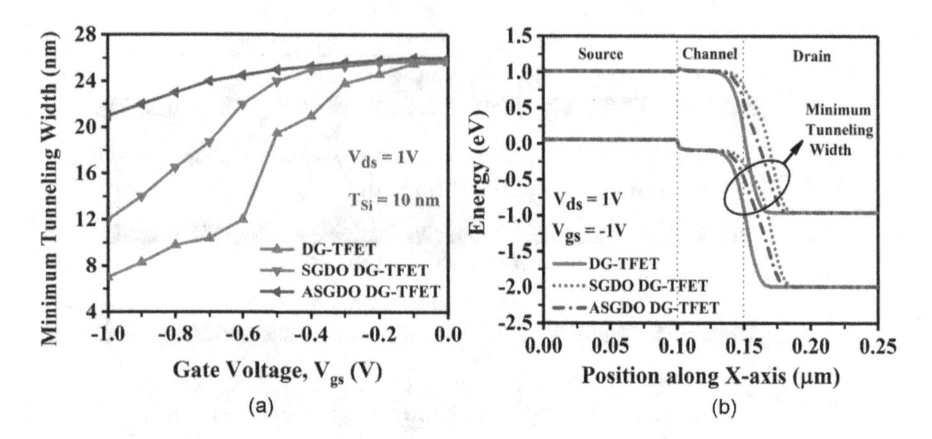

FIGURE 3.9 Comparison of (a) relative widening rate of tunneling barrier at the C-D interface and (b) energy band diagram between conventional, SGDO, and ASGDO DG-TFET [41].

Furthermore, it can be seen from the figure that I_{amb} is approximately the same in both DG-TFET with ASGDO and SGDO up to a gate voltage of −0.4 V, and it is reduced at a faster rate in ASGDO DG-TFET if the negative gate bias is increased further. This is because of the larger widening of the lateral tunneling barrier (W_{min}) at the C–D interface in ASGDO DG-TFET compared to SGDO DG-TFET when negative gate bias is increased beyond −0.4 V as shown in Figure 3.9a. The relative widening of the lateral tunneling barrier in the proposed device is found to be approximately 67% more than that of SGDO DG-TFET at $V_{gs} = -1$ V. The energy band profile of all three devices in the lateral direction are shown in Figure 3.9b, and the horizontal cutline is set at the center of the device. As shown in the figure, the tunneling width at the C–D interface is found to be maximum in ASGDO DG-TFET while the conventional DG-TFET shows a minimum value for it. The reason for this is that the rate of widening of the lateral tunneling width with applied gate voltage is higher in the proposed device than that in SGDO DG-TFET, as shown in Figure 3.9a.

Figure 3.10a compares C_{gs} of ASGDO DG-TFET with those of conventional and SGDO DG-TFET at different drain biases for channel length varying from 30 to 50 nm. It can be observed that there is no significant impact of GDO on C_{gs} even if the channel length is scaled down to 30 nm. Again, the impact of GDO on C_{gd} is shown in Figure 3.10b in which C_{gd} is found to increase due to overlap, thus degrading the RF performances.

3.4.2 Dielectric-Engineered Tunnel FETs

TFETs with a high-*k* dielectric pocket (DP) on the drain side (as shown in Figure 3.11) can reduce the ambipolarity by a huge order number. Through 2-D numerical simulations using Synopsys Sentaurus [42], it has been demonstrated that Dielectric Pocket SOI-TFET (DP SOI-TFET) eliminates the ambipolar conduction completely even at a large negative gate bias of −1.0 *V* [43]. The presence of the dielectric pocket above the partially scaled drain region enhances the underlying depletion region at

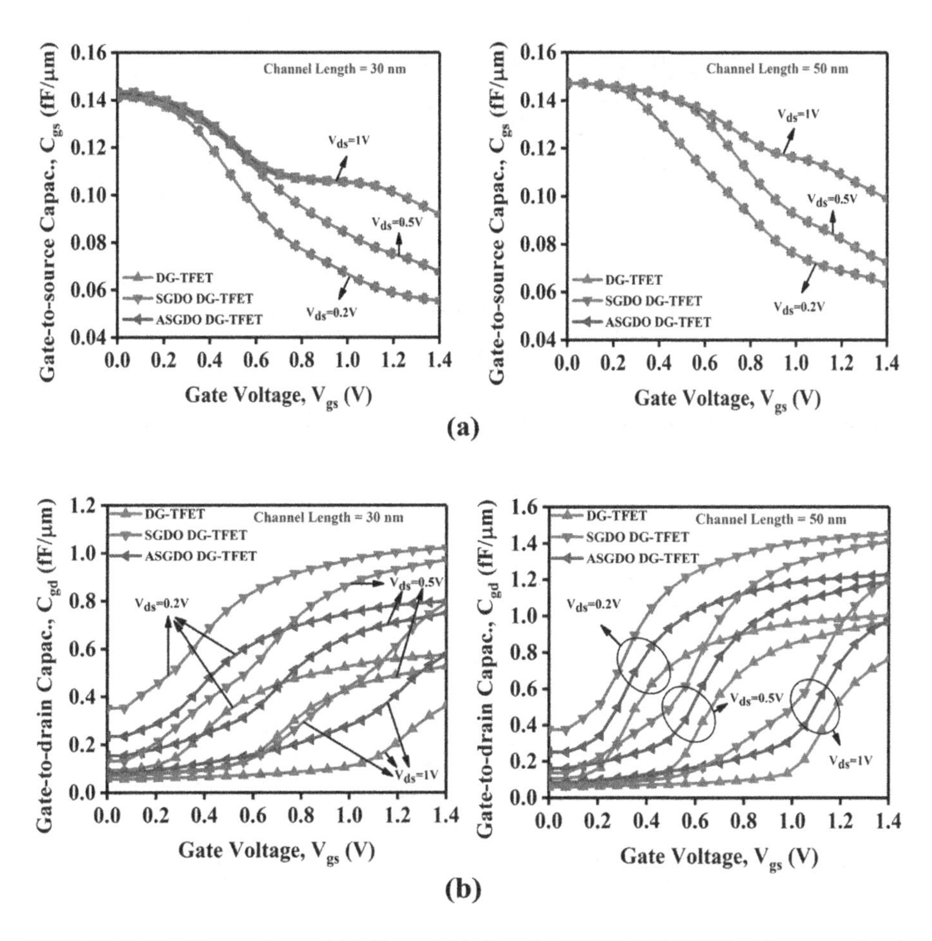

FIGURE 3.10 Comparison of (a) C_{gs} and (b) C_{gd} of ASGDO DG-TFET with conventional and SGDO DG-TFET [41].

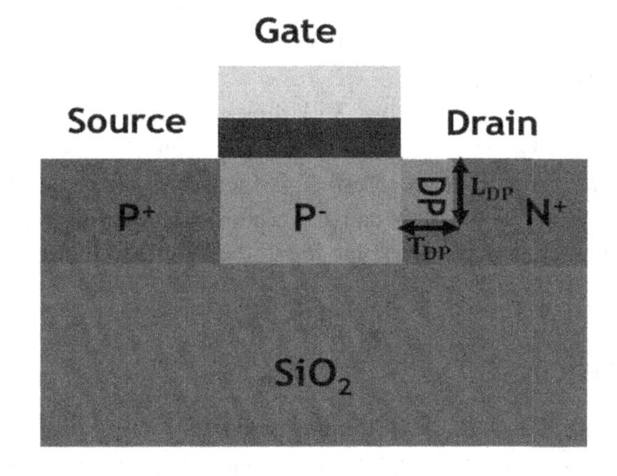

FIGURE 3.11 Cross-sectional view of DP SOI-TFET [43].

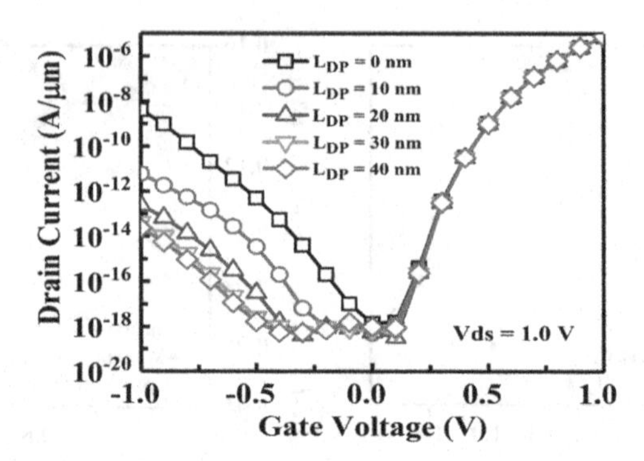

FIGURE 3.12 Transfer characteristics of the proposed device for varying L_{DP} [43].

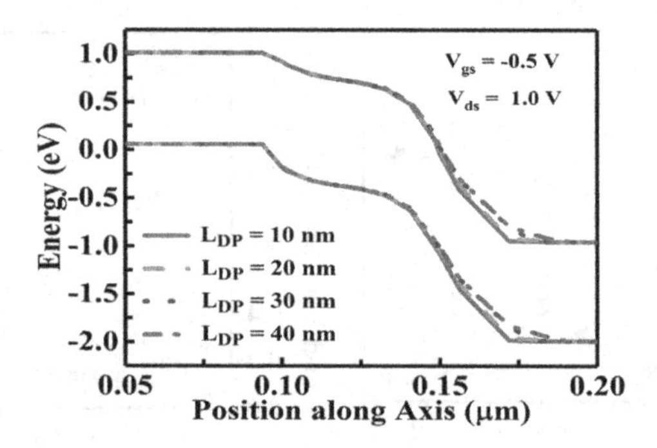

FIGURE 3.13 Energy band profiles of the proposed device for varying L_{DP}.

the channel–drain junction, which, in turn, modulates the energy band structure of channel and drain regions at the interface. The impact of the varying length of dielectric pocket on the device performance is analyzed and shown in Figure 3.12. For the simulations, T_{DP} is fixed at the optimum value of 5 nm. It can be noticed from the figure that ambipolar current decreases from 4.16×10^{-9}A to 4.15×10^{-14} A when L_{DP} is increased from 0 to 30 nm.

For a better understanding, Figure 3.13 shows the energy band diagram of the proposed structure for varying L_{DP}. It can be clearly seen from the figure that the tunneling width at the channel–drain interface is increased when the length of the DP is increased from 10 to 30 nm, thus causing a consistent decrease in I_{amb}. Since, the tunneling width gets the maximum value for L_{DP} = 30 nm, no further decrease in I_{amb} is observed in TFETs for L_{DP} > 30 nm.

3.5 CONCLUSIONS

In this chapter, first, the fundamental issues associated with MOSFETs, when technology node enters into the nanometer regime, have been discussed in detail. Then, in order to meet today's demand of low-power consumption, the possible devices that

YEAR (20-)	13	14	15	16	17	18	19	20	21	22	23	24	25	26	27	28
Nano-wire MOSFETs																
Alternate channels: III-V or Ge																
Alternate Channels: Carbon CNTs/Graphene																
Alternate Channels: 2-D crystals (MoSi2, BN...)																
Tunnel FET																
Non-CMOS Technology																

Research Required
Development Underway
Qualification/Pre-production
Continuous Improvement

FIGURE 3.14 Future prediction for tunnel FET mentioned in the ITRS Report, 2013 [44].

can replace MOSFETs from the current IC technology have been briefly discussed. Since TFETs have been considered as the best alternative device to MOSFETs due to their ability to have steeper SS, less than 60 mV/decade, and better immunity to SCEs, the working principle of this device has been explained in detail. Furthermore, two techniques such as asymmetric gate-drain overlap and dielectric pocket, which can overcome a major drawback of TFETs, that is, ambipolarity, have been analyzed and show that a remarkable reduction in ambipolar current can be achieved along with an improvement in the HF performances. Even though a many studies have already been performed to mature TFETs, this device is yet to be commercialized and replace the MOSFET from current IC technology, which is mainly due to its low ON-state current as compared to MOSFETs, ambipolarity, and, most importantly, inability to practically achieve steeper SS as expected and claimed in the literature. Even in ITRS 2013, it was predicted that TFETs will remain in the research only till the beginning of the 2020s before they can be commercialized and used in integrated circuits (as shown in Figure 3.14) [44].

REFERENCES

1. Lilienfeld JE. Device for Edgar LJ, inventor. Device for controlling electric current. United States patent US 1,900,018. 1933 Mar 7.
2. Kahng D. Silicon-silicon dioxide field induced surface devices. In *the Solid-State Device Research Conf., Pittsburgh, PA. June 1960.*
3. Moore GE, Cramming more components onto integrated circuits. *Electronics*. 1965 Apr 19;38(8):114–116.

4. Dennard RH, Gaensslen FH, Yu HN, Rideout VL, Bassous E, LeBlanc AR. Design of ion implanted MOSFET's with very small physical dimensions. *IEEE Journal of Solid State Circuits*. 1974 Oct;9(5):256–268.
5. Tsividis Y, *Operation and Modeling of the MOS Transistor.* McGraw-Hill, Boston; 1998. pp. 248–310.
6. Vandamme EP, Jansen P, Deferm L. Modeling the subthreshold swing in MOSFET's. *IEEE Electron Device Letters*. 1997 Aug;18(8):369–71.
7. Godoy A, López-Villanueva JA, Jiménez-Tejada JA, Palma A, Gámiz F. A simple subthreshold swing model for short channel MOSFETs. *Solid-State Electronics*. 2001 Mar 1;45(3):391–7.
8. Datta S. Exclusion principle and the Landauer-Büttiker formalism. *Physical Review B*. 1992 Jan 15;45(3):1347.
9. Waser R, editor. *Nanoelectronics and Information Technology: Advanced Electronic Materials and Novel Devices*. John Wiley & Sons, Weinheim, Germany; 2012 May 29.
10. Liu CW, Hsieh TX. Analytic modeling of the subthreshold behavior in MOSFET. *SolidState Electronics*. 2000 Sep 1;44(9):1707–10.
11. Datta S, Das B. Electronic analog of the electro-optic modulator. *Applied Physics Letters*. 1990 Feb 12;56(7):665–7.
12. Dreslinski RG, Wieckowski M, Blaauw D, Sylvester D, Mudge T. Near-threshold computing: Reclaiming Moore's law through energy efficient integrated circuits. *Proceedings of the IEEE*. 2010 Jan 22;98(2):253–66.
13. Sreenivasulu VB, Narendar V. Performance improvement of spacer engineered n-type SOI FinFET at 3-nm gate length. *AEU-International Journal of Electronics and Communications*. 2021 Jul 1;137:153803.
14. Bennett HS. International Technology Roadmap for Semiconductors 2015 Edition.
15. Wong HS. Beyond the conventional transistor. *IBM Journal of Research and Development*. 2002 Mar;46(2.3):133–68.
16. Yousefi R, Shabani M. A model for carbon nanotube FETs in the ballistic limit. *Microelectronics Journal*. 2011 Nov 1;42(11):1299–304.
17. Elmessary MA, Nagy D, Aldegunde M, Seoane N, Indalecio G, Lindberg J, Dettmer W, Perić D, Garcia-Loureiro AJ, Kalna K. Scaling/LER study of Si GAA nanowire FET using 3D finite element Monte Carlo simulations. *Solid-State Electronics*. 2017 Feb 1; 128:17–24.
18. Yang FL, Lee DH, Chen HY, Chang CY, Liu SD, Huang CC, Chung TX, Chen HW, Huang CC, Liu YH, Wu CC. 5nm-gate nanowire FinFET. In *Digest of Technical Papers. 2004 Symposium on VLSI Technology*. 2004 Jun 15 (pp. 196–197). IEEE.
19. Yoon JS, Kim K, Rim T, Baek CK. Performance, and variations induced by single interface trap of nanowire FETs at 7-nm node. *IEEE Transactions on Electron Devices*. 2016 Dec 28;64(2):339–45.
20. Song YS, Kim S, Kim G, Kim H, Lee JH, Kim JH, Park BG. Improvement of self-heating effect in Ge vertically stacked GAA nanowire pMOSFET by utilizing Al_2O_3 for high performance logic device and electrical/thermal co-design. *Japanese Journal of Applied Physics*. 2021 Mar 25;60(SC):SCCE04.
21. Song YS, Kim JH, Kim G, Kim HM, Kim S, Park BG. Improvement in self-heating characteristic by incorporating hetero-gate-dielectric in gate-all-around MOSFETs. *IEEE Journal of the Electron Devices Society*. 2020 Nov 17;9:36–41.
22. Sreenivasulu VB, Narendar V. Characterization and optimization of junctionless gate-all-around vertically stacked nanowire FETs for sub-5 nm technology nodes. *Microelectronics Journal*. 2021 Oct 1;116:105214.
23. Sreenivasulu VB, Narendar V. A comprehensive analysis of junctionless tri-gate (TG) FinFET towards low-power and high-frequency applications at 5-nm gate length. *Silicon*. 2021 Feb 23;14:1–3.

24. Nirschl T, Wang PF, Hansch W, Schmitt-Landsiedel D. The tunnelling field effect transistors (TFET): the temperature dependence, the simulation model, and its application. In *2004 IEEE International Symposium on Circuits and Systems (IEEE Cat. No. 04CH37512)*. 2004 May 23 (Vol. 3, pp. III-713). IEEE.
25. Rahi SB, Asthana P, Gupta S. Heterogate junctionless tunnel field-effect transistor: future of low-power devices. *Journal of Computational Electronics*. 2017 Mar 1;16(1):30–8.
26. Rahi SB, Ghosh B, Asthana P. A simulation-based proposed high-k heterostructure AlGaAs/Si junctionless n-type tunnel FET. *Journal of Semiconductors*. 2014 Nov;35(11):114005.
27. Nirschl T, Wang PF, Hansch W, Schmitt-Landsiedel D. The tunnelling field effect transistors (TFET): the temperature dependence, the simulation model, and its application. In *2004 IEEE International Symposium on Circuits and Systems (IEEE Cat. No. 04CH37512)*. 2004 May 23 (Vol. 3, pp. III-713). IEEE.
28. Jiang C, Liang R, Xu J. Investigation of negative capacitance gate-all-around tunnel FETs combining numerical simulation and analytical modeling. *IEEE Transactions on Nanotechnology*. 2016 Nov 10;16(1):58–67.
29. Jo J, Shin C. Impact of temperature on negative capacitance field-effect transistor. *Electronics Letters*. 2015 Jan;51(1):106–8.
30. Salahuddin S. Review of negative capacitance transistors. In *2016 International Symposium on VLSI Technology, Systems and Application (VLSI-TSA)*, 2016 Apr 25 (pp. 1–1). IEEE.
31. Sugahara S, Tanaka M. A spin metal-oxide-semiconductor field-effect transistor using half-metallic-ferromagnet contacts for the source and drain. *Applied Physics Letters*. 2004 Mar 29;84(13):2307–9.
32. Shen M, Saikin S, Cheng MC. Spin injection in spin FETs using a step-doping profile. *IEEE Transactions on Nanotechnology*. 2005 Jan 17;4(1):40–4.
33. Knoch J, Mantl S, Appenzeller J. Impact of the dimensionality on the performance of tunneling FETs: Bulk versus one-dimensional devices. *Solid-State Electronics*. 2007 Apr 1;51(4):572–8.
34. Sze SM, Li Y, Ng KK. *Physics of Semiconductor Devices*. John Wiley & Sons, Weinheim, Germany; 2021.
35. Knoch J, Appenzeller J. A novel concept for field-effect transistors-the tunneling carbon nanotube FET. In *63rd Device Research Conference Digest, 2005. DRC'05*. 2005 Jun 20 (Vol. 1, pp. 153–156). IEEE.
36. Sandow C, Knoch J, Urban C, Zhao QT, Mantl S. Impact of electrostatics and doping concentration on the performance of silicon tunnel field-effect transistors. *Solid-State Electronics*. 2009 Oct 1;53(10):1126–9.
37. Bardon MG, Neves HP, Puers R, Van Hoof C. Pseudo-two-dimensional model for double-gate tunnel FETs considering the junctions depletion regions. *IEEE Transactions on Electron Devices*. 2010 Feb 17;57(4):827–34.
38. Sylvia SS, Khayer MA, Alam K, Lake RK. Doping, tunnel barriers, and cold carriers in InAs and InSb nanowire tunnel transistors. *IEEE Transactions on Electron Devices*. 2012 Sep 17;59(11):2996–3001.
39. Brahma M, Kabiraj A, Saha D, Mahapatra S. Scalability assessment of Group-IV monochalcogenide based tunnel FET. *Scientific Reports*. 2018 Apr 16;8(1):1–10.
40. Abdi DB, Kumar MJ. Controlling ambipolar current in tunneling FETs using overlapping gate-on-drain. *IEEE Journal of the Electron Devices Society*. 2014 May 30;2(6):187–90.
41. Pandey CK, Singh A, Chaudhury S. Effect of asymmetric gate–drain overlap on ambipolar behaviour of double-gate TFET and its impact on HF performances. *Applied Physics A*. 2020 Mar;126(3):1–2.

42. TCAD Sentaurus Device Version J-2014.09. [Online]. Available: http://www.synopsys.com.
43. Pandey CK, Dash D, Chaudhury S. Approach to suppress ambipolar conduction in Tunnel FET using dielectric pocket. *Micro & Nano Letters*. 2019 Jan 17;14(1):86–90.
44 "International Technology Roadmap for Semiconductors", http://public.itrs.net/, 2013 Edition.

4 Analytical Modeling of Surface Potential of a Double-Gate Heterostructure PNPN Tunnel FET

K. Baruah and S. Baishya
National Institute of Technology

CONTENTS

4.1 INTRODUCTION

Tunnel FETs (TFETs) have been present in most semiconductor research laboratories in low-power applications for some years due to their improved electrical properties in the device [1,2]. The main benefits of TFETs are their low subthreshold swing (less than 60 mV/decade) and very small leakage current. Carriers that tunnel from source to channel cause current to flow in TFETs [3]. Because of this, TFETs have a very low ON-current. In addition, TFETs have an ambipolar current that is undesired and should be reduced for improved performance. Researchers have employed a range of structural/physical approaches to improve the overall performance of a TFET device during the last few decades [4–6]. One of these is the double-gate TFET structure, which can increase the device's ON-state performance by increasing the number of channels for current conduction [4]. The study of analytical modeling is critical for correctly understanding the physics of a device. Several studies have previously been published in this sector, utilizing various modeling methodologies to predict surface potential, vertical and lateral electric fields, and drain current of TFETs [7–12].

This work represents the analytical modeling of a proposed heterostructure double-gate pnpn TFET with a dual dielectric spacer and gate oxide of a high-dielectric-constant material. To obtain the solution of surface potential at various

DOI: 10.1201/9781003240778-4

places along the channel, the model employs Poisson's equation and parabolic approximation, as well as various boundary conditions. When examining the characteristics, several parameter factors such as gate and drain bias, body thickness, gate-dielectric thickness, and gate oxide materials are taken into account. The depletion zones at both the source and the drain side are considered in this technique. Also, the analytical expressions are extracted and validated using the Sentaurus TCAD 2D-device simulator [13].

4.2 STRUCTURE OF THE DEVICE AND THE MODELING APPROACH

The TFET suggested in this paper is depicted in Figure 4.1 as a cross-sectional view. The proposed structure is a double-gate structure with germanium and silicon-germanium (mole fraction, 0.3) as the source and pocket material, respectively. High-k HfO_2 ($\epsilon = 22\epsilon_0$) followed by low-k SiO_2 ($\epsilon = 3.9\epsilon_0$) is used as the spacer material. The dimensions of device parameters are shown along with the device structure in Figure 4.1. The proposed TFET is n-TFET and it has a p^+ source (doping: $10^{20}\,cm^{-3}$), n^+ pocket (doping: $5 \times 10^{18}\,cm^{-3}$), intrinsic (i) channel, and n^+ drain (doping: $10^{18}\,cm^{-3}$) regions. On both sides of the device, a metal of 4.25 eV work function is utilized as the electrode of the gate terminal. The material used in the source–channel junction pocket is $Si_{i\text{-}x}Ge_x$ with a germanium composition of 30% (i.e., $x = 0.3$).

In order to perform analytical modeling, the recommended device is first partitioned into four areas, namely, Region 1, Region 2, Region 3, and Region 4 (Figure 4.1). The depletion regions in the source and drain side are denoted by Region 1 (length L_1) and Region 4 (length L_4), respectively. The channel is divided into two parts, Region 2, which is the pocket (length L_2), and the remainder of the channel is Region 3 (length L_3). For each of the abovementioned regions, the potential is computed by solving the Poisson equation [14,15].

The 2D Poisson's equation can be written as:

$$\partial^2\psi_i(x, y)\Big/\partial x^2 + \partial^2\psi_i(x, y)\Big/\partial y^2 = -qN_i\Big/\epsilon_i \tag{4.1}$$

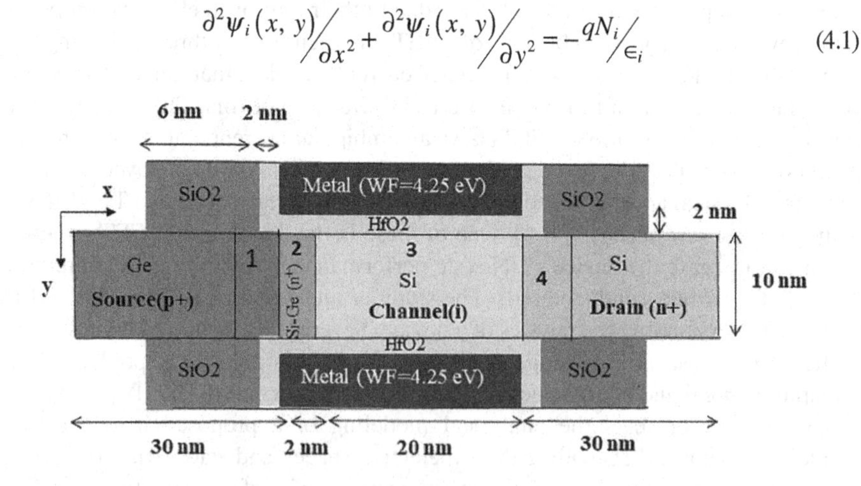

FIGURE 4.1 Cross-sectional view of the suggested DG heterostructure pnpn TFET.

where $\psi_i(x, y)$ the surface potential of the regions is $R_i(1, 2, 3, 4)$, q is the electron charge, N_i is the doping amount of each region, and ϵ_i is the dielectric permittivity of the respective regions of $i = 1, 2, 3, 4$.

In the vertical direction, the potential profile of a double-gate TFET structure is parabolic. Thus, using the parabolic approximation, the potential function is

$$\psi_i(x, y) = a_{1,i}(x) + a_{2,i}(x)y + a_{3,i}(x)y^2 \tag{4.2}$$

where $a_{1,i}(x)$, $a_{2,i}(x)$, *and* $a_{3,i}(x)$ are the coefficients of the polynomial in y.

To evaluate $\psi_i(x, y)$ in equation (4.2), first, we have to calculate the value of all coefficients present in equation (4.2). These coefficients can be evaluated using the below-mentioned boundary conditions in the y-direction:

$$\psi_i(x, 0) = \psi_{s,i}(x) \tag{4.3}$$

$$\psi_i(x, t_{si}) = \psi_{s,i}(x) \tag{4.4}$$

$$\left.\partial\psi_i(x, y)\middle/\partial y\right|_{y=0} = \eta_i \,\psi_{s,i}(x) - \psi_{G,i} \middle/ t_{si,i} \tag{4.5}$$

$$\left.\partial\psi_i(x, y)\middle/\partial y\right|_{y=t_{si,i}} = -\eta_i \,\psi_{s,i}(x) - \psi_{G,i} \middle/ t_{si,i} \tag{4.6}$$

where $\eta_i = C_{ox,i}/C_{si,i}$ *for*, $i = 1, 2, 3, 4$, $C_{ox,i} = \epsilon_{ox,i}/t_{ox,i}$ *for*, $i = 2$, and 3, $C_{ox,i} = \left(\frac{2}{\pi}\right)\epsilon_{ox,i}/t_{ox,i}$ *for*, $i = 1$, and 4. Also, $C_{si,i} = \epsilon_{si,i}/t_{si,i}$ *for*, $i = 1, 2, 3$, and 4 and $\psi_{G,i} = V_{G,i} - V_{fb,i}$, $V_{fb,i} = \phi_M - \phi_{s,i}$, $\phi_{s,i} = \chi_i + E_{g,i}/2q + \phi_{f,i}$ where $C_{ox,i}$ $C_{si,i}$, $\epsilon_{ox,i}$, $\epsilon_{si,i}$, $t_{ox,i}$, $t_{si,i}$, $\psi_{G,i}$, $V_{fb,i}$, ϕ_M, $\phi_{s,i}$, χ_i, $E_{g,i}$, and $\phi_{f,i}$ represent the oxide capacitance per unit area, substrate capacitance per unit are, permittivity of oxide, permittivity of substrate, thickness of oxide, thickness of substrate, gate potential, flat band voltage, metal work function, semiconductor work function, electron affinity, energy band gap, and fermi potential for regions $i = 1, 2, 3, 4$, respectively. Now using the above boundary condition equations (4.3–4.6) and equation (4.2), we obtain the coefficient of equation (4.2) as follows:

$$a_{1,i}(x) = \psi_{s,i}(x), \quad a_{2,i}(x) = \eta_i \frac{\psi_{s,i}(x) - \psi_{G,i}}{t_{si,i}}, \quad a_{3,i}(x) = -\eta_i \frac{\psi_{s,i}(x) - \psi_{G,i}}{t_{si,i}^{\;2}}$$

Substituting the above coefficients in equation (4.2) and solving for equation (4.1), we get

$$\partial^2\psi_{s,i}(x)\big/\partial x^2 - \frac{2\eta}{t_{si}^2}\psi_{s,i}(x) = -qN_i\big/\epsilon_i - \frac{2\eta}{t_{si}^2}\psi_{G,i}$$

Or

$$\partial^2\psi_{s,i}(x)\big/\partial x^2 - k_i^2\psi_{s,i}(x) = -k_i^2\psi_{d,i} \tag{4.7}$$

where $k_i = \sqrt{2\eta_i \big/ t_{si}^2}$

$$\psi_{d,i} = qN_i \big/ k_i^2 \in_i + \psi_{G,i}$$

where $1\big/k_i$ is known as the characteristic-length of the surface potential in each zone and $\psi_{d,i}$ is the region-dependent parameter.

The equation mentioned above equation (4.7) is a second-order differential equation with the following generic solution:

$$\psi_{s,i}(x) = C_i e^{-k_i x} + D_i e^{k_i x} + \psi_{d,i},\ for\ i = 1,\ 2,\ 3,\ 4 \tag{4. 8}$$

The following set of x-directional boundary conditions may be used to determine the value of coefficients $C_1,\ C_2,\ C_3,\ C_4,\ D_1,\ D_2,\ D_3,\ and\ D_4,$

$$\psi_{s1}(0) = -\frac{kT}{q} ln\left(\frac{N_1}{n_{i,1}}\right) = -V_{bis} \tag{4.9}$$

$$\psi_{s1}(L_1) = \psi_{s2}(L_1) \tag{4.10}$$

$$\partial\psi_{s1}(L1)\big/\partial x = \partial\psi_{s2}(L1)\big/\partial x \tag{4.11}$$

$$\psi_{s2}(L1+L2) = \psi_{s3}(L1+L2) \tag{4.12}$$

$$\partial\psi_{s2}(L1+L2)\big/\partial x = \partial\psi_{s3}(L1+L2)\big/\partial x \tag{4.13}$$

$$\psi_{s3}(L1+L2+L3) = \psi_{s4}(L1+L2+L3) \tag{4.14}$$

$$\partial\psi_{s3}(L1+L2+L3)\big/\partial x = \partial\psi_{s4}(L1+L2+L3)\big/\partial x \tag{4.15}$$

$$\psi_{s4}(L1+L2+L3+L4) = V_{DS} + \frac{kT}{q} ln\left(\frac{N_4}{n_{i,2}}\right) = V_{DS} + V_{bid} \tag{4.16}$$

where $L1 = \sqrt{\frac{2\epsilon_i\left|\psi_{d,2} - \psi_{s,1}\right| N_2}{qN_1(N_1+N_2)}}$, $L4 = \sqrt{\frac{2\epsilon_i\left|\psi_{s,4} - \psi_{d,3}\right| N_3}{qN_4(N_3+N_4)}}$ are the calculated depletion region lengths on the source side and drain side, respectively.

And $n_{i,1}$ and $n_{i,2}$ represent the intrinsic carrier concentration of SiGe and silicon, respectively. Applying the above boundary condition equations (4.9–4.16) in equation (4.8), eight new equations can be obtained. These equations can be solved to obtain the eight different coefficients of equation (4.8).

$$C_1 = t_1 - D_1,\ D_1 = \frac{t_3}{t_2} + D_2\frac{t_8}{t_2} + \frac{C_2}{t_8 t_2},\ C_2 = -\frac{t_7}{t_{10}} - D_2\frac{t_9}{t_{10}},\ D_2 = C_3 t_{19} + D_3 t_{18} + t_{17},$$

$$C_3 = \frac{t_{24}}{t_{22}} - D_3\frac{t_{23}}{t_{22}}, D_3 = C_4 t_{31} + D_4 t_{32} + t_{33}, C_4 = \frac{t_{36}}{t_{34}} - D_4\frac{t_{35}}{t_{34}}, D_4 = \frac{t_{49}}{t_{41}}$$

where

$$t_1 = V_{bis} - \psi_{d1},\ t_2 = 2\sinh(k_1 L_1), t_3 = \psi_{d2} - \psi_{d1} - t_1\exp(-k_1 L_1), t_4 = 2\cosh(k_1 L_1)$$

$$t_5 = t_1 k_1 \exp(-k_1 L_1), t_6 = \coth(k_1 L_1), t_7 = t_3 k_1 t_6 - t_5, t_8 = \exp(k_2 L_1),$$

$$t_9 = t_8(k_1 t_6 - k_2), t_{10} = \frac{k_1 t_6 + k_2}{t_8}, t_{11} = \exp(k_2(L_1 + L_2)),$$

$$t_{12} = t_{11} - \frac{t_9}{t_{10} t_{11}}, t_{13} = \frac{t_7}{t_{10} t_{11}}, t_{14} = \psi_{d2} - \psi_{d3}, t_{15} = \frac{1}{t_{12}}$$

$$t_{16} = \exp(k_3(L_1 + L_2)), t_{17} = t_{15}(t_{13} - t_{14}), t_{18} = t_{15} t_{16}, t_{19} = \frac{t_{15}}{t_{16}},$$

$$t_{22} = \frac{t_9 t_{19} k_2}{t_{10} t_{11}} + t_{11} t_{19} k_2 + \frac{k_3}{t_{16}}, t_{23} = \frac{t_9 t_{18} k_2}{t_{10} t_{11}} + t_{11} t_{18} k_2 - (t_{16} k_3),$$

$$t_{24} = -k_2\left(\frac{t_9 t_{17}}{t_{10} t_{11}} + \frac{t_7}{t_{10} t_{11}} + t_{11} t_{17}\right), t_{25} = \exp(k_3(L_1 + L_2 + L_3)),$$

$$t_{26} = t_{25} - \frac{t_{23}}{t_{22} t_{25}}, t_{27} = \psi_{d4} - \psi_{d3}, t_{28} = t_{27} - \frac{t_{24}}{t_{22} t_{25}},$$

$$t_{29} = \exp(k_4(L_1 + L_2 + L_3)), t_{30} = \frac{1}{t_{26}}, t_{31} = \frac{t_{30}}{t_{29}}, t_{32} = t_{30} t_{29}, t_{33} = t_{28} t_{30},$$

$$t_{34} = \frac{t_{23} t_{31} k_3}{t_{22} t_{25}} + t_{25} t_{31} k_3 + \frac{k_4}{t_{29}}, t_{35} = \frac{t_{23} t_{32} k_3}{t_{22} t_{25}} + t_{25} t_{32} k_3 - k_4 t_{29},$$

$$t_{36} = \frac{t_{24} k_3}{t_{22} t_{25}} - t_{25} t_{33} k_3 - \frac{t_{23} t_{33} k_3}{t_{22} t_{25}}, t_{37} = \exp(k_4(L_1 + L_2 + L_3 + L_4)),$$

$$t_{38} = V_{bid} + V_{DS} - \psi_{d4}, t_{39} = \frac{t_{36}}{t_{37} t_{34}}, t_{40} = t_{38} - t_{39}, t_{41} = t_{37} - \frac{t_{35}}{t_{37} t_{34}}$$

4.3 DISCUSSION ON THE RESULTS

Here, the solved model is analyzed in graphical form and compared with the Synopsys TCAD simulated graph. The validation of surface potential is also determined for varying the device's body thickness, gate oxide thickness, gate oxide materials, gate voltage, and drain voltages. For TCAD simulation, we have used various models such as the nonlocal band-to-band tunneling model, SRH recombination, bandgap narrowing model, doping dependence mobility model, and Fermi Dirac statistics [13,16–18].

The characteristic of surface potential with gate voltage variation is presented in Figure 4.2, at a constant V_{DS} of 0.7 V. The figure shows that there is a sudden change of potential in the junction between source and channel and drain and channel due to a change of charge carrier numbers. On the other hand, it is almost constant throughout the channel. It can also be visualized that the potential rises as the gate voltage rises, as the channel's gate control improves with higher gate voltage. In addition, as the gate voltage rises, the depletion charge rises, causing the surface potential to increase.

Figure 4.3 depicts the dissimilarity in surface potential with drain bias, and it can be seen that the fluctuation is only substantial at the drain edge. The potential curve is almost invariant in the source and channel regions. With increased drain voltage, its value marginally increases near the drain edge. As a result, the drain voltage is less at the source–channel tunneling junction. However, in the TCAD simulated graph, the variation of potential with V_{DS} is also observed in some parts of the channel, which may be due to the effect of the spacer dielectric included in the simulated device. The impact of the spacer dielectric material is not included in the modeling process. As demonstrated in Figure 4.4, variations in body thickness also affected the device's performance. Slight changes occur at the source–channel and channel–drain junction, showing the highest value at 10 nm body thickness. The junction potential has a decreasing behavior with increasing body thickness because of the lowering of capacitive coupling between the gate and the substrate. The good agreement between the modeled and simulated plots demonstrates the modeling approach's validity.

The type and thickness of gate oxide material have a major influence on device characteristics. If the gate-dielectric thickness is varied, the behavior of potential

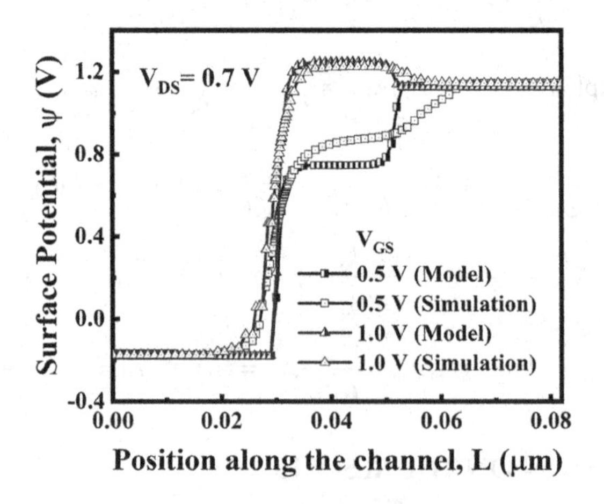

FIGURE 4.2 Characteristics of surface potential along x-axis for different V_{GS} at constant V_{DS}.

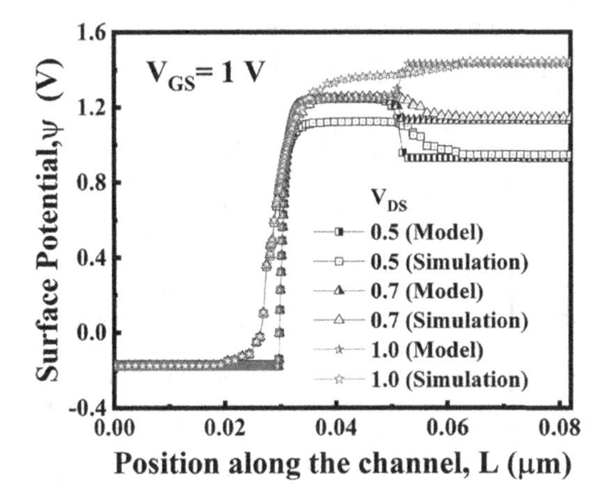

FIGURE 4.3 Characteristics of surface potential along the channel for various V_{DS} at constant V_{GS}.

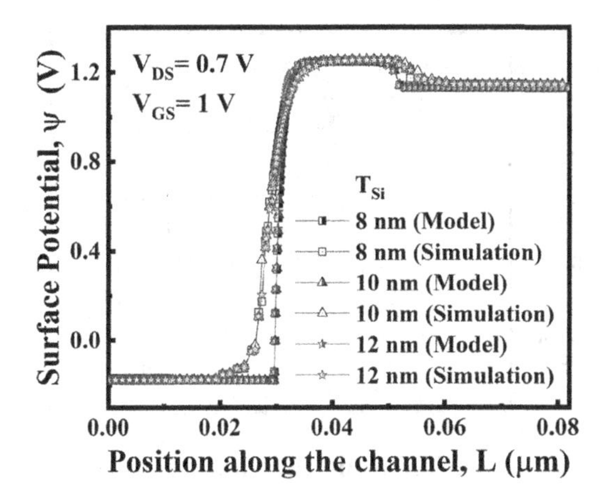

FIGURE 4.4 Characteristics of surface potential along the channel x-axis for different values of body thickness.

also varies, which can be perceived from Figure 4.5. The difference is more significant in source–channel and channel–drain junctions. The slope of the potential curve slightly decreases at both junctions as we increase the oxide thickness due to the degradation of control of the gate over the channel. As the gate oxide materials vary, the dielectric constant also varies, and the potential variation occurs from source to drain throughout the channel (Figure 4.6). It is easy to see how the potential value rises when the gate oxide material's dielectric constant rises; the highest value observed was for HfO_2 with a dielectric constant of 21, and the lowest for SiO_2 (dielectric constant of 3.9). This is because the capacitive connection between the channel surface and the gate is increased, as a result of which the dielectric value is increased.

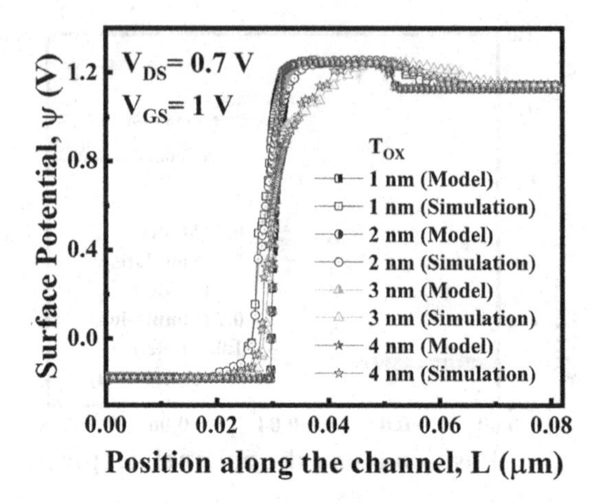

FIGURE 4.5 Characteristics of surface potential along the channel for different values of gate oxide thickness.

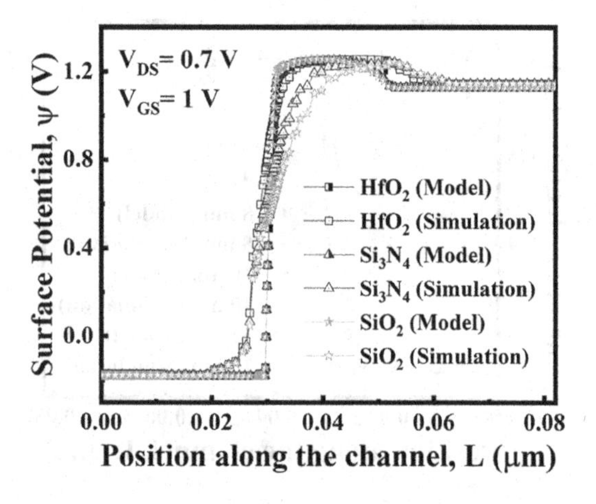

FIGURE 4.6 Characteristics of surface potential along the channel x-axis for different values of gate oxide material same V_{DS}.

4.4 CONCLUSIONS

An analytical model of surface potential for a suggested double-gate TFET is established in this paper. Sentaurus TCAD software is used to validate the proposed work's outcomes. The modeled and simulated plots are in good agreement. The source depletion area and drain depletion area are considered in all cases of analytical modeling. The surface potential is explored for adjusting several factors such as body thickness, gate-dielectric thickness, gate-dielectric material, gate bias, and drain bias, among others, to determine the correctness of analytical modeling. The proposed model in this paper is validated by the close proximity of the modeled and simulated graphs.

REFERENCES

[1] M. Ionescu and H. Riel, "Tunnel field-effect transistors as energy-efficient electronic switches", *Nature*, vol. 479, pp. 329–337, 2011.

[2] J. K. Mamidala, R. Vishnoi, and P. Pandey, *Tunnel Field-Effect Transistors (TFET): Modeling and Simulation*, John Wiley & Sons, 2016.

[3] E. O. Kane, "Theory of tunneling", *Journal of Applied Physics*, vol. 32, no. 1, pp. 83–91, 1961.

[4] K. Boucart, and A. M. Ionescu, "Double-gate tunnel FET with high-k gate dielectric", *IEEE Transactions on Electron Devices*, vol. 54, no. 7, pp. 1725–1733, 2007.

[5] A. Chattopadhyay, and A. Mallik, "Impact of a spacer dielectric and a gate overlap/underlap on the device performance of a tunnel field-effect transistor", *IEEE Transactions on Electron Devices*, vol. 58, no. 3, pp. 677–683, 2011.

[6] A. S. Verhulst, B. Sorée, D. Leonelli, W. G. Vandenberghe, and G. Groeseneken, "Modeling the single-gate, double-gate, and gate-all-around tunnel field-effect transistor", *Journal of Applied Physics*, vol. 107, no. 2, p. 024518, 2010.

[7] W. G. Vandenberghe, A. S. Verhulst, G. Groeseneken, B. Sorée, and W. Magnus, "Analytical model for a tunnel field-effect transistor", *MELECON 2008-The 14th IEEE Mediterranean Electrotechnical Conference*, pp. 923–928, May 2008.

[8] M. G. Bardon, H. P. Neves, R. Puers, and C. Van Hoof, "Pseudo-two-dimensional model for double-gate tunnel FETs considering the junctions depletion regions", *IEEE Transactions on Electron Devices*, vol. 57, no. 4, pp. 827–834, 2010.

[9] R. Goswami, B. Bhowmick, and S. Baishya, "Physics-based surface potential, electric field and drain current model of a δp+ Si 1–x Ge x gate-drain underlap nanoscale n-TFET", *International Journal of Electronics*, vol. 103, no. 9, pp. 1566–1579, 2016.

[10] R. Dutta, and S. K. Sarkar, "Analytical modeling and simulation-based optimization of broken gate TFET structure for low power applications", *IEEE Transactions on Electron Devices*, vol. 66, no. 8, pp. 3513–3520, 2019.

[11] R. Das, and S. Baishya, "Analytical modeling of electrical parameters and the analog performance of cylindrical gate-all-around FinFET", *Pramana*, vol. 92, no. 1, pp. 1–10, 2019.

[12] S. Kumar, E. Goel, K. Singh, B. Singh, M. Kumar, and S. Jit, "A compact 2-D analytical model for electrical characteristics of double-gate tunnel field-effect transistors with a SiO_2/High-k stacked gate-oxide structure" *IEEE Transactions on Electron Devices*, vol. 63, no. 8, pp. 3291–3299, 2016.

[13] Sentaurus Device User Guide, Version G-2012.06, 2012.

[14] N. Bagga, and S. K. Sarkar, "An analytical model for tunnel barrier modulation in triple metal double gate TFET", *IEEE Transactions on Electron Devices*, vol. 62, no. 7, pp. 2136–2142, 2015.

[15] J. Talukdar, G. Rawat, B. Choudhuri, K. Singh, and K. Mummaneni, "Device physics-based analytical modeling for electrical characteristics of single gate extended source tunnel FET (SG-ESTFET)", *Superlattices and Microstructures*, vol. 148, p. 106725, 2020.

[16] K. Baruah, R. Das, and S. Baishya, "Impact of trap charge and temperature on DC and analog/RF performances of heterostructure overlapped PNPN tunnel FET", *Applied Physics A*, vol. 126, no. 11, pp. 1–12, 2020.

[17] D. Kumar, "Performance evaluation of double gate tunnel FET based chain of inverters and 6-T SRAM cell", *Engineering Research Express*, vol. 1, no. 2, p. 025055, 2019.

[18] D. Kumar, S. B. Rahi, and P. Kuchhal, "Investigation of analog parameters and miller capacitance affecting the circuit performance of double gate tunnel field-effect transistors", In Choudhury, S., Gowri, R., Sena Paul, B., and Do, D. T. (eds.), *Intelligent Communication, Control and Devices*, Springer, Singapore, pp. 335–349, 2021.

5 Impact of Semiconductor Materials and Architecture Design on TFET Device Performance

M. Saravanan
Sri Eshwar College of Engineering

Eswaran Parthasarathy
SRM Institute of Science and Technology

J. Ajayan
SR University Warangal

D. Nirmal
Karunya Institute of Technology and Sciences

CONTENTS

DOI: 10.1201/9781003240778-5

5.1 INTRODUCTION

Similar to metal oxide-semiconductor field-effect transistor (MOSFET), the tunnel field-effect transistor (TFET) has the same structure, but the key mechanism of switching varies. In distinct typical MOSFET, the TFET switching mechanism is accomplished by manipulating the quantum tunnel flowing through the barrier, rather than modifying thermionic emission at the barrier. The transistor consists of three or four poles whose components are built using silicon. This type of transistor operates on the idea of a gate with a tunnel, and its basic structure is a closed PIN diode. It offers several benefits over MOSFET [1]. Because of the reduced output current, the subthreshold swing (SS) is not limited to 60 mV/decade, and there is improved resistance to short-channel effects (SCEs), suitability for low-power applications, faster tunnel working speed, and considerably lower threshold voltage and current. The gate of a well-designed TFET must efficiently alter the width of the tunnel barrier. To satisfy this criterion, the tunnel junction must obviously be built with the gate modulation electric field component in the same direction [2]. Controlling the carrier injection mechanism optimally enables improved regulation of the drain current level. Based on this idea, many TFET designs may be constructed, and an attempt to employ oblique source coupling in this approach has recently been made. For low-power and high-speed applications, consider TFET as a viable alternative to MOSFET. When compared to other alternative device designs, TFET is built by the reverse-biased gate, which have the benefit of being applicable for the processing of CMOS in a normal condition [2]. These types of transistors do not depend on any other processes such as the impact of ionisation, and so they are known as unreliable sources. Silicon-based TFETs have been investigated as electrostatic discharge (ESD) protection devices and proposed for use in ESD protection networks. It has been established that utilising TFETs in place of conventional ESD diodes in on-chip ESD circuits can enhance the reliability of on-chip ESD protection by introducing secondary discharge pathways.

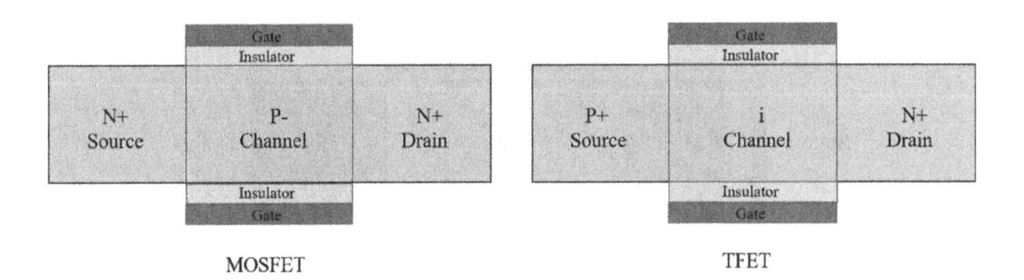

FIGURE 5.1 Comparing MOSFET and TFET.

5.2 OPERATING PRINCIPLE OF TFET

A TFET component's construction is similar to that of a MOSFET. Although the source and depletion of a MOSFET are doped with the same type of dopant, the source and depletion of a TFET are doped in opposite directions as show in Figure 5.1. The trapezoidal range profile is formed by the potential at the intersection of p_i and n_i. Built-in potential barriers prevent electron and hole currents from flowing, when a marginal V_{DS} is applied to the steady state. The channel's bands increase when a negative gate bias is introduced. The charge carries tunnel across the band gap, when the system valence band increases over the conduction band (CB) source, thus also resulting in the negative V_{DS}. Holes accumulate in this channel, the working mode of P channel at ON state in the TFET. A positive V_{GS} and a negative V_{DS} can be used to generate the n-channel ON-state.

For a large negative V_{DS} value, saturation occurs similarly to the situation of a MOSFET. The combination of a large V_{DS} value and only a moderate V_{GS} may even produce a potential distribution of two transitions at same time tunnelling, which leads to the operation of V_{DS}–I_D. Devices such as n-channel and p channel TFETs that work in the same device are called bipolar devices [3,4]. For a large enough positive V_{DS}, the connection configuration is forward biased, which implies that in the absence of a gate voltage, since electron–hole pair currents can both propagate, an increasing diode characteristic results in barriers, but not at the same time. A forward-biased tunnel diode operates with no net current for electrons on both sides of the junction with equal amount of energy levels. A great current flow, on the state of electron flow, occurs in the region of n-type to the p-type of the device [4], but when electrons enter the tunnel from the north side, the current reduces to a very small amount. Electrons are expelled through the barrier as a result of thermal emission, and the current rises once again.

5.3 STRUCTURE OF TFET

TFET is mainly used for low-power devices, and for its application, we employ band-to-band tunnelling (BTBT). By comparing with the structure of MOSFET, it contacts surfaces showing variation in the opposite doping polarity. In the source region of TFET based on different polarities, in nTFET, the source region is the p-type, and for pTFET, the source region is the n-type. There are several variants of the

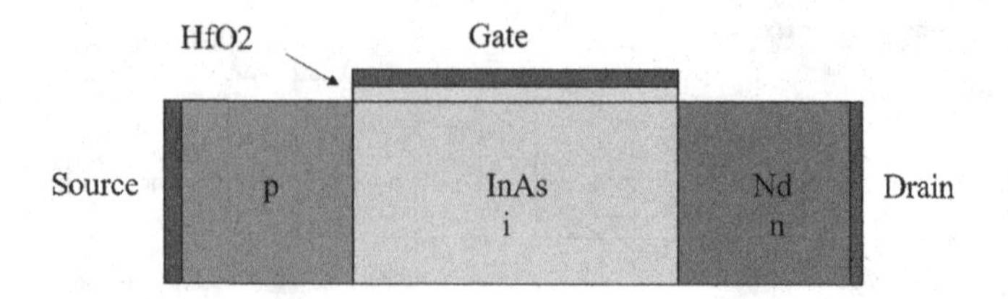

FIGURE 5.2 p-i-n TFET structure.

fundamental design, such as various gate overlaps or doping profiles. In doping the TFET, the Fermi level is associated with the valence band edge of the source region [5]. The TFET works by controlling the electrostatic potential at these places via the gate contact, permitting and opposing BTBT tunnelling among source and channel regions. The connection is given as reversed as in the source and drain, and it stimulates the diode to work on reverse as shown in Figure 5.2. In the diode, current flows in reverse leakage in the OFF condition, and this occurs often due to the diffusion current of the minority charge carrier.

The source is depleted as V_{GS} increases; then, the electric current near the channel connection increases. It causes an increased bending of band at the source–channel junction. At a certain V_{GS}, known as the onset voltage, the tunnel CB crosses over with the source valence band, allowing for tunnelling transitions between these two channels.

These transitions can occur directly between the maximum valence band value and minimum CB value in the section or indirectly between the highest valence band value and one of the CB values. In the second situation, the conversion is supported by phonons. The TFET is essentially a band pass filter, with only the statuses in the overlapping band contributing to the total current; as a result, the Fermi distribution's high-energy and low-energy states are taken off [5]. This state is similar to the conserving of traditional equipment: The regions above the Fermi level have a lower population density at lower temperatures, which reduces the output current and leads to an abrupt SS increase.

5.4 DEVICE DESIGN AND IMPLEMENTATION

The tunnelling FET design has been modified and efficiently designed in 30 nm technology using various materials by modifying the materials used for the drain, source, and channel. In particular, we designed a TFET heterostructure and a simple silicon TFET PIN structure in this work. SOI wafers are used in the development of the PIN-TFET workflow. The workflow utilises a high-quality metal gate technique based on the CMOS technology. Materials with a higher relative dielectric resistance are utilised to minimise leakage current and enhance productivity. The function of the diode between the source and drain must always be monitored while testing these contact design TFETs to ensure that the source/drain implantation half-masks are aligned. When measuring the characteristics of I_D–V_{GS}, the total current at the terminals must be checked to measure the actual tunnel current [6]. Managing germanium offers a number of difficulties, including obtaining a suitable gate dielectric surface and dopant initiation. However, because of its better mobility and several solutions to these processing issues, germanium has been actively researched as a channel-replacement material for MOSFETs.

Because of such benefits, the use of germanium to increase TFET performance is being investigated. It has been proven possible to create TFETs utilising silicon/germanium heterostructures. Because the tunnelling effect occurs mostly at the PIN-TFET structure's source–channel interface, reducing the band gap in this location will be linked to the investigation of utilising silicon-germanium heterostructures at the interface of the source and channel. Additionally, to understand the basics of silicon, the structure of silicon is used as an additional advantage [7]. Germanium provides the advantage of the tunnelling method having the lowest germanium band gap, and it also has the benefit of a good dielectric silicon gate contact. The tunnelling FET with excessive-k substances may be used as excessive semiconductor devices with high overall performance. Even after continuous functioning of transistors, high-k dielectric substances are still rare. The usage of high-k material provides sufficient electrical stability [6,7] and the quantity of rate within the excessive-k. The material employed must be scalable because it exhibits an appropriate stage of electron and hollow mobility even at reduced thickness.

5.5 CIRCUIT DESIGN FOR TFET PERFORMANCE

At the device level, TFET is superior to conventional TFET design. However, the characteristics of such circuit-level devices have yet to be studied. This section studies the enactment factors of TFET-based circuits and compares them with their Fin-FET equivalents. In terms of performance, the two devices channel lengths are maintained constant at 14 nm, and there are several circuits that emphasise performance, such as mirror circuits.

5.5.1 Current Mirror Circuit

The current TFET-based mirrors are analysed and compared to their Fin-FET equivalents. The basic structure of the current mirror circuit is shown in Figure 5.3. Since

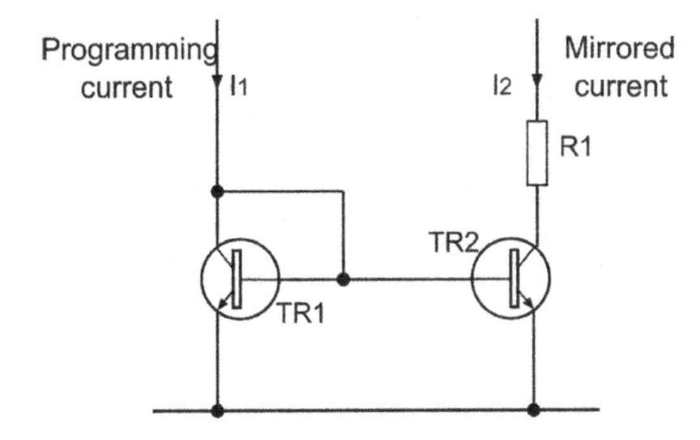

FIGURE 5.3 Current mirror circuit.

the resultant impedance of TFET is higher than the Fin-FET current mirror, the TFET current mirror can provide accurate output current. In addition, both current mirrors provide better I_{out} accuracy at an input current of 10 μA compared to an input current of 50 μA. Characteristics of conventional and cascaded current mirrors are as follows. The cascade transistor increases the output impedance of the current mirror, thereby obtaining improved saturation performance in the saturation range [8]. In FET tunnelling, the output impedance is quite high, and these devices exhibit excellent I_{DS} saturation characteristics in the saturation band.

The above diagram shows the architectural structure of a current mirror circuit. In addition, the cascaded stage delay I_{out} is saturated; this reduces the working functionality of TFET-based current mirrors. As a result, in situations where extremely exact replication is not required, traditional current mirrors based on the TFET may be preferable to cascaded current mirrors. Current TFET mirrors have better performance than Fin-FET mirrors.

5.5.2 TFET-Based SRAM

To store the data for forever in the memory circuit, we need a static state. The information storage cell, which is the one-bit memory cell within the static RAM arrays, consists of an easy latch circuit with two stable operative points. Counting on the preserved state of the two electrical converter latch circuits, the data being command in the memory cell are going to be understood either as logic 0 or logic 1 [9]. The 6-T SRAM's browse and write performance is mainly limited by the size of the transistors in the read stability and write capabilities of a 6-T SRAM cell as shown in Figure 5.4.

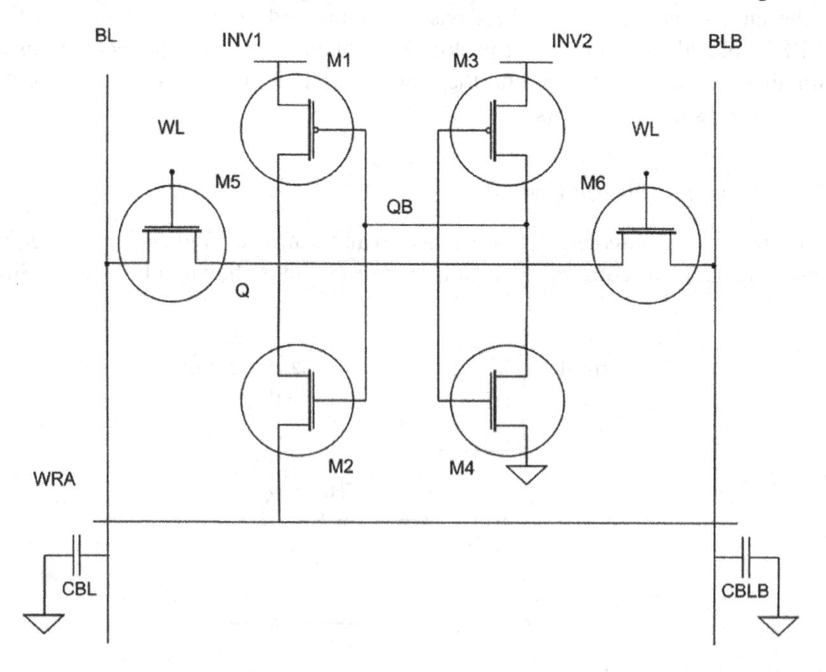

FIGURE 5.4 Circuit design of SRAM.

The N-curve simulation approach is employed. The N-curve may explain the entire practical analysis of the SRAM cell in terms of performance-moving parameters determined by static voltage noise margin, static current noise margin, and write trip current WTI. TFET SRAM cells use internal read access transistors; therefore, they suffer greater read damage and less than 50mV at $V_{dd} = 0.6\,V$. TFET SRAM cells 7T and 8T use read buffers to read the current path. A storage node is added from the cell to reduce read interference, despite the fact that the SRAM-TFET-7T cell utilises one transistor for reading rather than the two stacked transistors in the SRAM-8T-TFET cell. It should be noted that because TFETs have unidirectional current flow characteristics [10], the transistors in the 7T TFET cell can be read, and this prevents current from passing into the input node from the unselected cell in the read transistor of the specified bit line. The 6T TFET SRAM design is depicted in the diagram below.

Write Mode: Because TFETs can only conduct current in one way, all TFET SRAM cells are utilised in the writing procedure. As a result, the WSNM of the TFET SRAM cell is much lower than that of the 8T-MOSFET-SRAM cell.

Writing Assist: To improve the writing capability and performance of TFET SRAM cells without the need of extra cell circuits (for example, using full conduction gates in place of single pass transistors to improve the writing process), circuit technology that allows for writing in the form of virtual ground Pin and floating power (can be written using the FP line) can be used [11]. Using the FP Write Assistant has significantly improved the WSNM of the SRAM-TFET-7T/8T cell.

The structure of the SRAM cell and the resultant to the read and write path: The SRAM 4T DL Fin-FET cell presented use for the independent gate control capability. The fundamental concept is to merge the accessible transistor and the extraction transistor into a gate extraction device that can be operated separately, with the back gate acting as the access transistor.

5.6 TFET FABRICATION AND CHARACTERISTICS

The concept of using planar transistors in silicon-on-insulator substrates in TFET: TFET fabrication follows a manufacturing flow that is quite similar to that of normal CMOS manufacture, making the TFET a highly appealing device for the semiconductor industry to adopt. Molecular beam epitaxy (MBE) is used to produce a 100nm intrinsic silicon layer using an n+ doped silicon substrate, which acts as the channel area (source). The intrinsic layer is produced first, and then the silicon epitaxy layer is doped, which works as the drain. On the sidewall, the gate oxide is produced, and then the n+ doped polysilicon is coated as the gate interface. Diffusion alloying can also be used to create TFETs [12]. The SOD layer's external diffusion might first create the highly doped n or p regions. The surface doping concentration p achieved via diffusion is quite high when compared to the MPE doping technique. This manufacturing method begins with silicon with its own additives, which is utilised as a substrate. To restrict the diffusion of

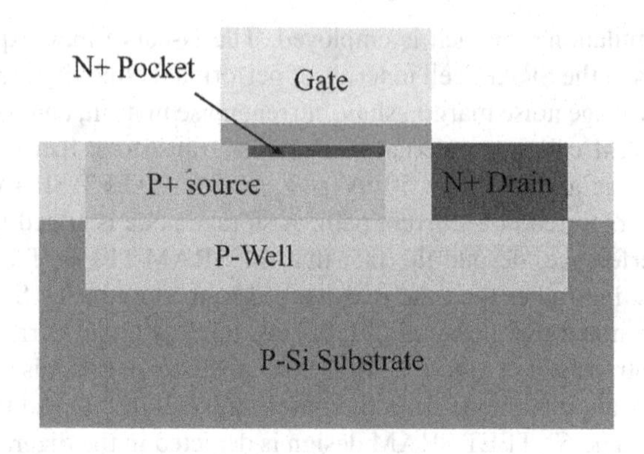

FIGURE 5.5 Schematic of the fabricated Si line tunnelling TFET.

p dopants, n + diffusion occurs before the p + diffusion [13]. Following all diffusion processes, the oxidation process occurs. The length is 1 m, and the thickness of gate oxide is about 4.2 nm, in order to compare with the experimental gadget.

A high doping level and the thin gate oxide are the primary requirements for excellent performance of tunnel effect transistors. In order to meet these technical requirements, production technology must improve. An improvement required for vertical TFETs, vertical group diodes, self-levelling gates, and shallow trench in isolation (STI). Based on the etching profile, the etching process may be divided into two types: isotropic etching and anisotropic etching [14,15]. The etching of the substrate material is independent of direction in isotropic etching, resulting in lateral etching of the substrate as illustrated in Figure 5.5. The dimensional changes caused by side etching during lateral pulling should be considered. The etching technique may be divided into two types based on the etching environment: wet and dry etching. Initially, wet etching was most commonly used in the IC industry. Dry etching is becoming more common in today's semiconductor production lines.

5.6.1 Homojunction Vertical TFET Fabrication

The InAs TFET is constructed using the same parameters as those for Sib-based TFET. It is a homojunction TFET device because all its source, gate, and drain contain InAs material. To enhance the device's surge current, a material with a smaller band gap can be used instead of Sib-based TFET devices. To maximise the tunnelling effect and boost surge current and oscillation below the threshold of TFET devices, low-band-gap materials (such as InAs) are utilised for the source, drain, and the channel.

The first-generation transistor is used because it is not a flip gate and does not set itself. Figure 5.6 shows the homojunction schematic of InAs TFET. This section describes the use of $In_{0.53}Ga_{0.47}$ As to fabricate a vertical homojunction tunnel transistor [16]. The gate dielectric is deposited using a vertical TFET fabrication process mentioned above, the atomic layer deposition (ALD). The basis of ALD is self-limiting atomic layer-by-layer growth and very conformal coating. ALD is distinguished by the pulse of chemical precursors that combine to produce one atomic layer and one cycle.

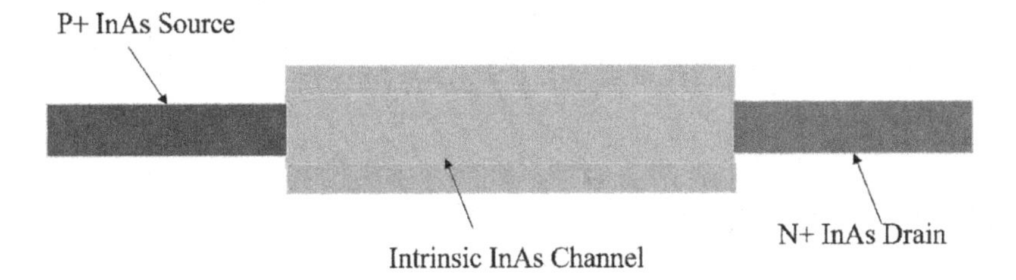

FIGURE 5.6 InAs homojunction TFET schematic.

As a consequence, conformal coating with no pinholes is accomplished in deep pores, trenches, and cavities. In the forward bias domain, gate-controlled negative differential resistance characteristics may be detected, following which, a standard diode switch at a greater negative drain is turned on to provide voltages. Upon oxide-semiconductor contact in the NDR region, in the p + source area, direct BTBT activity occurs in the CB to valence band channel. The BTBT current is temperature-independent in the pre-NDR region, the valley current or surplus current is temperature-dependent, increasing with increasing temperature and decreasing ultimately to the valley current ratio.

Compared with Si TFET, InAs TFET shows improved surge current and subthreshold fluctuation. As in InAs material, its energy gap (0.45 eV) is smaller than that of silicon (1.1 eV) [16]. This shows a significant improvement in the subliminal fluctuation value of 42.16 mV/decade. Figure 5.7 shows the comparison between Si-based and InAs-based TFET. Therefore, this device configuration shows promising performance, but it is ineffective for low-power, high-performance applications.

According to the findings of the experiments, the InAs TFET device configuration exhibits good SS. The TFET device has made significant progress and is now a viable option to replace MOSFETs in low-power, high-performance applications.

5.6.2 Heterojunction TFET Design and Fabrication

At room temperature, tunnel FET transistors may attain switching edges of less than 60 mV/decade, allowing for the power supply voltage to be increased. The heterojunction TFETs based on III-V semiconductors are gaining popularity due to their maximum ON current and OFF current ratio (I_{ON}/I_{OFF}) and high ion limitation. Using AsSb-based heterojunction hybrid TFETs, a wide range of adjustable effective barrier heights can be achieved. A heterojunction TFET n-channel FET with a suitable band gap design is believed to produce 135 μA/μm MOSFETs with a high I_{ON}/I_{OFF} of 104 at 0.5 V V_{DS} [17].

Instead of using a p+ source as part of a traditional TFET, the tunnel connection is located among the p+region and a thin n-layer with full drain below the gate as in Figure 5.8, thus reducing the width of the tunnel and causing adjacent tracks to bend. Compared with the traditional gate oxide TFET, its pocket improves the tunnelling effect by achieving steep subthreshold conductivity while also providing high ion concentration. In a heavily doped (P^+) N-channel TFET substrate, the device design can be optimised to reduce design costs, improve design efficiency, and achieve the best device and technology development. Simulation can be used to predict the electrical performance

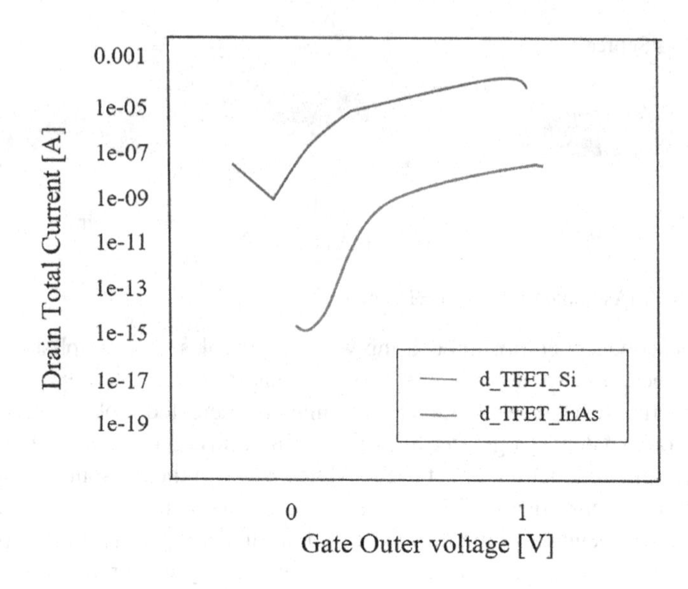

FIGURE 5.7 Combined curve for both Si-based TFET and InAs TFET configuration.

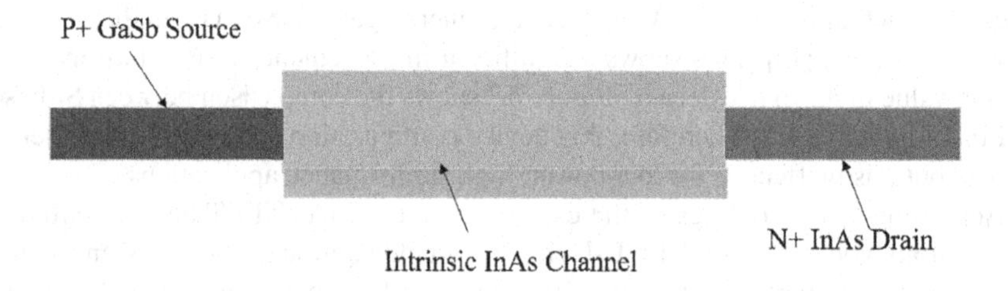

FIGURE 5.8 GaSb-InAs heterojunction TFET schematic.

of equipment. The selection of source/channel materials with a smaller (effective) band gap is critical for enhancing tunnelling current while decreasing OFF current.

A tunnelling FET's ON-state current might be significantly boosted by using materials with lower band gaps. Electric current passes from the source to the channel region, and then vice versa. When current travels from a drain to a channel and then back to the source, the tFET uses a MOS gate to drive the band-to-band tunnel through the resulting pn junction. An n-channel TFET's cross-section and band diagram in its OFF and ON modes, which prevents band-to-band tunnelling, are shown. When the gate movement is small, the bulk of the channel valence band may flow better than the drain CB's bottom [17], which it connects the channel to the drain via an electron tunnel. The channel conductivity varies from one kind of carrier to another in this situation, and the transient response is bipolar. This is a rather typical occurrence, using TFET geometry.

5.6.3 Effect of Doping Concentration

The intensity of doping has a significant impact on the TFET's performance. The integrated $V_{bi,}$ pi is determined by the doping concentration, which also specifies the

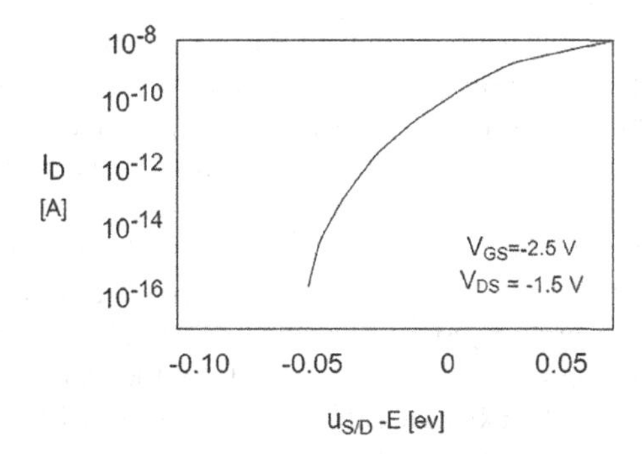

FIGURE 5.9 Concentration of doping affects the potential of the p-i-n structure and band bending at junctions.

potential with offset required for tunnel building, as well as the V_{DS} and V_{GS} used. The biggest V_{bi} is the smallest V_{DS} that produces a similar band bending. All electric fields that enter these regions are filtered by the charge carriers at the region of source and drain. A higher doping concentration decreases parasitic series resistance in a component construction based on doped lines because the doped semiconductor resistance decreases as the doping concentration increases [18]. Despite the reality that this impact may be shown in experimental data, it is observed in the current study. At V_{DS} 0.1 V, the distance between the chemical capacity and the related band side varies up to 50–350 meV. The shifting internal potential is represented by a change in the distance between the n and p branches. The higher the level, the greater the bipolar influence. At the lowest possible doping concentration, V_{DS} is relatively low because the intrinsic potential cannot be compensated; therefore, the transmission is blocked.

The current level rises when doping concentration increases; however, there is another negative effect that offsets the advantage of high doping concentrations. At about $V_{GS} = 1.5$ V, the slope below the threshold declines rapidly at high doping concentrations as shown in Figure 5.9. Increased alloy concentration [18]: The extended Fermi distribution shifts in the direction of the source CB as doping levels increase.

This suggests that there must be a trade-off for doping concentration. On the other hand, excessive levels of high current need excessive doping degrees. A high slope, on the other hand, restricts the number of permitted doping degrees. The current change is influenced not only by the potential attached but also by the screening length. In addition to the band bending at the intersections, the doping has an effect on the p-i-n structure's inherent potential. The output current is standardised as a consequence of VDS [19]. Because of the curvature of band increases, the nonlinearity of the onset decreases as the doping level increases. Doping concentration is a significant element that influences TFET performance in various aspects. The optimum choice is quite narrow, but based on the research given, a Fermi level of 50–100 meV in the case of silicon, the band looks to be the most capable.

5.6.4 Subthreshold and Output Characteristics

When compared to traditional silicon MOSFETs, the exponents are generally recognised to have high contact resistance and contact Schottky barriers; TFETs are predicted to exhibit this behaviour even with the best connections. We tested the forward current of the p-i-n structure under comparable bias settings to exclude any pin-related effects in our device in 100 mA mode. This suggests that the index can initially be explained by the large number of subvolumes of the return channel when the device is turned on, rather than by the effects related to contact. For reversible loads, the energy band in the channel becomes very weak compared to the external gate voltage [19]. However, the degree of channel potential build-up is determined at the drain contact in the Fermi level energy region. The channel potential varies as you travel, causing a nonlinear performance. Only when the state of the channel strip corresponds exactly to the voltage of the external gate can the linear start be expected. It is too large for a device with quantum capacitance because SOI-TFET is a two-dimensional system with density solid conditions ón the channel. As a result, lowering the density of states in the channel is one technique for obtaining a linear start. A further in-depth study of the output characteristics reveals that the total current intensity is lower than that of standard silicon MOSFETs, which is related to the lower tunnelling probability. Ion limit compared to classic MOSFET: To obtain the upper limit, the gate length depends on the saturation current shown in the illustration.

The measurement error is 1%, and the output current is independent of L, which indicates that BTBT is the main leakage event. Doping has no effect on the minimal inverse subthreshold slopes. However, when the doping concentration increases, the average inverse subthreshold slope improves. This implies that there is no evidence of an excessive doping concentration, which might worsen the anticipated subthreshold slopes [20]. There is no noticeable trend for the n-branch at positive V_{GS}, most likely due to activation difficulties in the case of boron. Dopant cluster formation and out-diffusion are two examples of effects, particularly at high doses, which may result in an unintended dependence of active doping concentration on implantation dosage that is not even accounted for in process models. When the n and p branches are compared, the finding of variations in doping concentration cannot account for the significant difference in reverse subthreshold increases seen in both branches. Figure 5.10 shows the output characteristics of TFET. Saturation is unique for all dosages and follows at

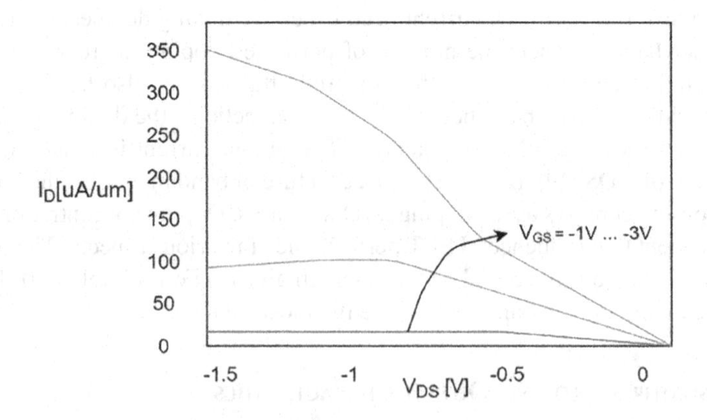

FIGURE 5.10 Output characteristics of TFET.

the same V_{DS}, indicating that the threshold voltage fluctuation across models is minimal. As predicted, the present levels correspond to the implantation dosages.

5.7 TFET CIRCUIT EVALUATION METHODS

5.7.1 TFET Circuits and Current Mode Logic

Current mode logic (CML) employs a set of differential digital logic. The CML gate has a tail current source, a current control core, and a differential load. CML's gate operating approach involves changing the DC current via the input transistor's differential circuit and causing the decreased voltage swing on the output on both the load devices. Because CML is rarely utilised in fundamental circuit design, its distinct characteristics, such as low latency and constant power consumption, may be applied in specialised applications such as DPA measures. CML gates based on TFETs have also lately been developed. Early CML gate circuit designs relied on a recently discovered GaSbInAs heterojunction TFET, which increased ON-state current with heteroband alignment. When compared to CMOS counterparts [21], TFET-based CML designs consumed less power. However, TFET-CML gates had been displayed, and TFET-CML is now no longer used in any respect within the hardware safety domain. To thoroughly assess TFET-based logic not just in terms of old metrics such as latency and power but also in terms of novel criteria such as security, the schematic diagram of CML circuit is shown in Figure 5.11.

The pull-up network for TFET-CML is made up of two resistors or two P-type TFETs (PTFET). Because contemporary technology's power consumption and resistance area are much larger than those of field-effect transistors, pull-up networks based on field-effect transistors are predominant. The pull-up network functions

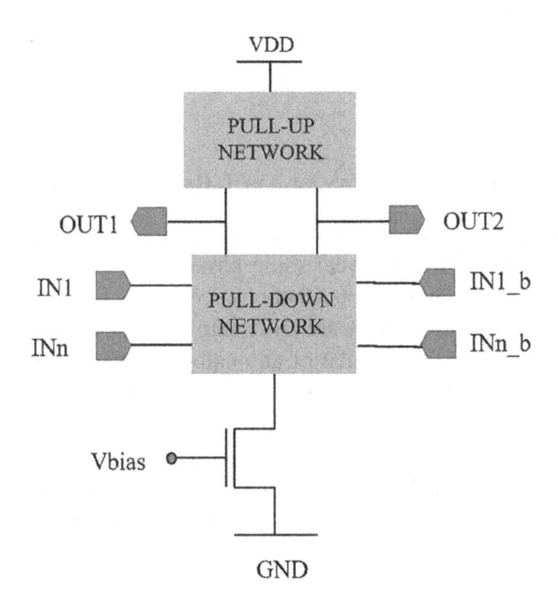

FIGURE 5.11 Schematic diagram of CML circuit.

primarily as a load device, controlling the voltage drop on the output DC [21]. The resistance of the PTFE may be changed to adjust the output voltage simply by modifying the grid bias of the P-type FET.

5.7.2 Design Optimisation

The TFET I_{ON} increases as the oxide thickness decreases. The narrower crawler tractor opens a vertical tunnel on the lower V_{GS}. Gate greater than 1.2 nm significantly increases gate leakage current. It has been perceived that TFET provides the best performance when the oxide thickness is in the 2–4 nm range. The capacitive coupling between source and gate is improved. The gate oxide layer thickness and the dielectric constant are complementary, and the combination of the two may characterise the sensitivity of the gate voltage across the tunnel junction, which affects the device's performance. When the effective channel length is scaled [21], the TFET performance does not change. This is because the source and drain are not aligned in a plane. Effective channel length and scaling of 12 nm increases the device's turn-off current while decreasing the TFET's overall performance. However, the gate-source capacitance also increases at the same time, which limits the high-frequency applications of the device. This is a compromise between I_{ON} and C_{GS}. The equipment design optimisation is relevant to the respective material and alloy concentration. However, it is dependent on the substance and the concentration of doping. Any escalation in the epitaxial layer's doping concentration will result in an increase in I_{ON} and I_{OFF}.

Understanding the device's usefulness necessitates investigating the operational characteristics of the TFET circuit. The operational parameters of the inverter circuit were briefly studied in order to understand the circuit characteristics of the TTFET construction. The inverter circuit is implemented in the Cadence Virtuoso environment using the Verilog-A model of the planned construction of the TFET device. Due to the sharp drop below the threshold [22], the TFET inverter turns on and off, and then turns on and off very suddenly. The inverter circuit delays energy as a function of supply voltage.

5.7.3 Evaluation of the Security of TFET-Based CML Gates

Although the hardware implementation of the encryption algorithm can provide better performance and speed, it may reveal some physical information to retrieve sensitive data. CML can perform different functions according to different settings. When stacking planes and various pairings are taken into account, the latency of a gate with more than three planes surpasses that of an analogous static multiplexer with three planes; in order to accomplish optimisation, the number of stages of differential pairs is restricted to three. Four CML functions based on two-input TFETs with a two-layer structure are introduced. As inputs, each gate contains three differential pairs. The power supply voltage and peak-to-peak voltage characteristics serve as the foundation for other factors such as transistor size and bias voltage. To reduce the area, the TFET width is equal to the technical length. The TFET-CML gateway is used for lightweight encryption implementation [23], and we will first carefully study TFET-CML from a hardware security perspective. The primary idea of differential energy analysis, as we know, is primarily based totally at the energy intake within the circuit transition.

Static XOR TFET gate and differential XOR TFET gate power supply line: Obviously, the XOR-TFET-CML valve provides almost constant power, rather than a significant power surge from a static XOR element. The static XOR element filters extra information for attackers to discover the cryptographic system's internal operations. However, the TFET-CML-XOR gate's nearly constant power consumption offers little information about the conversion of the data. Furthermore, as previously mentioned, the shift from zero to one is clearly represented in the transition from one to zero. On the CML gateway, attackers can recover some information after a power outage, but it is difficult for them to determine the value of the processing logic. CML is not commonly used to construct cryptographic hardware due to its huge size and high-power consumption, particularly in lightweight cryptographic systems. Researchers often use other methods to protect encryption schemes from DPA attacks. It already contains a large amount of computing power and consumes relatively large energy and space. Therefore, when considering devices for the Internet of Things, WSN nodes, etc., CML based on low-power TFETs is particularly valuable. The energy advantages of TFET-based CML valves also promise that they would continue to optimise our circuit specs and develop standard CML libraries. As stated earlier, the beauty of creating a standard cell library in the current mode is that you can use standard logic gates to generate additional logic gates that follow the patterns in the CML design pattern. In addition, various configurations of the driving force of the logic element can be realised by changing the tail power supply. TFET can help improve the circuit's resistance to CPA attacks while yet consuming lesser power than comparable CMOS devices.

5.8 PERFORMANCE AND RELIABILITY ISSUES

FETs are distinguished by their low I_{OFF}, SS, resistance to short-channel impacts, and compatibility with CMOS technology, which are all advantages, making them excellent for accurate low-power applications. Due to the tunnelling phenomenon, it is unable to meet the growing demand for very high-speed and low-power applications (BTBT). Another drawback of TFETs is their bipolar conductivity, which can lead to significant issues such as inverter logic circuit failure. Device performance has mostly been studied for several new device architectures and TFET material designs in an attempt to improve device operating functionality [24]. In order to evaluate the device's dependability and suitability for applications covering a wide temperature range, its performance on the device impact must be evaluated. Furthermore, it has been claimed that the dependability of TFETs is a more significant issue than that of MOSFETs because of field tunnelling variations produced by interface trap charge (ITC). A larger electric field is necessary near the tunnel junction (source–channel junction) in FET tunnelling to minimise the width of the tunnel barrier; however, this high lateral field might contribute to trap formation. Local charge and interface (donor and acceptor) (positive and negative): These charges of deception may have significant consequences for equipment dependability and longevity. The device's performance, on the other hand, is constantly affected by the operating temperature. The operating temperature of the microcircuit and the device rises significantly as the microcircuit increases heat dissipation.

5.8.1 Radiation Reliability of Low-Power Devices

Radiation-induced single interference, also known as soft error, has become a serious concern for data centre applications as the number of computer nodes increases. If the soft error rate (SER) of each state bit of each generation of technology increases by 8%, the error rate of each chip is expected to increase by 100 times from the 180-nm node technology to the 16-nm node technology, which may lead to significant system reliability. This is lowered by reducing the load on the circuit nodes, and scaling V_{DD} poses a greater challenge in keeping SER low [25]. As we know, they are good candidates for realising TFET to lower the tunnel barrier. However, because of their low ionisation energy, these narrow-band-gap materials are more susceptible to radiation than silicon.

5.8.2 Models of Threshold Voltage

One of the most important aspects of a MOSFET is its threshold voltage. A MOSFET's threshold voltage has a straightforward physical description. It is the gate voltage when the quantity of charge in the channel's conductive barrier layer is the same. A conductive inversion layer is created however due to tunnel barrier modulation, and this definition cannot be applied to the TFET. Another contrast between TFETs and MOSFETs is the existence of two threshold voltages at the gate and drain. Even though the gate bias is maximum in the OFF state of a TFET, the channel potential is made equal to the drain potential when the drain bias is low. When the drain bias voltage rises, the TFET flips from OFF to ON, and the drain threshold voltage is produced as a second threshold value. As a result, the threshold voltage requirements for MOSFETs cannot be directly transferred to TFETs. The constant current technique, which is based on practical factors, is one of the oldest methods for determining the threshold voltage of a TFET [26]. As the gate voltage increases, the threshold voltage consumed by this approach reduces. This is because a higher gate voltage creates a stronger electric field at the source–channel junction, resulting in a bigger current even when the current is limited.

Once calculated, the threshold tunnelling length is the same as the tunnelling length as a function of the bias provided by the analytical model. The device's threshold voltage is the gate bias at which the tunnelling factor is equal to the threshold tunnelling distance. As observed in the evolution of the threshold voltage model through time, the constant current approach is utilised to compute the threshold voltage by equating the current with an arbitrarily set value in Figure 5.12. Because the threshold voltage values generated by this method have no physical basis, they do not provide information about the true behaviour of the TFET. Therefore, to calculate

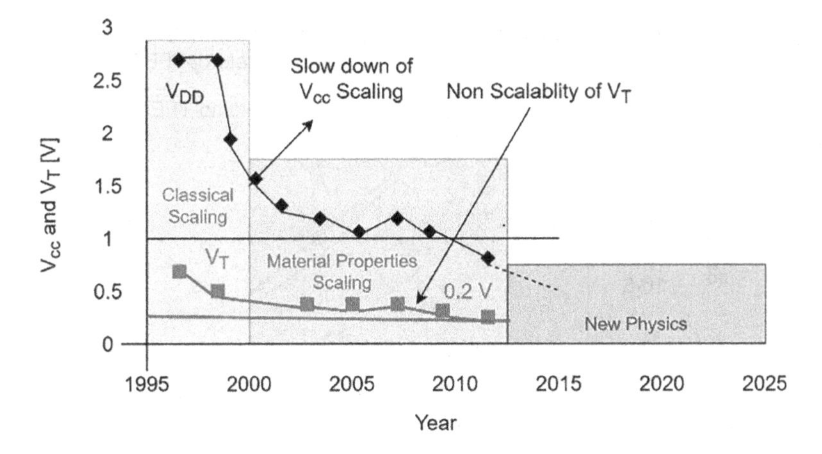

FIGURE 5.12 Threshold voltage model.

the threshold voltage of TFET, the transconductance change technique, which is frequently utilised in MOSFETs, was created.

The aim of TFET optimisation is to obtain the greatest possible I_{ON} while also achieving the lowest possible I_{OFF}. The objective of TFET is to outperform CMOS transistors by having an I_{ON} of hundreds of milliamps. To produce large tunnel currents and steep slopes, the transmission source tunnel barrier probability, you must approach the device closely to obtain a small V_G change.

5.8.3 Noise Performance in TFET and Characteristics

Electrical noise is becoming a major dependability concern for the most gratifying device designs at scaled generation nodes. At low-frequency V_{DD}, the overall performance and sign diversity are reduced due to the circuit's higher sensitivity. In analogue signal and RF circuits, as well as semiconductor memory, the noise parent is an important layout feature [27]. As a result, assessing the electric noise characteristics of TFETs as part of the strength reduction process is critical.

The major source of low-frequency noise in the gate oxide is flicker noise, which is generated by carrier trapping and detrapping in various trap states. The source–channel tunnelling barrier (E_{beff}) design has a considerable influence on the flicker noise characteristics of III-V TFETs as the BTBT identified the features of current transfer. The normalised drain current noise levels of homojunction and heterojunction TFETs are similar as shown in Figure 5.13, demonstrating that both BTBT and trap-assisted tunnelling (TAT) affect transfer qualities.

Heterojunction TFET reveals notably much less flicker noise than homojunction TFET on the equal drain cutting-edge at 77 K, while most effective BTBT dominates. The propagation of charge carriers generated by BTBT in the channel in TFETs is significantly shorter than the channel length [28].

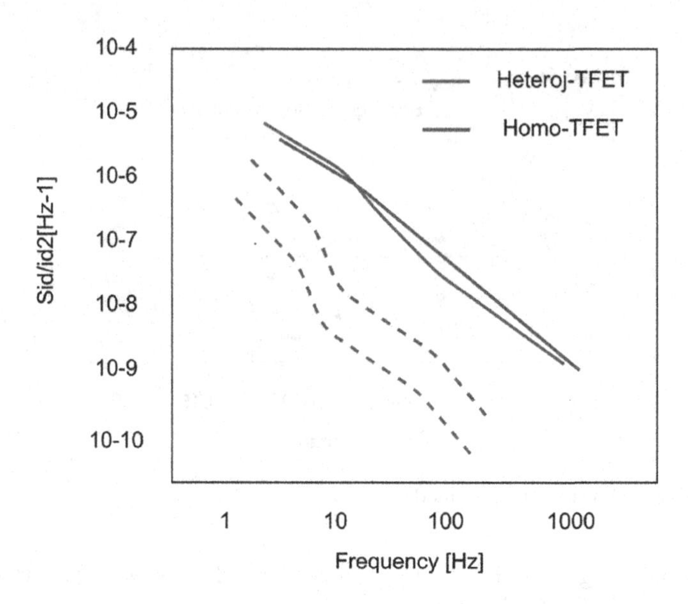

FIGURE 5.13 Flicker noise power comparison of homojunction TFET and heterojunction TFET.

The noise contribution from the tunnelling process is studied exclusively in this thorough research of tunnel FET noise modelling. Thus, for frequency-independent noise, the analytic noise models used are pure shot noise, and for frequency-dependent noise, Kane's model is used. However, for typical TFETs, drift-diffusion processes, in addition to tunnelling and ballistic transport [29], may play a major role in TFET conduction. White noise compositions differ due to substantial differences in conduction processes. The ballistic transport nature of tunnelling leads to a shot noise, while that of drift-diffusion transport is better expressed as thermal noise in equilibrium state.

5.8.4 Shot Noise Tunnel Junction

Shot noise is the statistical fluctuation of current associated with ballistic transport, such as electron emission in vacuum tubes, ballistic field-effect transistors, and tunnel junctions. The load that arrives at a specific terminal with a data packet of a specific size q at random and independent arrival times will produce start-up noise.

5.8.5 Thermal Noise in Channel

The thermal noise formula is true when the device is in quasi-thermal equilibrium, in other words, assuming a drift-diffusion model; in general, shot noise is reduced since the arrival times of various payloads are no longer independent occurrences. It influences the arrival of subsequent electrons in the source channel. As the device length drops below the carrier's mean free path, the transmission mechanics shifts from drift diffusion to quasi-ballistic and eventually ballistic transmission [29]. During the transition, the noise expression shifts from thermal noise to recording noise.

Since TFET can form tunnel junctions in the source and drain channels, it exhibits undesirable bipolar currents. Tunnelling at the channel–drain junction in a negative gate voltage mode generates bipolar current in nTFETs. Proposed by CG-TFET: Because the tunnel barrier height at the channel–drain junction is high, tunnelling rates are low. The transfer characteristics for various lengths of gate–drain underlap. The main framework of electrical noise evaluation was initially established by Shockley et al. and introduced as the impedance area technique, which was later extended for many semiconductor devices. To begin, noise sources (flicker, shot, and thermal noise) are assessed on a mesoscopic stage within the tool and are considered to be independent [28,29]. Then, each noise supply at a specific role is modelled as a (small signal) Langevin pressure riding a noiseless PDE primarily based complete semiconductor model. The resulting fluctuations on the tool terminals are then computed. For a Langevin pressure of current and a drain terminal reaction of open circuit voltage, the role-based ratio between the two is referred to as "impedance area," which may be used to describe any localised noise supply to the drain terminal reaction. This technique may be expanded to various forms of input and output, and the gain can be converted to other units. For example, if Langevin is chosen as the current and the drain terminal response is chosen as the short-circuit current, the gain becomes dimensionless [30–34].

5.9 DISADVANTAGE OF SILICON-BASED TFETs

The disadvantage of Si TFETs is that their ON current is extremely low, making them incompatible with contemporary CMOS-based circuits. As a result, the Si body was replaced with a compound material with a narrow band gap, as a result of which the tunnelling distance at the source–channel junction was lowered. When evaluating the results of Si TFETs, the use of smaller and direct-band gap III-V materials in the fabrication of TFETs results in higher I_{ON}, lower I_{OFF}, higher f_T, and f_{max}.

5.10 III-V-TFET STRUCTURES

Indirect-band gap progress is much slower, and it requires the collision of two things in order to proceed, an electron and photon. It is related to chemical processes. A reaction between two molecules will occur at a considerably faster rate than a process involving three molecules at a certain reaction stage. The same is true for recombinant electrons and holes in order to generate photons. For a direct-band semiconductor, the recombinant approach is far more effective than for an indirect-band semiconductor. Gallium arsenide and other direct-strip semiconductors are employed in TFETs due to these factors. A list of different semiconductor properties can be listed in Table 5.1.

Because of the narrower band gap and increased electron mobility, III-V semiconductors are preferred over silicon and germanium to increase ON current and reduce OFF current.

TABLE 5.1
Different Semiconductor Elements' Material Properties [4,35]

Semiconductor	Dielectric Constant	Effective Electron Mass	Effective Hole Mass	Electron Affinity	Band Gap	Electron Mobility
Si	11.7	0.98 m_0	0.49 m_0	4.05 eV	1.12 eV	1,400 cm²/V-S
Ge	16.2	1.6 m_0	0.33 m_0	4.0 eV	0.661 eV	3,900 cm²/V-S
$In_{0.53}Ga_{0.47}As$	13.9	0.041 m_0	0.45 m_0	4.5 eV	0.72 eV	12,000 cm²/V-S
InAs	15.15	0.023 m_0	0.41 m_0	4.9 eV	0.36 eV	40,000 cm²/V-S
InSb	16.8	0.014 m_0	0.43 m_0	4.59 eV	0.18 eV	77,000 cm²/V-S

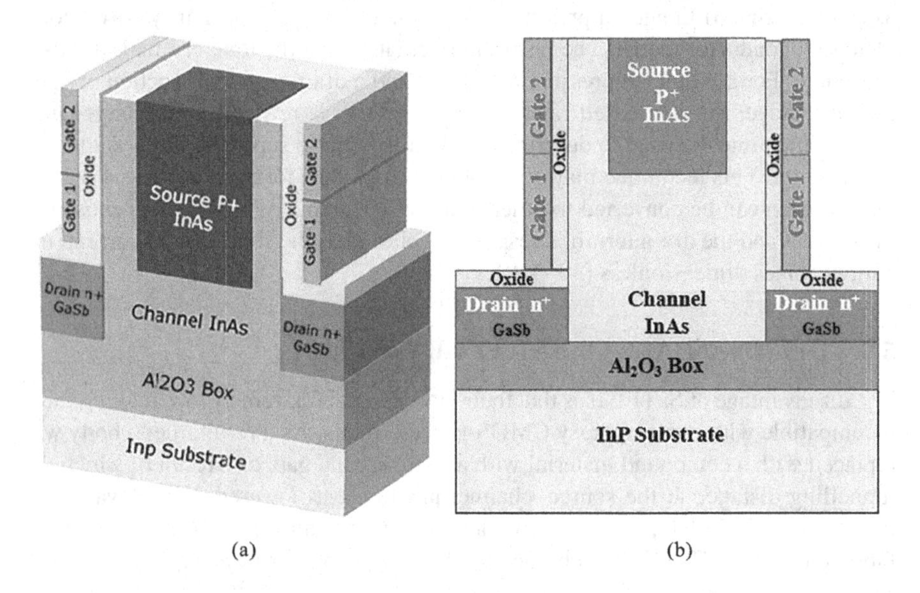

FIGURE 5.14 T-Shaped TFET structure. (a) single material gate and (b) dual material gate.

5.10.1 T-Shaped Tunnelling Field-Effect Transistor

Figure 5.14 shows a T-shaped tunnelling field-effect transistor (TTFET). InAs is the source material, while GaSb is the drain material. In addition to the single material gate construction, a dual material gate structure is presented here. Because of the enhanced tunnelling cross-sectional area and superior electrostatics, the TTFET has a higher ON current. The TTFETs reduced ambipolar current is owing to its gate–drain overlapping topology. According to the results, the TTFET has a better performance parameter. Furthermore, the TTFET structure's source and channel are spread vertically. As a result, the device offers greater potential for aggressive scaling than ordinary TFETs for future technologies beyond CMOS.

5.10.2 Vertical TFET

A TFET made on vertically grown InAs/GaSb (V-TFET) is shown in Figure 5.15. A source pocket (InAs) is introduced to increase the device's performance. In this

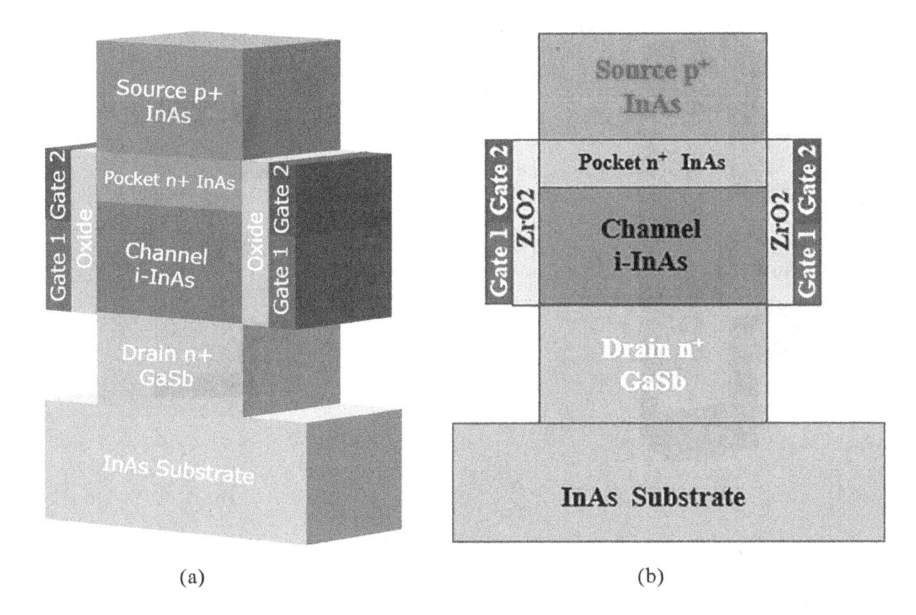

FIGURE 5.15 Vertical TFET structure.

scenario, ZrO_2 is employed as a dielectric layer in addition to SiO_2 to boost scalability. The DC and analogue/RF performances of a V-TFET with a pocket are demonstrated to be superior than those of a V-TFET without a pocket and other reported TFET designs. Because of its decreased propagation delay, V-TFET-WP is well suited for high-speed digital circuits. The V-TFET with pocket based on GaSb/Si heterojunction is suitable for low-power, low-voltage, high-speed analogue/RF/digital circuit applications.

5.10.3 Z-Shaped TFET

Figure 5.16 depicts a Z-shaped tunnel field-effect transistor (ZS-TFET). ZS-TFET is used to decrease ambipolar behaviour and increase RF performance in TFETs. In comparison to traditional TFETs, the ZS-TFET is more scalable and delivers greater ON-state current (I_{ON}), bigger ON/OFF current ratio (I_{ON}/I_{OFF}), and smaller SS. These advantages lead to the conclusion that the tunnelling junction in the ZS-TFET is perpendicular to the channel direction, allowing for the formation of a comparatively large tunnelling junction area. The ZS body employs both vertical and horizontal fields to control the lateral parasitic tunnelling current. Furthermore, the proposed device demonstrated increased RF performance due to greater transconductance and decreased gate–drain parasitic capacitance.

5.11 MILLER CAPACITANCE IN TFET

To simulate the tunnel current, a nonlocal tunnel model is used, depicting the actual space charge transfer across the tunnel barrier while taking into consideration the potential profile throughout the whole tunnel route [34]. The gate-to-drain capacitance clearly displays the total gate capacitance due to the presence of the source-side tunnel barrier (C_{gg}) for TFETs, when the gate-to-source capacitance (C_{gs}) stays low. C_{gd} increases with positive gate voltages because the channel-to-drain potential barrier decreases.

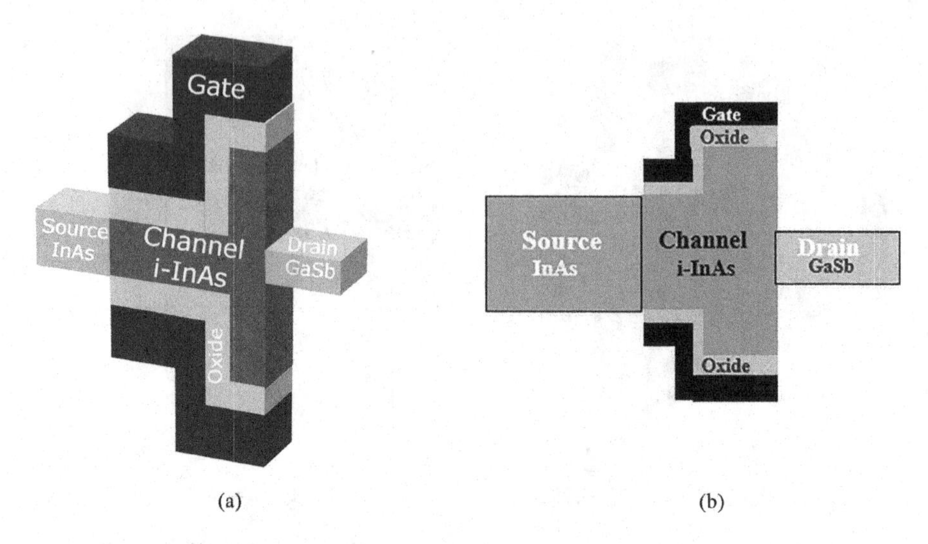

FIGURE 5.16 Z-shaped TFET.

Figure 5.17 compares the inverters constructed using TFET and MOSFET based on the normalised value in the input to output capacitance as the function in the input voltage. The gate-to-drain capacitance (C_{gd}) of both n- and p-type transistors contributes to the overall Miller capacitance (CM). The regions of A, B, D, and E are the place where the transistors remain linear, which results in the same value of 0.5 C_{gg} in both CM and C_{gd}. The two transistors flowing into the saturation zone during the 0–1 V input ramp create the drop in Miller capacitance at point C. InAs TFETs have significant driving current (I_{ON}) at lower supply voltages due to their reduced tunnel barrier height and

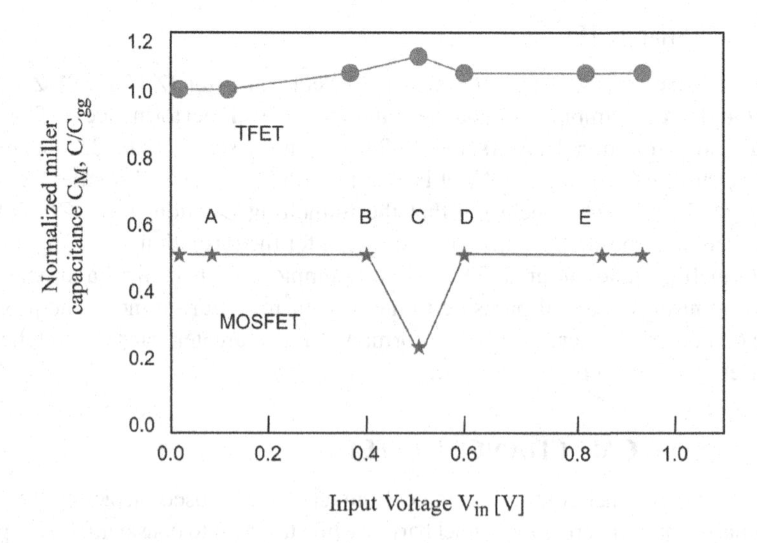

FIGURE 5.17 Miller capacitance for TFET/MOSFET inverters as a function of inverter input voltage [34].

width, as well as their lower tunnelling mass, and their gate capacitance C_{gg} is limited by the quantum capacitance caused by their lowered density of states.

The capacitance–voltage characteristics of InAs TFETs are depicted in Figure 5.18, demonstrating that the overall gate capacitance is only 10% of the gate oxide capacitance. Based on the transistor design, the C_{gd} again acts as a primary contributor to C_{gg} [34], attributed to the fact that the capacitance value is much lower than that of Si TFETs, V_{dd} = 1v. Furthermore, the ON-resistance of the InAs TFET is much lower than that of the Si TFET. Peak overshoot voltages in InAs TFET inverters are less than 20% of the peak input voltage as shown in Figure 5.19. The peak overshoot is reduced in both TFET inverters as the capacitive load increases as expected, and because of its lower switching resistance and Miller capacitance, the overshoot in the InAs-based TFET inverter is substantially lower [34].

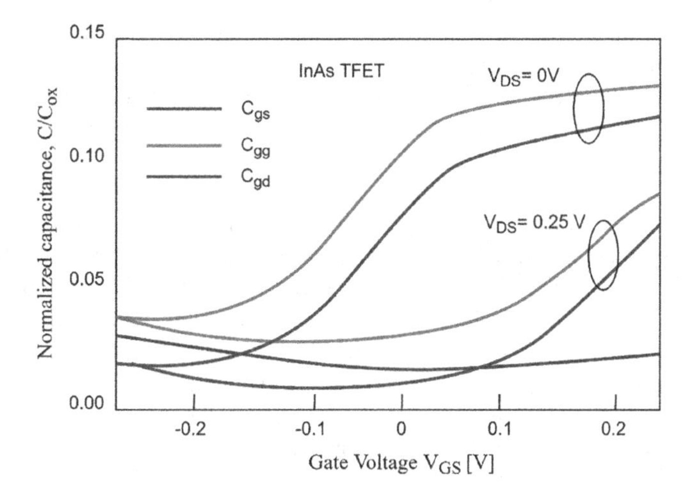

FIGURE 5.18 InAs TFET's capacitance–voltage characteristics [34].

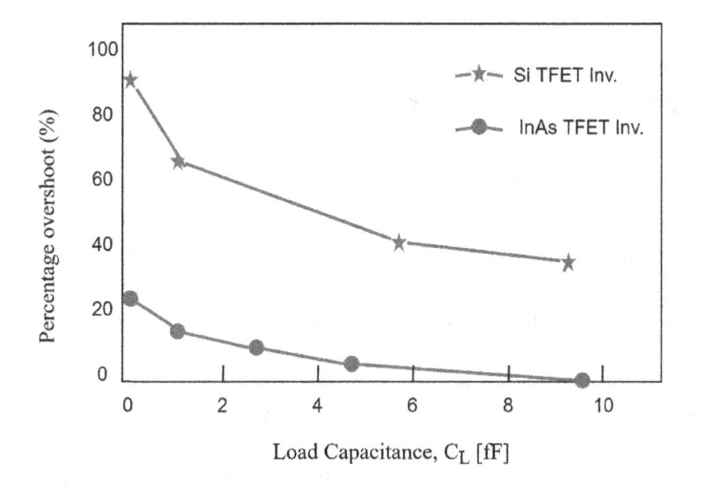

FIGURE 5.19 Si and InAs TFET inverter percentage overshoot of load capacitance [34].

TABLE 5.2
Inverter Delay Is Compared to Benchmark Methodologies [34]

Delay [ps]	$C_{OX}\, V_{DD}/2I_{ON}$	$C_{gg}\, V_{DD}/2I_{ON}$	Q_{ON}-Q_{OFF}/I_{ON}	Inverter Fall Delay
Si MOSFET	1	0.8	0.63	1.15
Si TFET	18.5	15	8	48
InAs TFET	3.5	0.38	0.3	1.1

5.12 CV CHARACTERISTICS AND MATERIAL PROPERTIES

The generally used measurements for today's scaled CMOS technology with scaled threshold voltages differ from those of actual MOSFET inverter fall latency by an unreasonable amount. Because the real switching current of a MOSFET inverter might be substantially lower than the saturation current of a single transistor, an effective driving current (I_{EFF}) should be utilised to anticipate the actual delay instead of I_{ON} [36]. Analytical approaches for calculating the average or effective driving current (I_{EFF}) while taking the real inverter switching current trajectory into account were also presented.

Table 5.2 shows that the benchmarking measures now in use stray greatly from the performance of the TFET inverter, highlighting that a proper quantification of the actual output capacity and the actual switching current is necessary to anticipate the performance of the TFET. TFET has been employed in two distinct material systems, Si and InAs, and only Si MOSFET findings are presented for comparison. Silicon and InAs were chosen because they have a high density of states (DOS) and a low DOS [34].

The excessive gate-to-drain capacitance inherent within the functioning of a TFET tool has huge implications for its brief responsiveness. The reaction of Si TFET and MOSFET inverters is short with a 1 V entry step voltage and a 5 ps upward push time. Si TFETs have an output voltage overshoot of around 90% because of the large Miller feedthrough capacitance resulting from its crucial tool feature combined with its low pressure as compared to MOSFETs.

5.13 VARIATION IMPACT ON TFET MODELLING AND LOW-VOLTAGE EFFECT

On the steep intuitive slope and high-slope potential in low-voltage power supplies, tunnel FETs have emerged as options for low-power and high-frequency embedded electronic components. With near/subthreshold CMOS, scalable transistors and energy-saving applications are possible. Because of the exponential relationship of I_{ON} with the width of the tunnel barrier, the source of fluctuation that might cause the tunnel barrier's width to alter will cause significant I_{ON} fluctuations in the TFET. We simulate the change of TFET and its effect on the signal-to-noise ratio (SNR) of TFET SRAM for low-voltage applications [31]. Because of their low-slope (SS) and high-density low-power slopes, tunnel FETs have become a viable alternative for low-power, high-frequency inputs. The source of variation that might vary the width of the tunnel barrier will generate large I_{ON} fluctuations in the TFET due to the exponential dependency of ions on the width of the tunnel barrier. We simulate the effect of TFET modification on the SNR of a TFET-SNR SRAM for applications with lower voltage.

The thin channel mitigates the effect of the short channel by shielding the source junction from leakage voltage; nevertheless, the thin channel decreases the accessible area for tunnelling current, lowering the maximum possible current. It is made up of a stack of two transition metal dichalcogenide (TMD) elements. When the two layers are appropriately biased, the van der Waals gap between them functions as a tunnelling gap. Several researchers have investigated such structures as tunnelling diodes and discovered that they exhibit BTBT properties. A source area with electrically regulated carrier density might eliminate dopant-induced band tails, improving the steep slope. Furthermore, improved electrostatics may be achieved in lateral tunnel construction resulting in excessive gate efficiency to all monitored locations, as opposed to vertical tunnelling with lesser gate control in the channel's centre.

5.14 APPLICATION OF TFET DEVICES

In TFET, I_{ON}/I_{OFF} ratio is further improved when the ON current is doubled, while the OFF current remains of the order of picoampere even if it is increased by the same factor [21,31–34,36]. The TFET is part of the family of steep slope devices, which operate at intermediate frequencies and are currently being investigated for the below-mentioned ultralow power applications:

- RF Switches.
- Digital ICs: Adder, Multiplier, Flip-Flop, Inverter, etc.
- SRAM.
- Biosensors.
- Energy-Harvesting Circuits (DC-DC Converters).
- Analogue/Mixed-Signal System-on-Chip Applications.

5.15 OPTIMISATION AND FUTURE DESIGN OF TFETs

The aims of TFET enhancement are to obtain the greatest I_{ON}, the lowest SS throughout multiple times higher drain current, and the lowest I_{OFF} all at the same time. The objective characteristics for TFETs to surpass CMOS transistors are I_{ON} in the hundreds of milliampere range and SS significantly below 60 mV/decade for five decades of current [21,37,38]. TFETs are designed to operate at low voltages. To obtain a large tunnelling current and a steep slope for a small change in V_G, the transmission probability of the source tunnelling barrier should approach unity.

According to researchers, the power consumption of this new form of transistor is 90% less than that of regular transistors, allowing it to operate well beyond the theoretical limit of electronic devices. These results, according to the researchers, might one day lead to circuits that are ultradense and low-power, as well as biosensors and gas sensors that are extremely sensitive. The continual downsizing of field-effect transistors, the building blocks of most microchips, has driven the growth of computer power during the last half-century. Transistors function similarly to switches in that they may be switched on and off to represent data as 0 and 1. At room temperature, the switching properties of FETs are currently limited to a theoretical limit of 60 mv/decade [32]. This constraint is known as subthreshold oscillation, and it implies that every 60 mV rise in voltage causes a 10-fold increase in current. Reducing oscillations improves channel control, using less power for adjustments.

Scientists attempted to circumvent this restriction by employing TFETs. The electrons are being used in the devices to the cross barriers, a process known as quantum tunnelling. Traditional field-effect transistors turn off at considerably lower voltages. Fin-FET technology took approximately 20 years to become a reality in semiconductor goods, replacing planar components. Fin-FETs [32], on the other hand, employ planar nodes, such as high metal gates, boost/epiphyseal sources, deformed silicon, and the newest gate technology, due to technical development that does not allow for significant modifications. Although it is commonly acknowledged that Si or SiGe Fin-FETs may be replaced by TFETs, it remains unclear if reducing nodes below 7 nm would result in significant technological advances. We must implement the TFET idea in Fin-based technology to enable a seamless transition from Fin-FET to Fin-based vertical tunnel FET while using Fin-FET architecture.

The TFET has decreased I_{OFF} in the present day compared to conventional MOSFET, and all TFETs have comparable relatively low I_{ON}, but increasing the I_{ON} is massive mission in the future [21,33,34,36–38]. For the TFET, lower SS and increasing I_{ON} may be an aggressive alternative for the traditional MOSFET to meet the requirements of I_{ON} in accordance with ITRS low band hole material for providing excessive doping awareness necessary. Similarly, a heteroshape may be utilised to get a steep subthreshold drop and with high I_{ON}. Using a thick buried oxide layer can reduce the ambipolar behaviour of TFETs.

5.16 CONCLUSIONS

Tunnel FETs are a new type of transistor technology whose investigation is ongoing. Numerous TFET designs are being suggested, tested, and improved over time. Existing TFET models have limitations, and modelling of TFETs frequently necessitates a trade-off between time and accuracy. Because of the core operating concept based on BTBT, TFET can reach SS greater than the 60 mV/decade at an ambient temperature, making it a feasible choice for overcoming the growing performance difficulties in scalable MOSFET technology. The lower average SS, low I_{OFF}, or high I_{ON} are the major problems, which is true for both nTFET and pTFET. To solve these difficulties, reliable predictive models are required, and quantum mechanics simulations are increasingly being utilised to augment the semiclassical models usually used for MOSFETs. These prototypes can assist you in selecting the appropriate material system for your needs, including Group IV, III-V, and 2D materials that can give intriguing silicon alternatives. Vertical TFETs' scattered physics produces a high capacitance value in saturation, a lengthy channel length, and a partially diminished drain barrier in saturation. In contrast to the lateral TFET design example, shorter gate length enhances RF performance by significantly reducing capacitance while maintaining transconductance. Even when extrinsic components are present, reducing the gate length can significantly increase the cut-off frequency, it is advantageous for low-power, high-frequency applications, and other speed gains can be obtained using dopant pockets or optimising the gate stack design. Finally, several device-level features of TFET operation have implications for its application in circuits, emphasising the necessity for TFET device and circuit co-optimisation.

REFERENCES

1. J.G. Webster, D. Verreck, G. Groeseneken, and A. Verhulst. *The Tunnel Field-Effect Transistor.* Wiley Encyclopedia of Electrical and Electronics Engineering, Weinheim, Germany; 2016.
2. H. Lu and A. Seabaugh. Tunnel field-effect transistors: State of-the-art. *IEEE J. Electron. Devices Soc.*; 2014; 2: pp. 44–49.
3. J. Appenzeller, Y.-M. Lin, J. Knoch, and P. Avouris. Band-to-band tunneling in carbon nanotube field-effect transistors. *Phys. Rev. Lett.*; 2004; 93: p. 196805.
4. M. Saravanan and E. Parthasarathy. A review of III-V tunnel field effect transistors for future ultra low power digital/analog applications. *Microelectron. J.*; 2021; 114: p. 105102.
5. M. Alioto and D. Esseni. Tunnel FETs for ultra-low voltage digital VLSI circuits: Part II evaluation at circuit level and design perspectives. *IEEE Trans. Very Large Scale Integr. (VLSI) Syst.*; 2014; 22(12): pp. 2499–2512.
6. G. Dewey, B. Chu-Kung, J. Boardman, J.M. Fastenau, J. Kavalieros, R. Kotlyar, W.K. Liu, D. Lubyshev, M. Metz, N. Mukherjee, and P. Oakey, Fabrication, characterization, and physics of III–V hetero junction tunneling field effect transistors (H-TFET) for steep sub-threshold swing. IEEE Electron Device Meeting (IEDM), Washington, DC; 2011; pp. 33.6.1–33.6.4.
7. B. Ganjipour, J. Wallentin, M. T. Borgström, L. Samuelson, and C. Thelander. Tunnel field-effect transistors based on InP-GaAs hetero structure nano wires. *ACS Nano*; 2012; 6 (4): pp. 3109–3113.
8. Dubey, P.K. and Kaushik, B.K. Evaluation of circuit performance of T-shaped tunnel FET. *IET Circuits Devices Syst.*; 2020; 14(5), pp. 667–673.
9. Y. Chen, M. Fan, V.P. Hu, P. Su, and C. Chuang. Design and analysis of robust tunneling FET SRAM. *IEEE Trans. Electron. Devices*; 2013; 60(3), pp. 1092–1098.
10. A. Tura and J. Woo. Performance comparison of silicon steep subthreshold FETs. *IEEE Trans. Electron. Devices*; 2010; 57(6): pp. 1362–1368.
11. D. Kim, Y. Lee, J. Cai, I. Lauer, L. Chang, S. J. Koester, et al. Low power circuit design based on heterojunction tunneling transistors (HETTs). *IEEE Trans. Very Large Scale Integr. (VLSI) Syst.*; 2013; 21(9): pp. 1632–1643.
12. P.F. Wang, T. Nirschl, D. Schmitt-Landsiedel, and W. Hansch. Simulation of the Esaki-tunneling FET. *Solid-State Electron.*; 2003; 47: pp. 1187–1192.
13. W. Hansch, P. Borthen, J. Schulze, C. Fink, T. Sulima, and I. Eisele. Performance improvement in vertical surface tunneling transistors by a boron surface phase. *Jpn. J. Appl. Phys.*; 2001; 40: pp. 3131–3136.
14. C. Sandow, J. Knoch, C. Urban, Q. Zhao, and S. Mantl. Impact of electrostatics and doping concentration on the performance of silicon tunnel field-effect transistors. *Solid-State Electron.*; 2009; 53: pp. 1126–1129.
15. W.Y. Choi, B. Park, and J.D. Lee. Tunneling field-effect transistors (TFETs) with subthreshold swing (SS) less than 60 mV/dec. *IEEE Electron. Device Lett.*; 2007; 28(8): pp. 743–745.
16. S. Badiger, K. Sujatha, and K. Bailey. Homojunction TFET device design and analysis for low power applications. *Int. J. Adv. Res. Electr. Electron. Instrum. Eng.*; 2018; 7(9): pp. 3675–3681.
17. M. Saravanan and E. Parthasarathy. Investigation OF RF/Analog performance of InAs/InGaAs channel-based nanowire TFETS. International Conference on Communication, Control and Information Sciences (ICCISc-2021). DOI: 10.1109/ICCISc52257.2021.9484973.
18. A. Varghese, C.S. Praveen, A.P. Mani, and A. Ravindran. InGaAs/GaAsSb heterojunction TFET. International Conference on Emerging Trends in Technology and Applied Sciences; 2015.
19. J. Knoch, S. Mantl, and J. Appenzeller. Impact of the dimensionality on the performance of tunneling FETs: Bulk versus one-dimensional devices. *Solid-State Electron.*; 2007; 51: pp. 572–578.

20. A.S. Verhulst, W.G. Vandenberghe, K. Maex, and G. Groeseneken. Tunnel field-effect transistor without gate-drain overlap. *Appl. Phys. Lett.*; 2007; 91: p. 053102.
21. W. Y. Choi, B.G. Park, J.D. Lee, and T.J.K. Liu, Tunneling field effect transistors (TFETs) with subthreshold swing (SS) less than 60 mV/dec. *IEEE Electron Device Lett.*; 2007; 28(8): pp. 743–745.
22. Y. Bi, K. Shamsi, J.S. Yuan, Y. Jin, M. Niemier, and X.S. Hu. Tunnel FET current mode logic for DPA-resilient circuit designs. *IEEE Trans. Emerg. Top. Comput.*; 2017; 5(3): pp. 340–352.
23. P.C. Kocher, J. Jaffe, and B. Jun. Differential power analysis. *J. Cryptogr. Eng.*; 2011; 1: pp. 5–27.
24. J. Madan and R. Chaujar. Temperature associated reliability issues of heterogeneous gate dielectric gate all around tunnel FET. *IEEE Trans. Nanotechnol.*; 2018; 17 (1): pp. 41–48.
25. S.J. Cho, M.C. Sun, G.R. Kim, T.I. Kamins, B.G. Park, and J.S. Harris Jr. Design optimization of a type-I heterojunction tunneling field-effect transistor (I-HTFET) for High performance logic technology. *J. Semicond. Technol. Sci.*; 2011; 11(3): pp. 182–189.
26. G. Soelkner, W. Kaindlb, H. Schulzea, and G. Wachutkab. Reliability of power electronic devices against cosmic radiation-induced failure. *Microelectron. Reliab.*; 2004; 44: pp. 1399–1406.
27. K. Boucart and A.M. Ionescu. A new definition of threshold voltage in tunnel FETs. *Solid State Electronics*; 2008; 52(9): pp. 1318–1323.
28. K.K. Bhuwalka, J. Schulze, and I. Eisele. Scaling the vertical tunnel FET with tunnel bandgap modulation and gate work function engineering. *IEEE Trans. Electron. Devices*; 2005; 52: pp. 909–917.
29. S. Datta, H. Liu, and V. Narayanan. Tunnel FET technology: A reliability perspective. *Microelectron. Reliab.*; 2014; 54 (5): pp 861–874.
30. L. Knoll, Q.-T. Zhao, A. Nichau, S. Trellenkamp, S. Richter, A. Schafer, et al. Inverters with strained Si nanowire complementary tunnel field-effect transistors. *IEEE Electron. Dev. Lett.*; 2013; 34(6): pp. 813–815.
31. L. Zhang, J. He, and M. Chan. A compact model for double-gate tunneling field effect transistors and its implications on circuit behaviors. 2012 International Electron Devices Meeting. DOI: 10.1109/IEDM.2012.6478994
32. R. Narang, M. Saxena, R. S. Gupta, and M. Gupta. Impact of temperature variations on the device and circuit performance of tunnel FET: A simulation study. *IEEE Trans. Nanotechnol.*: 2013; 12 (6): pp. 951–957.
33. A. R. Trivedi, S. Carlo, and S. Mukhopadhyay. Exploring tunnel-FET for ultra-low power analog applications: A case study on operational transconductance amplifier. IEEE Design Automation Conference (DAC); 2013. DOI: 10.1145/2463209.2488868.
34. H. Liu, S. Datta, M. Shoaran, A. Schmid, X. Li, and V. Narayanan. Tunnel FET-based ultra-low power low-noise amplifier design for bio-signal acquisition. ACM International Symposium on Low Power Electronics and Design (ISLPED); 2014. DOI: 10.1145/2627369.2627631.
35. http://www.ioffe.ru/SVA/NSM/Semicond/
36. S. Mookerjea, R. Krishnan, S. Data, and V. Narayanan. Effective capacitance and drive current for tunnel FET (TFET) CV/I estimation. *IEEE Trans. Electron. Devices*; 2009; 56(9): pp. 2092–2098. DOI: 10.1109/TED.2009.2026516.
37. S Saxena, SL Tripathi, SK Sinha, GS Patel, C Pravalika. Review on performance evaluation of TFET structures & its applications. *THINK INDIA J.*; 2019: 22: pp. 220–227.
38. M. Saravanan, and E. Parthasarathy. Investigation of RF/Analog performance of Lg=16nm Planner In0.80Ga0.20As TFET; Fourth International Conference on Electrical, Computer and Communication Technologies (ICECCT); 2021. DOI: 10.1109/ICECCT52121.2021.9616769.

6 Performance Analysis of Emerging Low-Power Junctionless Tunnel FETs

G. Lakshmi Priya
VIT University

M. Venkatesh
REVA University

S. Preethi
Sri Krishna College of Technology

T. Venish Kumar
Nadar Saraswathi College of Engineering & Technology

N. B. Balamurugan
Thiagarajar College of Engineering

CONTENTS

DOI: 10.1201/9781003240778-6

6.1 INTRODUCTION

Device scaling is the most challenging task in the expanding microelectronics industry. Solid-state devices can be scaled to reduce cost, power consumption, and speed. Drain-induced barrier lowering (DIBL), channel length modulation (CLM), hot carrier effects (HCEs), reduced subthreshold slope (SS), high OFF state leakage current, and low transconductance efficiency are all realistic restrictions to device scaling. Circuit designers are seeking alternatives to standard MOSFETs in order to overcome these concerns. Tunnel field-effect transistor (TFET) is being examined as a possible candidate for improving a device's subthreshold properties. SS less than 60 mV/decade can be achieved via band-to-band tunneling (BTBT). With work function designed and heterogate-dielectric-based low-power junctionless TFETs (JLTFTs), the subthreshold performance of these nanodevices has risen tremendously. The main focus of this chapter is on studying the short-channel behavior of tunneling devices using significant parameters such as surface potential, electric field, threshold voltage, drain current, transconductance-to-drain-current ratio, and subthreshold current.

6.2 MOS TRANSISTOR: EVOLUTION

Julius Edgar Lilienfeld and Oskar Heil patented the first FET in 1926 after which the practical semiconducting transistor called JFET was developed by William Shockley at Bell Labs in 1947. As an expansion to the original FET design, at Bell Laboratories, Dawon Kahng and Martin (John) developed the metal oxide semiconductor field-effect transistor (MOSFET) in 1959. MOSFET differs from the bipolar junction transistor (BJT) by its design, construction, and working.

The MOSFET is a semiconductor device widely used for switching and amplification of signals in electronic devices. The core of all integrated circuits is MOSFET because of its design and ease of fabrication. The MOSFET is basically categorized into two types, namely, N-channel MOSFET and P- channel MOSFET. Figure 6.1 shows an N-channel and P-channel MOSFET and two different types of MOSFETs called as enhancement (E)- and depletion (D)-type MOSFET.

MOSFET is a voltage-controlled device and uses an insulating layer between the gate metal region and the semiconductor substrate. The silicon wafer is oxidized to obtain the dielectric/insulating material for the transistor as silicon dioxide (SiO_2). Oxidizing SiO_2 from the silicon wafer is much easier as the best choice for manufacturing the semiconductor device is silicon. In recent years, some integrated circuit (IC) manufacturers

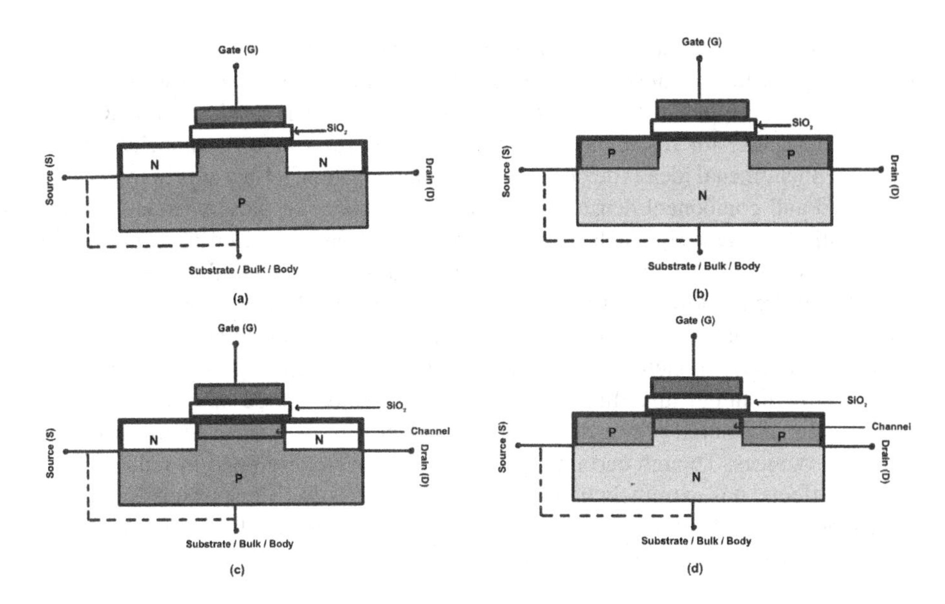

FIGURE 6.1 (a) N-type and (b) P-type enhancement MOSFET and (c) N-type and (d) P-type depletion MOSFET.

such as IBM and Intel have initiated the use of compound materials [1,2]. Most of the compound semiconductors are combinations of Group III and Group V elements such as GaAs, InP, GaN, and ZnS. It is also possible to combine different elements of the same group (Group IV) to make compound material such as SiC and SiGe. Though the cost of manufacturing compound semiconductors is still higher than the cost of silicon (Si), the spectral properties of these crystals plays a vital role in certain optoelectronic and microwave applications.

Advanced MOS structure popularly denoted as CMOS (complementary metal oxide semiconductor) is a notable invention in the semiconductor industry. In 1963, Frank Wanlass invented CMOS for constructing the integrated circuits. The most significant essence of CMOSs devices is that they are largely immune to noise and have limited static power consumption, which facilitates their usage in fabricating controllers, memories, and other digital circuits. MOSFETs can also be used in radio systems to transform frequencies as oscillators or mixers [3]. Recent audio-frequency power amplifiers deploy MOSFET semiconductors. For smaller sizes of less than 45 nm node technology, IBM and Intel have announced to exploit the benefits of using high-K gate dielectric materials. To overcome the limitations posed by conventional transistor scaling, silicon dioxide has been replaced by high-K gate dielectric material and gate stack architectures.

6.3 NANOELECTRONICS ERA

The rise of various electronic gadgets and devices is remarkable for the economic and technical advancement of the nanoelectronics age in the twentieth century. The advancement in the semiconductor industry has significantly contributed to the global economy growth. The most important salvation of this progress is due to the invention of different types of transistors and new materials. The fabrication of ICs has paved a new way for researchers to begin with the revolution in nanoelectronics. Another phenomenal idea is device scaling/miniaturization, which aids in increasing the speed and component density [4,5]. The performance of these nanodevices has exponentially increased, which was successfully anticipated by Gordon Earle Moore (1965). However, with continuous device downscaling, conventional transistors have contributed greatly toward adverse effects in the device. As the device dimensions are becoming smaller, the channel length reduces and the device becomes heedless to SCEs, quantum tunneling effects, and subthreshold leakage issues. In nanoelectronics, this quantum tunneling turns out to be a major source of current leakage resulting in substantial power drain and heating effects that affect high-speed systems and devices. Though certain techniques have been developed to reduce these adverse effects, it is extremely hard to nullify the effects. As a response, researchers have looked into alternative materials for electronic device applications, as well as new ways for fabricating electronic devices.

6.4 DEVICE SCALING

Since the inception of Moore's law, device scaling has been continuing over the past few decades. Aggressive scaling of MOSFET has led to enormous growth of the IC industry incorporating channel lengths to tens of nanometers. This ongoing reduction was possible after Dennard's work on scaling [6]. In late 2009, Intel began its production with feature size of be 32 nm. In April 2019, Samsung announced a 5 nm process to their potential customers. Transistor size smaller than 7 nm will experience quantum tunneling effect through the gate dielectric layer. But a research team from Berkeley National Laboratory has created a transistor with a working 1 nm gate length. They strongly believe that with proper choice of materials, there is huge room to shrink our electronics. The key element that changed the perception of many scientists is the replacement of silicon material with molybdenum disulfide (MoS_2). This material has immense potential for application in LASERs, LEDs, nanoscale transistors, solar cells, and more. The development of MoS_2 has kept Moore's law still alive and enhanced the performance of nanoelectronics devices.

In general, as we move beyond 14 nm, new Fin materials will be required like SiGe, SiC, GaAs, InP, and more. The other approach to overcome the pitfalls of scaling is by using multigate architectures. Gate-all-around (GAA) FET is expected to replace all FinFET structures in the near future as it is believed to be the most effective structure in mitigating the quantum tunneling effects. GAA FET was targeted for 3 nm technology node in 2020 as it provides 40% performance boost for the same amount of power. Monolithic 3D transistors are also on the verge of overruling the semiconductor industry with 50% reduction in area. It is a 3D sequential integration of a layer of transistors placed over the substrate. The core essence of Moore's law continues by proper engineering of the device architecture and semiconductor materials to meet the demands of modern computing.

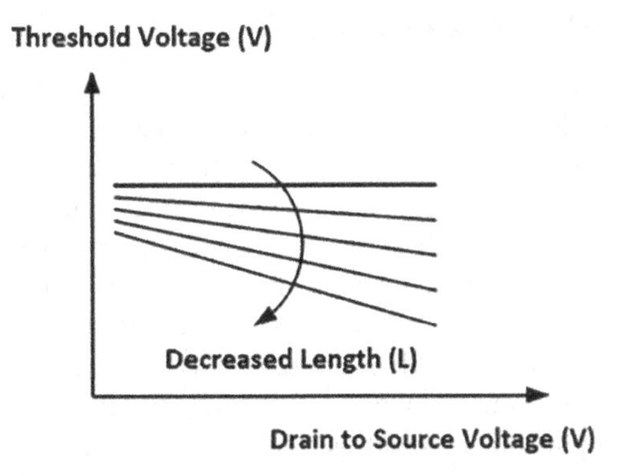

FIGURE 6.2 Variation of threshold voltage as a function of drain-to-source voltage (V_{ds}).

6.5 IMPACT OF SCALING ON DEVICE PERFORMANCE

Due to rigorous device downscaling, conventional transistors have contributed greatly toward degradation of device performance. The device performance is assessed by the minimization of SCEs. Short-channel risks come into play as the length of the channel becomes approximately equal to the sum of source and drain depletion layer widths [7,8]. These effects are associated with CMOS device scaling. It is well known that the primary motivation for device engineers to reduce the size of transistors, that is, their lengths, is to increase speed and reduce cost per IC [9].

Smaller gate lengths allow for higher drive current and speedier circuits. Additionally, when circuits become smaller, their capacitance decreases, boosting the working speed of the device. However, as device dimensions are miniaturized, many SCEs arise to degrade the device performance. These SCEs impose two fundamental limitations: one on the electron drift attributes in the channel and the other on the threshold voltage modification as shown in Figure 6.2.

6.6 SILICON-ON-INSULATOR (SOI) FET TECHNOLOGY

The adverse impacts of DIBL, CLM, punch-through effect, hot electron effect, threshold voltage roll-off, and subthreshold leakage have emerged to be strong barriers for fabricating under bulk substrates [10]. Also, when the gate length is reduced, the gate-oxide thickness is also reduced producing a significant quantum mechanical tunneling at high electric fields. Shortening the channel length has the biggest drawback from the drain bias, which can be reduced by enhancing the gate control over the channel with multiple gate and gate metal work function engineering techniques. Replacing silicon dioxide with high-K gate dielectric materials and using heterodielectric stack architecture can probably reduce the gate-oxide scaling issues. Therefore, we cannot compromise on having shorter channel lengths, as it reduces the cost per IC and produces faster switching transistors. Owing to all these facts, a long search for altering the bulk silicon technology has ended with the initiative of silicon-on-insulator (SOI) technology. SOI FET can be considered to be the best technology in providing high speed, low power, low SS, and low parasitic capacitance.

6.7 NOVEL LOW-POWER BTBT FET

Future generations of ICs demand low-power and energy-efficient transistors with less SS [11]. Semiconductor engineers have tried to squeeze a large number of transistors onto a single integrated chip. To accommodate such a huge number of transistors on smaller-size chips, the distance separating the source–drain and oxide thicknesses have been reduced. As a result, the electronic barrier that blocked the flow of current is now able to allow large current due to its thinner oxide thickness. This ability of an electron to penetrate the barrier is known as quantum tunneling. BTBT is the basic underlying principle of TFETs,which is observed to have SS less than 60 mV/decade and probably less subthreshold leakage current [12]. In MOSFET, the current flow is due to the diffusion phenomenon and thermal injection of carriers.

Figure 6.3 shows a simple TFET structure with its BTBT. An electron in the valence band tunneling across the band gap and entering the conduction band is BTBT. The band gap acts as a potential barrier for the electrons to tunnel across. This potential barrier between source and channel remains wider during the OFF state of the device. At this state, there is a very small amount of leakage current. During the ON state, once the gate bias is applied, the potential barrier becomes narrow, allowing for significant tunneling current to pass [13–15]. The tunneling of electrons can cause the transistor to move to ON and OFF states even at lower voltages compared to MOSFET's operational voltage. Thus, TFET has the capability to lessen the power consumption of the device.

6.8 MULTIGATE JLTFET

To overcome the SCEs encountered by the CMOS devices, many structural variations have been suggested. As discussed earlier, among the future device alternatives such as SOI FET, multigate architecture, and TFET has been the long-time awaited combination in combating SCEs and leakage issues. Now, another alternative best possible solution to transcend the shortcomings of TFET is using a novel device structure called junctionless transistor (JLT). JFET has evolved to be the most promising candidate in providing cheaper and denser microchips. The notable feature of JLT is that it has monolithic doping among the semiconductor channel, source, and drain regions [16]. The schematic of N-type JLT in OFF and ON states is shown in Figure 6.4. The gate electrode controls the mobility of electrons and thus the current

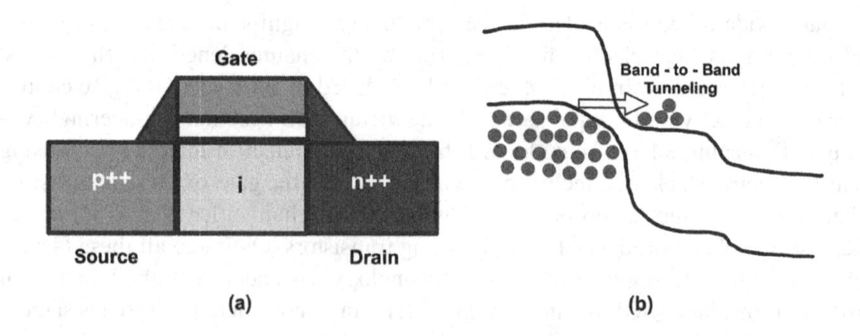

FIGURE 6.3 (a) Simple structure of n-TFET and (b) BTBT.

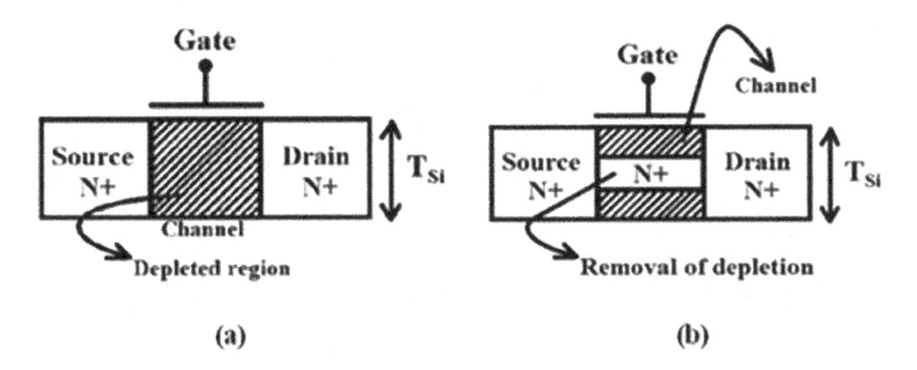

FIGURE 6.4 Schematic view of N-type JLT in the (a) OFF state and (b) ON state.

flow. An appropriate choice of gate metal work function depletes the area under the gate terminal leading to a "normally OFF" state device.

With the invention of JLTs, Bahniman Ghosh and Mohammad Waseem Akram (2013) investigated the switching characteristics of n-type heavily doped JLTFET [17–19]. This JLTFET deploys a high-K gate dielectric material (titanium oxide, TiO_2) to improve the transistor performance. Bal et al. (2014) incorporated gate work function engineering in JLTFET to further enhance the device characteristics in weak inversion regions. Baruah and Paily (2014) used a high-K spacer in dual material gate JLT to study its analog performance [20–22]. Recently, Shradya Singh et al. (2018) formulated an analytical model for JLT with a split gate for biosensor applications.

6.8.1 Dual Material Double-Gate Junctionless Tunnel FET (DMDG JLTFET)

With all key innovations emerging in JLTFETs, the development of the DMDG JLTFET is very similar to that of the n-type DMDG TFET, except for the uniformly doped regions. Figure 6.5 illustrates a cross-sectional view of the proposed DMDG JLTFET. The device dimensions and materials used are indexed in Table 6.1.

The band energy distribution of DMDG JLTFET in the cut-OFF and ON states is shown in Figure 6.6. The DMDG JLTFET remains in its OFF state without any bias, and the tunnel barrier width between source and channel is substantial. The control gate bias can be increased to alter the barrier width of the tunneling region. This barrier is reduced when bias ($V_{gs} = 1$ V) is given to the control gate, and the device is turned ON. The spacer thickness is set accordingly low to reduce the ambipolar behavior of TFETs. When homogeneous doping, high-K insulation, and gate metal engineering are combined, this device can be an appropriate choice for limited-power analog CMOS devices and rapid switching circuits.

6.8.2 Dielectric-Modulated DMDG JLTFET

Figure 6.7 represents a cross-sectional perspective of a DMDG JLTFET along with heterodielectric/stacked gate-oxide material. The overall gate-oxide thickness is expressed as $t_{ox} = t_{lk}(SiO_2) + t_{high-K}(TiO_2) = 2$ nm because the dielectric layer

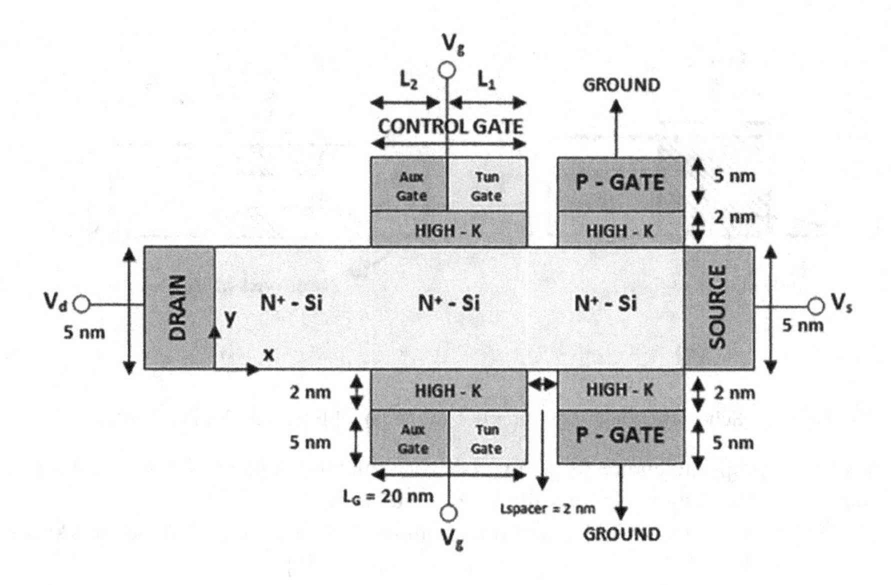

FIGURE 6.5 Cross-sectional view of DMDG JLTFET.

TABLE 6.1
Device Parameters of DMDG JLTFET

Parameters	Symbol	Values
Doping concentration	N_{ch}	$1\times10^{19}\text{cm}^{-3}$
Gate length	L_G	20 nm
Spacer thickness	L_{spacer}	2 nm
Free space permittivity	ε_0	$8.854\times10^{-12}\,F/m$
Permittivity of Silicon	ε_{si}	$11.9\,\varepsilon_0$
Permittivity of high-K dielectric (TiO_2)	ε_{ox}	$80\varepsilon_0$
Thickness of silicon substrate	t_{si}	5 nm
Front – gate dielectric thickness	t_f	2 nm
Back – gate dielectric thickness	t_b	2 nm
Control Gate Metal – 1 – Work Function (Cu)	ΦM1	4.7 eV
Control Gate Metal – 2 – Work Function (Al)	ΦM2	4.1 eV
P – Gate Metal – Work Function (Pt)	ΦP	5.93 eV

[23–28] includes a low-K (K = 3.9, SiO_2) dielectric material combined with the current high-K (K = 80, TiO_2) dielectric.

6.9 PERFORMANCE INVESTIGATION OF DMDG JLTFET

The device simulation incorporates mobility, bandgap narrowing, recombination, nonlocal BTBT, and Kane's tunneling models to illustrate the carrier transport mechanism in DMDG JLTFET [29]. Table 6.1 lists the device parameters that have been used for the proposed device.

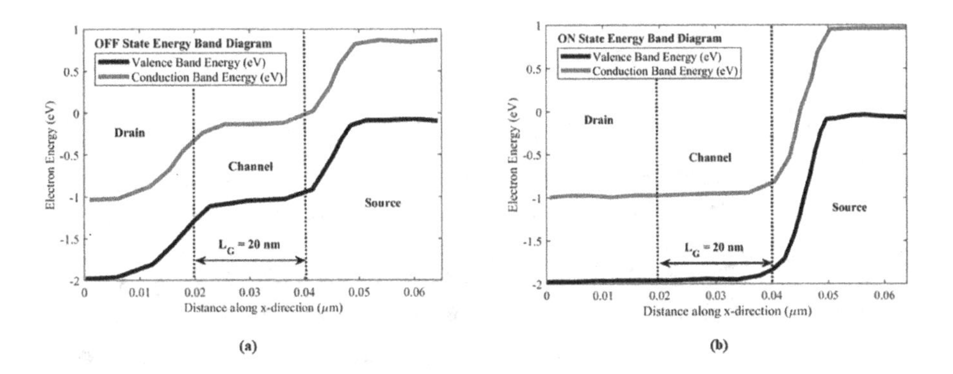

FIGURE 6.6 Energy band diagram of DMDG JLTFET in the (a) OFF state and (b) ON state.

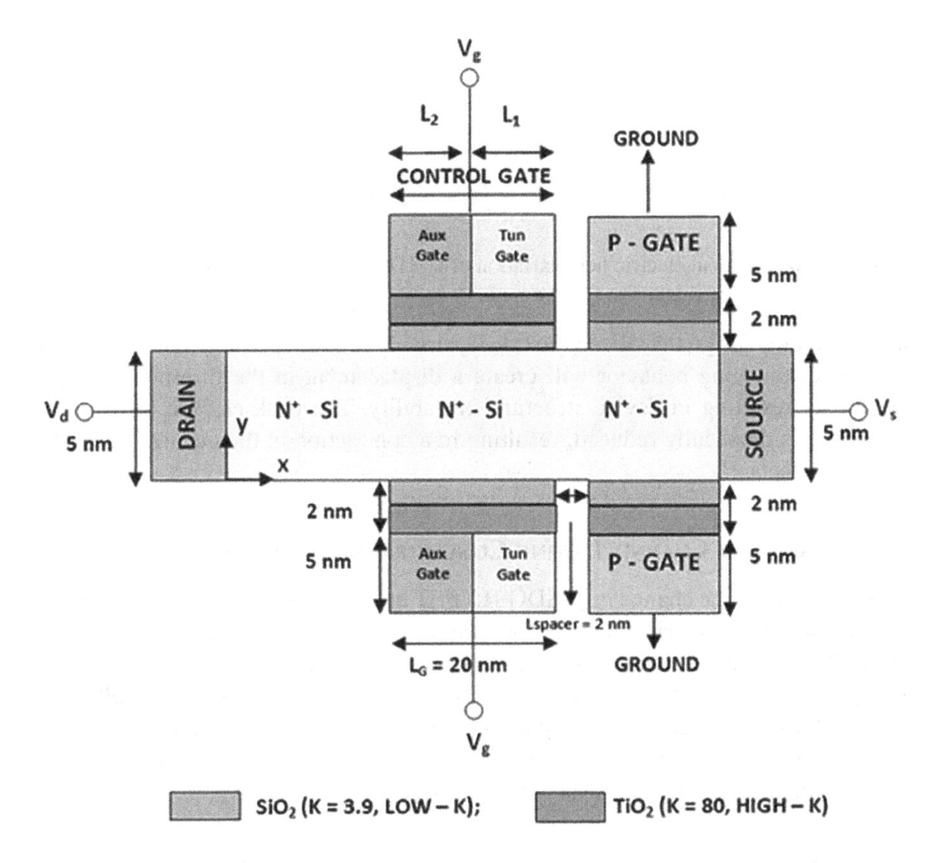

FIGURE 6.7 Cross-sectional view of DMDG JLTFET with stacked gate-oxide material.

6.9.1 Vertical Electric Field Profile

The vertical electric field patterns of DMDG JLTFET and DMDG TFET are compared in Figure 6.8. The vertical field component in both models is minimal for thinner oxide thickness. Electrons can seamlessly tunnel through the gate-oxide

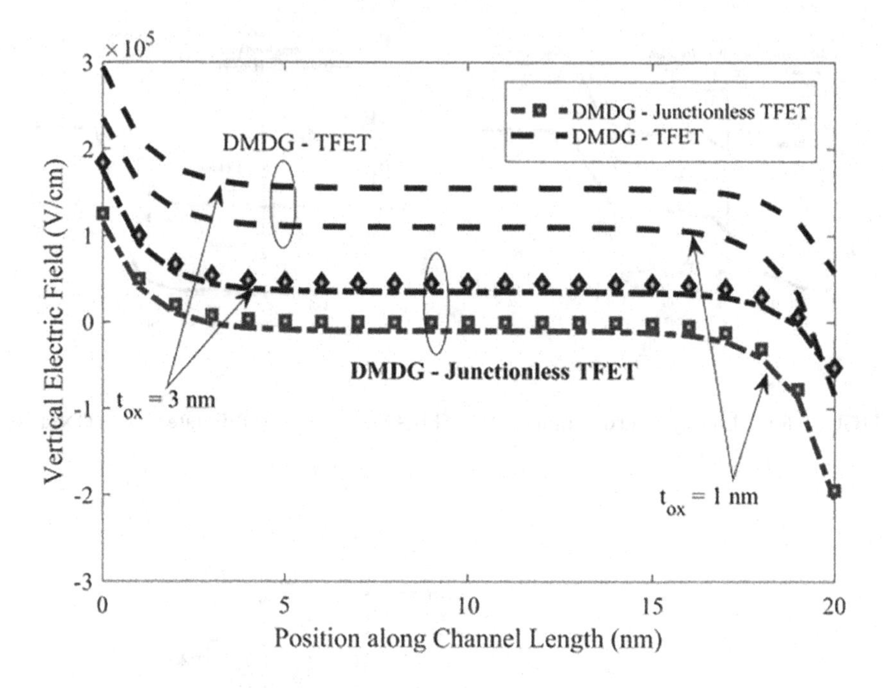

FIGURE 6.8 Vertical electric field variation of DMDG JLTFET and DMDG TFET for different values of oxide thickness: $t_{ox} = 1\,\text{nm}$ and $t_{ox} = 2\,\text{nm}$.

layer when we design the silicon dioxide layer with minimal possible thickness. This undesired tunneling behavior will create a displacement in the threshold voltage, eventually resulting in device structural instability. The peak electric field on the drain side is drastically reduced, resulting in a contraction in the occurrence of hot electron effects.

6.9.2 Effect of Channel Doping Concentration on Threshold Voltage

Figure 6.9 shows the change in DMDG JLTFET and DMDG TFET threshold voltage as a measure of dielectric thickness for multiple channel doping values. The DMDG JLTFET is observed with a threshold voltage of 0.33 V, whereas DMDG TFET's voltage threshold is 0.43 V. For DMDG JLTFET, the drop in threshold voltage is attributable to the consistent doping through the use of stronger gate dielectric material. Besides, a hard oxide layer demonstrates a decrease in threshold voltage for a given doping, indicating fewer short-channel effects.

6.9.3 Drain Current Profile

Designing nanoelectronics devices with minimal power consumption requires a good understanding of drain current and its effects. TFET's core idea of BTBT allows for a high ON state current. The evenly doped channel improves the BTBT mechanism thereby increasing the carrier tunneling generation rate.

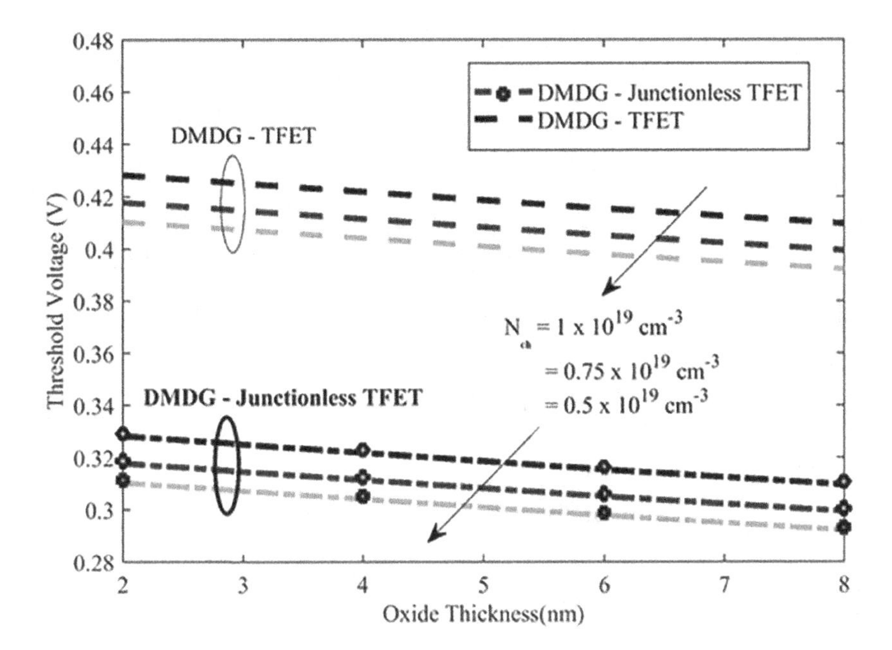

FIGURE 6.9 Threshold voltage of DMDG JLTFET and DMDG TFET as a function of different channel doping concentrations.

The drain current is estimated using the carrier generation rate from BTBT over the space of the proposed DMDG JLTFET structure.

$$I_{ds} = q\int TGR_{BTB}\,dv \tag{6.1}$$

The formulation for tunneling generation rate TGR_{BTB} using Kane's model is stated below:

$$TGR_{BTB} = A_{kane}\frac{|E|^2}{\sqrt{E_g}}\exp\left[-B_{kane}\frac{E_g^{3/2}}{|E|}\right] \tag{6.2}$$

where q is the electronic charge, E is the average electric field, E_g is the band energy gap, and A_{kane} $(4\times10^{14}\ \mathrm{V^{-5/2}s^{-1}cm^{-1/2}})$ and B_{kane} $(1.9\times10^{7}\ \mathrm{V/cm})$ are the distinct tunneling process parameters [13].

Figure 6.10 shows the drain current outputs of DMDG JLTFET and DMDG TFET for varied oxide thickness values. In addition, DMDG JLTFETs with strong gate dielectric and high work function toward the tunneling gate have predominant electric field lines along the source–channel interface, which reinforces gate control over the channel. Tunneling barrier toward the source is limited as the gate bias is increased. This induces electrons to tunnel into the drain's conduction band,

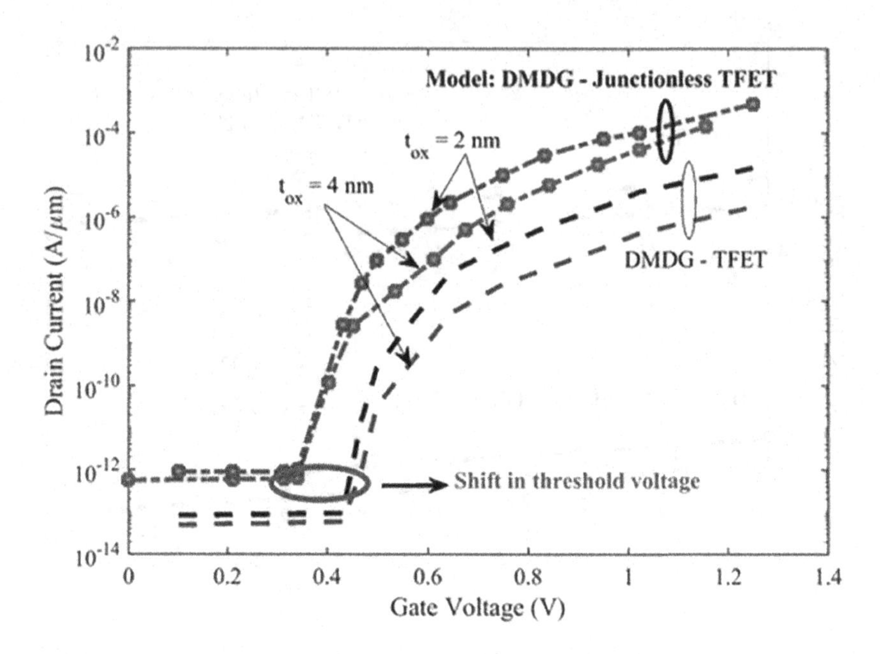

FIGURE 6.10 Drain current profile of DMDG JLTFET and DMDG TFET for different values of oxide thickness.

increasing the drain current exponentially. The proposed device's threshold voltage persists at 0.33 V with gate bias of 1 V, as illustrated in Figure 6.9, and the ON state drain current attributes commence thereafter. Relative to SiO_2- deployed DMDG TFET devices, embodiment of immense and strong gate dielectric material minimizes the threshold voltage for DMDG JLTFET devices, which also boosts the drain current [30].

6.10 SUBTHRESHOLD ANALYSIS OF DIELECTRIC-MODULATED DMDG JLTFET

The consistent doping in JLTs increases the density of carriers in the channel, resulting in a higher drain-to-source current. The drain current of junctionless tunneling transistors can be enhanced further with minimal equivalent oxide thickness (EOT) in layered gate-oxide architecture [31–34]. Figure 6.7 shows a dielectric- modulated dual material JLTFET that involves the integration of JLTs and TFETs through the use of heterogate dielectric materials.

6.10.1 Potential Distribution

Figure 6.11 contrasts the potential patterns of distinct TFET models: DMDG TFET, DMDG JLTFET, and DMDG stack JLTFET. The surface potential of the dielectric-modulated/gate-tack-based DMDG JLTFET is obviously superior to the outcomes of

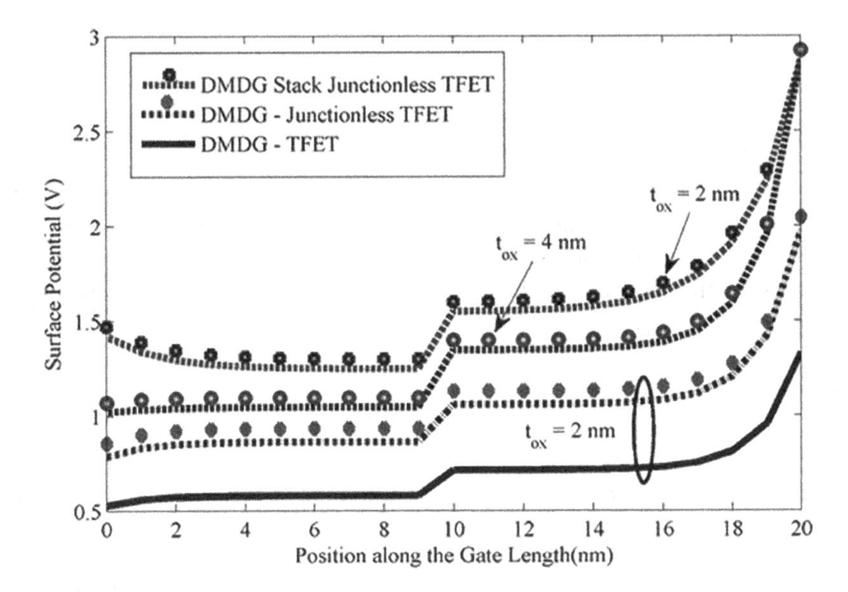

FIGURE 6.11 Surface potential distribution of different TFET device models for different values of oxide thickness.

the other TFET models. The proposed device's surface potential is presented for varied oxide thicknesses (t_{ox} = 2 nm and t_{ox} = 4 nm). The potential at the surface is more noticeable with a limited oxide thickness value, as shown in this plot.

6.10.2 Transconductance-to-Drain-Current Ratio

Transconductance-to-drain-current ratio (g_m/I_{ds}) is another important metric in assessing the strength of DMDG JLTFETs with heterogate-oxide material. The transition of drain-to-source current (I_{ds}) with applied gate-to-source bias (V_{gs}) is quantified by transconductance, as shown below:

$$\frac{g_m}{I_{ds}} = \frac{q}{K_B T}\left(\frac{\alpha^2 + \sqrt{2\cosh\ (\alpha L_1) - 2}\left(1 - \frac{1}{2}\sqrt{\frac{V_{ds}}{V_{gs}}}\right)}{\text{Sinh}\ (\alpha L_1)}\right) \tag{6.3}$$

where K_B is Boltzmann constant = 1.38×10^{-23}/JK,

T is temperature in K = 300 K,
Q is the charge of electron = 1.602×10^{-19}C,
L_l is the channel length of the gate metal region M_1,
V_{ds} is the drain-to-source bias,
V_{gs} is the gate-to-source bias,

and

$$\alpha - \text{constant} = \sqrt{\frac{4C_{ox}}{C_{si}t_{si}^2}} \tag{6.4}$$

where C_{si} is the channel capacitance, C_{ox} is the gate dielectric capacitance, ε_{ox} is the permittivity of silicon dioxide (3.9 ε_o), ε_{high-K} is the permittivity of titanium oxide (80 ε_0), and t_{oxeff} represents the effective gate-oxide thickness with t_{ox} and t_{high-K} indicating the thickness of silicon dioxide and high-K gate dielectric (titanium oxide) material, respectively.

$$C_{si} = \frac{\varepsilon_{si}}{t_{si}} \tag{6.5}$$

$$C_{ox} = \frac{\varepsilon_{ox}}{t_{oxeff}} \; and \; t_{oxeff} = t_{ox} + \frac{\varepsilon_{ox} t_{high-K}}{\varepsilon_{high-K}} \tag{6.6}$$

Figure 6.12 demonstrates the variation of the dielectric-modulated DMDG JLTFET's transconductance-to-drain-current ratio against the results of the other two TFET models. DMDG stack JLTFET outperforms DMDG JLTFET and DMDG TFET by 15.33% and 57.33%, respectively. This enhancement reflects the device's relatively high carrier generation efficiency and the proportion of gate voltage that can

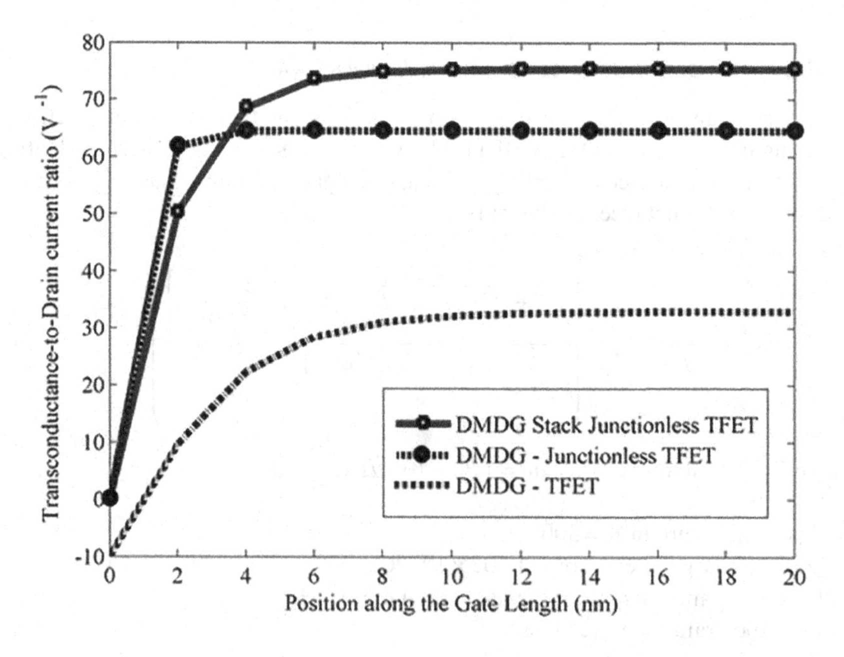

FIGURE 6.12 Transconductance-to-drain-current ratio of DMDG stack JLTFET, DMDG JLTFET, and DMDG TFET.

be turned into optimal ON current. To avoid OFF current leakage, the layered gate-oxide material works as a superior insulating layer.

6.10.3 Subthreshold Current Profile

The device's subthreshold conduction creates a dangerous scenario for SCEs to arise. Since the applied gate bias is much lesser than the device's threshold voltage, a situation like this emerges. During the OFF state, a confined amount of current runs across the source and drain terminals, allowing for the device to consume more power. Subthreshold current, represented by I_{ds}, is the current triggered by this conduction.

$$I_{ds} = \frac{\mu t_{si}^2 n_i^2 K_B T\left(1 - e^{-qV_{ds}} / K_B T\right)}{\left(e^{\frac{L_1}{q(\phi_{1,\min}/K_B T)}} e^{\frac{L_2}{q(\phi_{2,\min}/K_B T)}}\right) N_{ch}} \quad (6.7)$$

where n_i is the intrinsic carrier concentration of silicon = $1.5 \times 10^{10}/\text{cm}^3$,
μ is the electron mobility = 3,900 cm^2/(V-s), and
$\phi_{\min}$ is the minimum surface potential.

In examining the device's subthreshold properties, subthreshold current plays a significant role. Figure 6.13 shows even more clearly how the layered gate-oxide material suppresses the OFF state current (subthreshold current). It exhibits subthreshold current versus gate-to-source voltage for varied drain-to-source voltage values (V_{ds}).

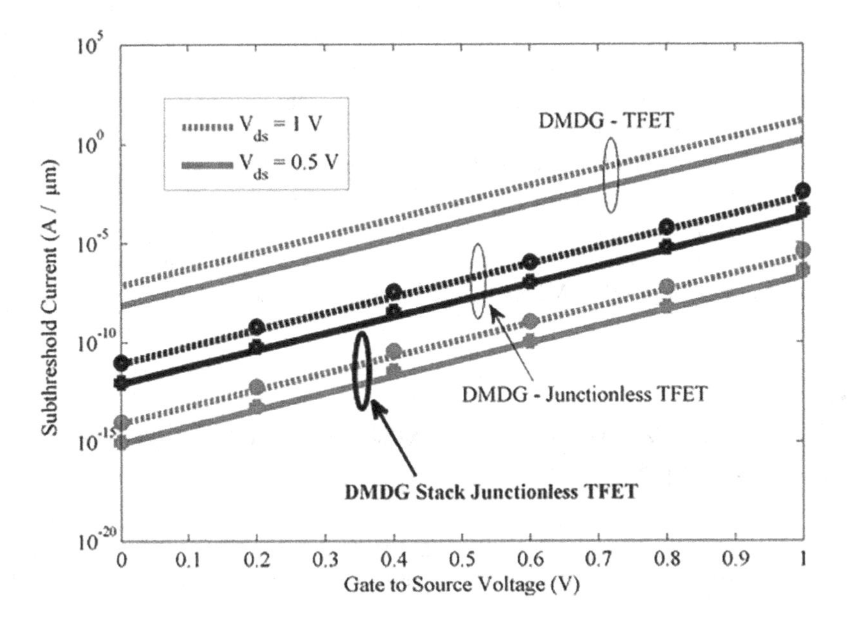

FIGURE 6.13 Subthreshold current of DMDG stack JLTFET, DMDG JLTFET, and DMDG TFET for various values of drain-to-source voltages.

DIBL is widely attributed to V_{ds}. When the gate loses control of the channel and the drain takes over, this effect typically occurs and degrades the device performance, lowering the threshold voltage. The superimposed gate dielectric material and gate metal engineering in DMDG stack JLTFETs restrict the drain from overpowering the channel. When examining the subthreshold leakage current of the DMDG stack-based JLTFET to the subthreshold current of the other two TFET models, it is noted that the DMDG stack JLTFET has a minimal subthreshold current of 10^{-15} (A/μm).

6.11 PERFORMANCE ANALYSIS OF LOW-POWER JLTFETs

With its major technological revolution in new materials and device structures, modern integrated circuit technology and the nanoscale industry have genuinely transformed human life. CMOS circuits were the most popular technological choice in the early 1980s, owing to their low static power consumption. MOSFETs have been used in the microelectronics industry for decades. MOS transistors, however, have begun to contribute second-order effects known as SCEs as a result of significant changes in scaling. TFETs have also emerged as a promising device to suppress SCEs, especially when gate lengths are below 100 nm, thanks to the invention of several unique transistors and constant scaling. The only shortcoming of the TFET is its ambipolar behavior, which results in a low ON state device current. Next-generation transistors with identically doped source, drain, and channel areas, known as JLTs/dopingless FETs [32], have evolved to combat this.

The most distinctive combination of incorporating the tunneling in JLTs was offered in the literature to resist second-order effects owing to device scaling. Furthermore, multigate TFETs were introduced to improve the potential of the gate metal to regulate the channel. Despite all of these advantages, we still used a substantial dielectric-modulated gate metal engineering approach to produce a greater resistance to SCEs. For various device parameters, Table 6.2 presents the results of DMDG TFET, DMDG JLTFET, and DMDG stack JLTFET. When each parameter is compared for all DMDG-based TFET models, it is clear that our proposed DMDG stack/dielectric-modulated DMDG JLTFET produces significantly better results.

In evenly doped TFETs, useful options of gate-oxide materials appear to be the optimal combination for significantly reducing short-channel issues such as HCE and DIBL. In a nutshell, DMDG JLTFET along with layered gate-oxide layer is

TABLE 6.2
Illustration Electrical Parameters for DMDG TFET, DMDG JLTFET, and DMDG Stack JLTFET

Electrical Parameters	DMDG TFET	DMDG JLTFET	DMDG Stack JLTFET
Potential at the surface region (V)	1.32	2.04	2.992
Transconductance-to-drain-current ratio (V^{-1})	33.43	65.12	75.52
Subthreshold current (A/μm)	7.58×10^{-8}	8.6×10^{-13}	9.8×10^{-16}

envisioned to strengthen switching characteristics and is found to be the best device for minimal power-consuming nanoscale applications.

6.12 SUMMARY

The JLTFET has demonstrated to be an effective device in dealing with the short-channel issues. Traditional FETs must be updated by adopting modern gate and channel engineering approaches in order to ensue Moore's law and avoid SCEs. Drain structure of the dual work function engineering is lightly doped, which weakens the peak electric field at the drain end. Then, consistent doping in channel aids in ensuring desired tunneling generation rate and higher carrier velocity. Employing a high-K gate dielectric material in the gate-oxide region is the most essential consideration in the design of emerging low-power JLTFETs. To increase the electrostatic regulation between gate and channel regions, silicon dioxide has been substituted with titanium oxide. The outcomes of the DMDG JLTFET showed a significant reduction in the device's electric field and threshold voltage. Also, an enhanced subthreshold device performance with stacked/layered gate-oxide/heterodielectric material is observed. The superior performance of the novel dielectric-modulated dual material double gate is validated in terms of increased transconductance/carrier generation efficiency and low subthreshold leakage current.

REFERENCES

1. Yang, ES 1988. *Microelectronic devices*. McGraw-Hill, New York, pp. 285–294.
2. Thompson, SE & Parthasarathy, S 2006. Moore's law: The future of Si microelectronics. *Materials Today*, 9(6): 20–25.
3. Skotnicki, T, Hutchby, JA, King, TJ, Wong, HS & Boeuf, F 2005. The end of CMOS scaling: Toward the introduction of new materials and structural changes to improve MOSFET performance. *IEEE Circuits and Devices Magazine*, 21(1): 16–26.
4. Suzuki, K, Tanaka, T, Tosaka, Y, Horie, H & Arimoto, Y 1993. Scaling theory for double gate SOI MOSFETs. *IEEE Transactions on Electron Devices*. 40(12): 2326–2329.
5. Kalra, S & Bhattacharyya, AB 2018. Scalable α-power law based MOSFET model for characterization of ultra-deep submicron digital integrated circuit design. *AEU - International Journal of Electronics and Communications*. 83: 180–187.
6. Moore, GE 1965. Cramming more components onto integrated circuits. *Electronics Magazine*. 3(8): 114–119.
7. Young, KK 1989. Short-channel effect in fully depleted SOI MOSFETs. *IEEE Transactions on Electron Devices*. 36(2): 399–402.
8. Hugo, MOSFET short channel effects, Available from <http://www.onmyphd.com/?p=mosfet.short.channel.effects#h2_dibl>
9. Veeraraghavan, S & Fossum, JG 1988. A physical short channel model for the thin film SOI MOSFET applicable to device and circuit CAD. *IEEE Transactions on Electron Devices*. 35: 1866–1875.
10. Lee MJ & Choi WY. 2011a. Analytical model of a single-gate silicon-on-insulator (SOI) tunneling field-effect transistors (TFETs). *Solid-State Electronics,* 63(1): 110–114.
11. Zhang, Q, Zhao, W & Seabaugh, A 2006. Low-subthreshold-swing tunnel transistors. *IEEE Electron Device Letters*, 27(4): 297–300.
12. Arun Samuel, TS, Balamurugan, NB, Bhuvaneswari, S, Sharmila, D & Padmapriya, K. 2013. Analytical modelling and simulation of single-gate SOI TFET for low-power applications. *International Journal of Electronics*, 101: 779–788.

13. Bagga, N & Dasgupta, S 2017. Surface potential and drain current analytical model of gate all around triple metal TFET. *IEEE Transactions on Electron Devices*, 64(2): 606–613.
14. Vanitha, P, Balamurugan, NB & Lakshmi Priya, G 2015. Triple material surrounding gate (TMSG) nanoscale tunnel FET-analytical modeling and simulation. *Journal of Semiconductor Technology and Science*. 15(6): 585–593.
15. Venkatesh, M, Priya, GL & Balamurugan, NB 2021. Investigation of Ambipolar conduction and RF stability performance in novel germanium source dual halo dual dielectric triple material surrounding gate TFET. *Silicon*. 13: 911–918.
16. Preethi, S, Venkatesh, M, Karthigai Pandian, M & Priya, GL 2021. Analytical modeling and simulation of gate-all-around junctionless Mosfet for biosensing applications. *Silicon*. 13(10): 3755–3764.
17. Ghosh, B & Akram, MW 2013. Junctionless tunnel field effect transistor. *IEEE Electron Device Letters*. 34(5): 584–586.
18. Ghosh, B, Bal, P & Mondal P 2013. A junctionless tunnel field effect transistor with low subthreshold slope. *Journal of Computational Electronics*. 12(3): 428–436.
19. Bal, P, Ghosh, B, Mondal, P, Akram, MW & Tripathi, BMM 2014. Dual material gate junctionless tunnel field effect transistor. *Journal of Computational Electronics*. 13(1): 230–234
20. Priya, GL, Venkatesh, M, Balamurugan, NB, & Samuel, TSA 2021. Triple metal surrounding gate junctionless tunnel FET based 6T SRAM design for low leakage memory system. *Silicon*. 13: 1691–1702.
21. Baruah, RK & Paily, RP 2014. A dual-material gate junctionless transistor with high-k spacer for enhanced analog performance', *IEEE Transactions on Electron Devices*. 61(1): 123–128.
22. Shradya S, Raj, B & Vishvakarma, SK 2018. Analytical modeling of split-gate junctionless transistor for a biosensor application. *Sensing and Bio-Sensing Research*. 18: 31–36.
23. Zhou, X 2000. Exploring the novel characteristics of hetero-material gate field-effect transistors (HMGFETs) with gate-material engineering. *IEEE Transactions on Electron Devices*. 47(1): 113–120.
24. Priya GL & Balamurugan NB 2018. Subthreshold modeling of triple material gate-all-around junctionless tunnel FET with germanium and high-K gate dielectric material. *Journal of Microelectronics, Electronic Components and Materials*. 48: 53–61.
25. Wallace, RM & Wilk, GD 2003. High-K dielectric materials for microelectronics. *Critical Reviews in Solid State and Materials Sciences*. 28(4): 55.
26. Darwin, S & Samuel, TSA 2019. A holistic approach on junctionless dual material double gate (DMDG) MOSFET with high k gate stack for low power digital applications. *Silicon*. 12: 393–403.
27. Ajayan, J, Nirmal, D, Prajoon, P & Charles Pravin, J 2017. Analysis of nanometer-scale InGaAs/InAs/InGaAs composite channel MOSFETs using high-K dielectrics for high speed applications. *AEU - International Journal of Electronics and Communications*. 79: 151–157.
28. Amin, S & Sarin, RK 2016. Enhanced analog performance of doping-less dual material and gate stacked architecture of junctionless transistor with high-k spacer. *Applied Physics A*. 122(380): 1–9.
29. Venkatesh, M. & Balamurugan, N.B. 2021. Influence of threshold voltage performance analysis on dual halo gate stacked triple material dual gate TFET for ultra low power applications. *Silicon* 13, 275–287.
30. Priya GL & Balamurugan NB 2019. New dual material double gate junctionless tunnel FET: Subthreshold modeling and simulation. *AEU - International Journal of Electronics and Communications*. 99: 130–8.

31. Venkatesh, M & Balamurugan, NB 2019. New subthreshold performance analysis of germanium based dual halo gate stacked triple material surrounding gate tunnel field effect transistor. *Superlattices and Microstructures*. 130: 485–498.
32. Priya GL & Balamurugan NB 2020. Improvement of subthreshold characteristics of dopingless tunnel FET using hetero gate dielectric material: Analytical modeling and simulation. *Silicon*. 12: 2189–2201.
33. Preethi S & Balamurugan NB 2020. Analytical modeling of surrounding gate junctionless MOSFET using finite differentiation method. *Silicon*. 13(9): 2921–2931.
34. Kumar D 2019. Performance evaluation of double gate tunnel FET based chain of inverters and 6-T SRAM cell. *Engineering Research Express*. 1(2): 025055.

7 Recent Developments in Schottky Diodes and Their Applications

S. Sreejith, B. Sivasankari,
S. Babu Devasenapati, and A. Karthika
SNS College of Technology

Anitha Mathew
IES College of Engineering

CONTENTS

7.1 INTRODUCTION

Low forward voltage and rapid switching speed are the major advantages of SBD. SBD, which is a majority carrier device, finds application in computers, power factor correction systems, high-speed microelectronic applications, optoelectronic devices, and transistors [1–4]. SBDs were also used in solar cells, industrial electronics, and radiofrequency detectors [5–7]. When a metal–semiconductor junction is created, a potential energy barrier is formed resulting in the formation of a SBD [8–13]. The major parameters used in the analysis of SBDs were R_s (series resistance), ϕ (barrier height), n (ideality factor), and G_p (shunt resistance) [14–24]. SBD current flow is expressed in equation (7.1) [15].

$$I = I_d + I_P \tag{7.1}$$

where I_d is the diode current at bias voltage V and I_p is the shunt current flowing through R_P (shunt resistance)

DOI: 10.1201/9781003240778-7

Diode current is given in equation (7.2) [15].

$$I_d = I_s\left\{\exp\left[\frac{\beta}{n}(V - R_s I\right] - 1\right\} \tag{7.2}$$

where I_s the saturation current and β is the inverse thermal voltage.

Shunt current is expressed in equation (7.3) [15].

$$I_p = G_p(V - R_s I) \tag{7.3}$$

Saturation current (I_s) is expressed in equation (7.4) [15].

$$I_s = AA^{**}T^2\exp(-\beta\phi) \tag{7.4}$$

where A is the diode area,

A^{**} is the modified Richardson constant, and
T is the temperature (Kelvin)

Also, β is expressed in equation (7.5) [15].

$$\beta = \frac{q}{kT} \tag{7.5}$$

where q is the electronic charge, k is the Boltzmann constant, and T is the temperature.

The basic parameter information of microelectronic devices was obtained from its voltage–current plot. SBD conduction analysis is helpful in obtaining parameters involving ϕ, n, and R_s. The ideality factor and barrier height of SBD can be calculated from the voltage–current plot's slope and intercept [25]. SBD has a cathode pin, anode pin, and substrate pin connected to VSS ground. SBD's conventional dc model used for its modeling is shown in Figure 7.1 [26].

In this article, we have studied recent developments in emerging SBDs such as GaN SBD, 4H-SiC SBD, ZnO SBD, organic SBD, and diamond SBD and their applications.

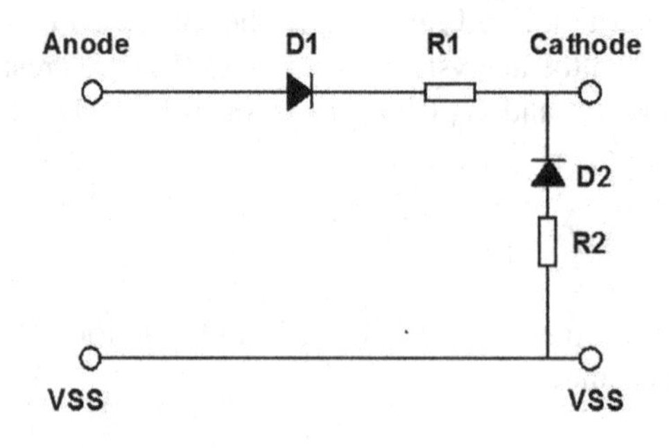

FIGURE 7.1 SBD conventional DC model [26].

7.2 GAN-BASED SBD

High saturation velocity, direct band gap, and high temperature durability are the advantages of GaN materials, which make them suitable for optoelectrical devices; high-speed electron devices; and in high-power, high-voltage applications [27–34]. The temperature dependence of SBD parameters, namely, barrier height and ideality factor, of a gold/n-type GaN SBD has been investigated [35]. The comparison of voltage–current density and voltage–capacitance measurements was done for the 165–480K temperature range. Metal–organic vapor phase epitaxy method is used to fabricate GaN on a sapphire C-plane substrate. It was observed that with an increase in temperature, there was a decrease in ideality factor (Figure 7.2a) and an increase in barrier height (Figure 7.2b). Similarly, barrier height and built-in voltage measured from voltage–capacitance measurement show a gradual increase with increase in temperature up to 400 K. At 400K, a sudden increase in both barrier height and built-in voltage has been reported (Figure 7.3).

GaN SBD is used as an effective hydrogen sensor [36]. Zhong et al. [37] reported the fabrication of a nano-SBD H_2 sensor that was able to detect H_2 gas in the range of 320–10,000 ppm at room temperature. The H_2 sensing ability of Pt-SBD fabricated on m-GaN has been investigated [38]. For various H_2 concentration exposures, Pt-SBD on m-GaN crystal reported reversible and rapid response. A maximum of 2×10^4% sensitivity was measured using a m-GaN SBD sensor at 0.1 V forward bias and 4% H_2 exposure (Figure 7.4). An increase in forward current was observed upon exposure to H_2 in the sweeping bias range. Also, reduction in forward turn-on voltage to 0.3 from 0.6 V was reported. Upon exposure to 4% H_2 in N_2, there was a decrease in SBD barrier height. Once H_2 gas is switched off, within a time period of 5 min, recovery in barrier height has been observed (Figure 7.5).

Humidity effect in GaN/AlGaN-based SBD hydrogen sensors can be eliminated by the PMGI (polydimethylglutarimide) water-blocking encapsulation layer [39]. Upon H_2 exposure of 500 ppm at 25°C, high responsivity of 5.97×10^7% was

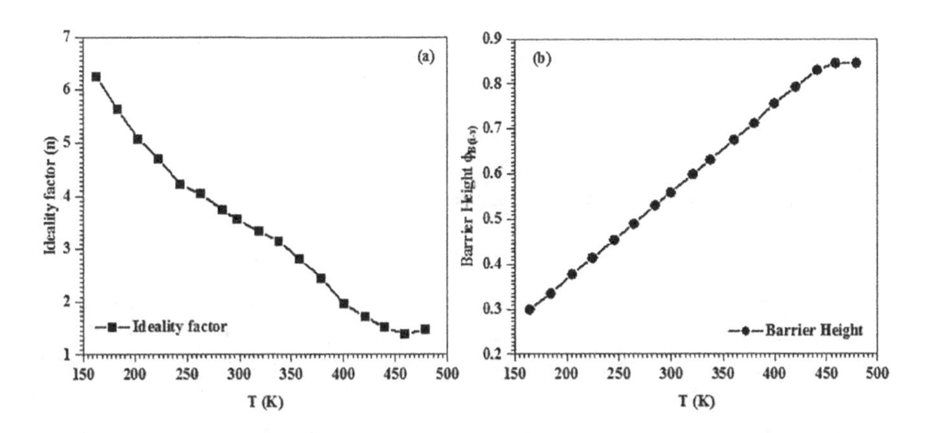

FIGURE 7.2 (a) Ideality factor vs. temperature of gold/n-type GaN SBD [35] and (b) zero bias barrier height vs. temperature of gold/ n-type GaN SBD [35].

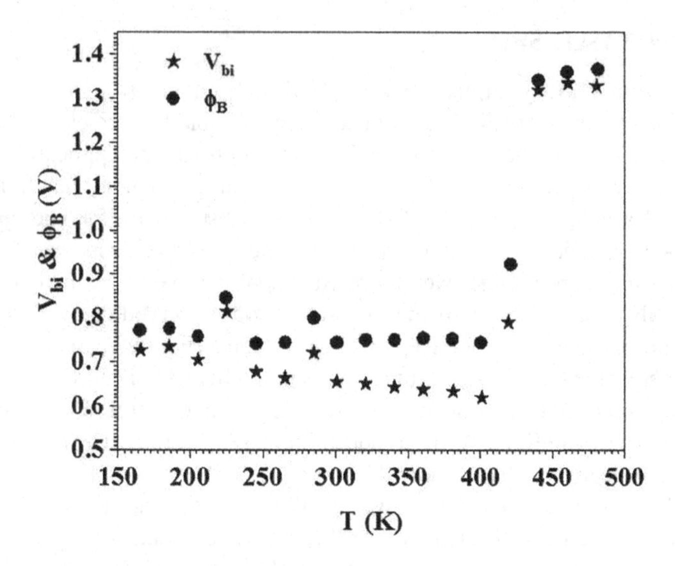

FIGURE 7.3 Built-in voltage and barrier height vs. temperature of gold/ n-type GaN SBD [35].

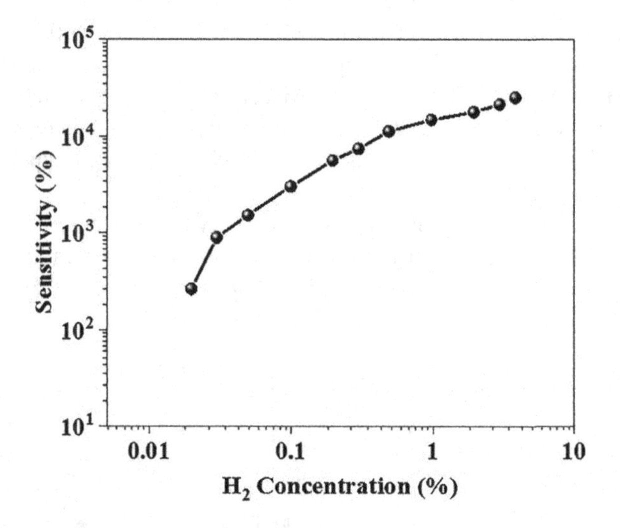

FIGURE 7.4 Sensitivity vs. hydrogen concentration plot of Pt-SBD on m-GaN wafer at 0.1V forward bias [38].

exhibited by PMGI-encapsulated SBD (Figure 7.6). Stable operation was reported for both wet and dry H_2 up to a temperature of 300°C in the PMGI-encapsulated device.

Reverse leakage current is a major factor affecting the reliability of GaN SBD [40–46]. In 2019, Li et al. [47] fabricated nickel nitride n-GaN SBD with a Ni target by a magnetron reactive sputtering process. The barrier height of 1.2 eV was reported in NiN SBD and barrier height of 1.0 eV was reported in Ni SBD from the voltage–capacitance measurement method. Figure 7.7 shows the voltage–capacitance plot of

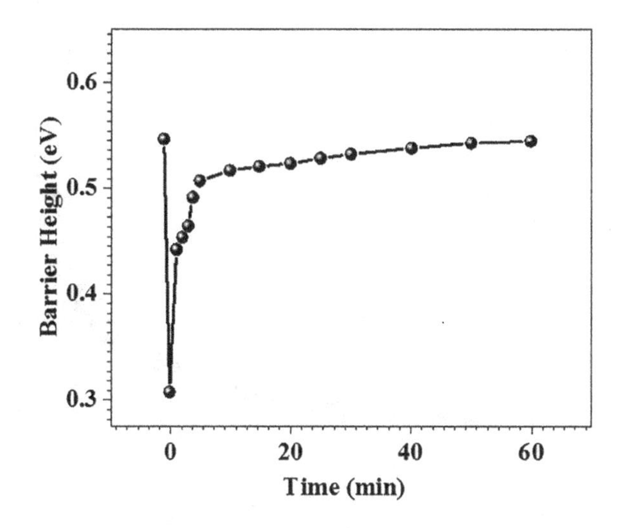

FIGURE 7.5 Barrier height vs. time plot of SBD on m-GaN after exposure to 4% hydrogen in N_2 [38].

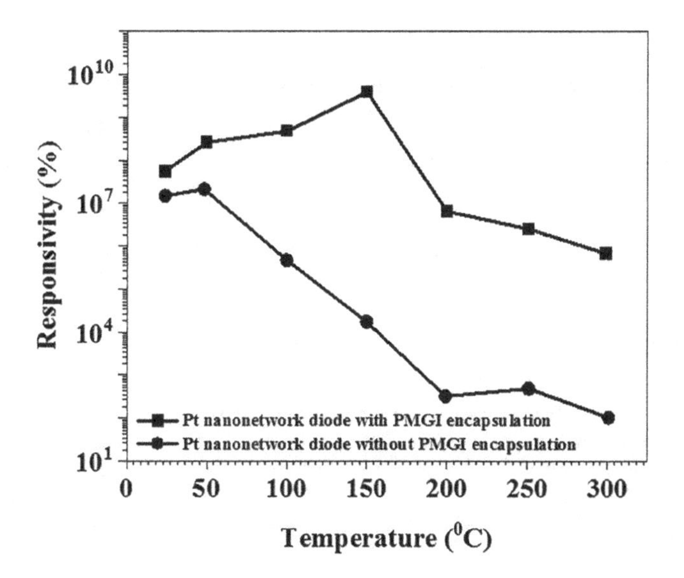

FIGURE 7.6 Hydrogen responsivity vs. temperature plot of PMGI-encapsulated and -unencapsulated SBD upon 500 ppm dry H_2 exposure in the temperature range of 25°C–300°C [39].

Ni SBD and NiN SBD. Good rectification characterization is reported in NiN SBD at various temperatures. Reduction by 2 orders of magnitude in reverse leakage current is also reported for NiN SBD. Better temperature stability observed in NiN SBD makes them ideal for temperature-sensing applications.

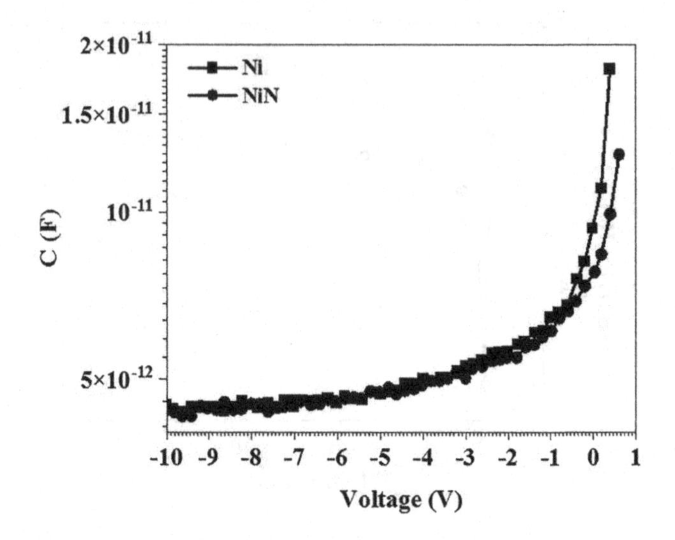

FIGURE 7.7 Voltage–capacitance plot of Ni SBD and NiN SBD [47].

TABLE 7.1
Ideality Factor and Barrier Height of Various GaN-Based SBDs

Ref.	SBD	Ideality Factor (n)	Barrier Height (ϕ) (eV)
[48]	GaAsN SBD with 1.2% N	1.16	1.15
[49]	Ag/InGaN/n-Si	2.84	0.79
[50]	n-GaN SBD	1.3	1.01
[51]	GaN NW SBD(CVD growth method)	1.65	0.80
[52]	InAlN/GaN SBD after O_2 plasma treatment	1.59	0.94
[53]	Vertical GaN SBD on Ammono GaN Substrate (HVPE growth method)	1.65	1.05
[54]	Ni_xN-GaN SBD($P(N_2)=0.069$ Pa)	1.09	1.21
[55]	Cu/AlGaN/GaN SBD	1.3	1.66
[56]	Vertical GaN SBD on Ge-doped GaN Substrate	1.27	0.99
[57]	Au/GaN/n-GaAs SBD	2.84	0.44
[58]	Au/AZO/n-GaN SBD	1.57	0.90
[47]	NiN-GaN SBD	1.13	1.20
[59]	Pt/n-type GaN SBD (100K)	1.88	0.44

An overview of electrical parameters, namely, ideality factor and barrier height of different GaN-based SBDs, is presented in Table 7.1.

7.3 4H-SiC-BASED SBD

4H SiC SBD finds wide applications in high-temperature sensors, power factor correction of uninterruptable power supplies, wind energy, traction system, motor-driven power converters, IGBT antiparallel diode, and also in same-class switching devices [60–68]. In 2018, Rao et al. [69] investigated the performance of proportional to absolute temperature (PTAT) sensor based on V_2O_5/4H-SiC SBD. The proposed PTAT sensor exhibited good reproducibility and high linearity. PTAT sensor based on V_2O_5/4H-SiC SBD also reported small sensitivity in output for variations in bias current. A sensitivity of 307 μV/K was reported. The thermal annealing process can enhance the electrical parameters of 4H-SiC SBD. Annealing helps in reducing the reverse leakage current in 4H-SiC SBD. Metal–semiconductor interfaces have a strong influence on the electrical properties of SBD. The thermal annealing process decreases the resistance between Schottky metal and SiC interface, thereby improving the contact quality [70–75]. Temperature dependence of ideality factor and barrier height of Pd/n-4H SiC SBD were investigated by varying the temperature to 800 from 300 K [76]. Both parameters were observed to have strong temperature dependence. An increase in temperature causes an increase in barrier height and a decrease in ideality factor. At 300 K, 1.42 eV barrier height and at 800 K, 2.27 eV barrier height were reported. Similarly, at 300 K, ideality factor of 1.62 and at 800 K, ideality factor of 1.20 were reported. The variation of ideality factor and barrier height as a function of temperature ranging from 300 to 800 K for Pd/n-4H SiC SBD is shown in Figure 7.8a and b, respectively.

In the year 2020, Zeghdar et al. [77] investigated voltage (V_D), current (I_D), and temperature characteristics of Mo/4H-SiC SBD. The temperature is varied to 498 from 303 K, and an increase in barrier height and a decrease in ideality factor are observed with an increase in temperature (Figure 7.9a and b). The temperature-sensing performance of Mo/4H-SiC SBD is investigated by forward-biasing the SBD and varying the current level to 10 nA, 100 nA, 1 μA, 10 μA, 100 μA, and to 1 mA, respectively (Figure 7.10). For current level ID = 1 μA, high sensitivity of 1.92 mV/K was reported.

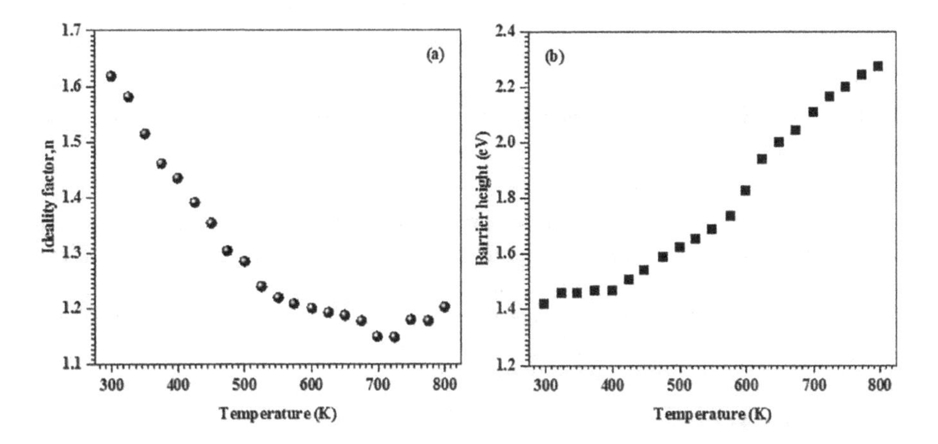

FIGURE 7.8 (a) Ideality factor vs. temperature of Pd/n-4H SiC SBD [76] and (b) barrier height vs. temperature of Pd/n-4H SiC SBD [76].

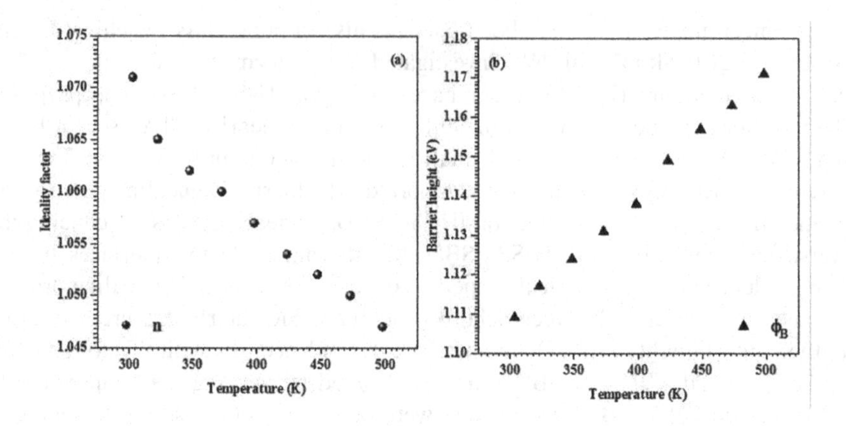

FIGURE 7.9 (a) Ideality factor vs. temperature of Mo/4H-SiC SBD [77] and (b) barrier height vs. temperature of Mo/4H-SiC SBD [77].

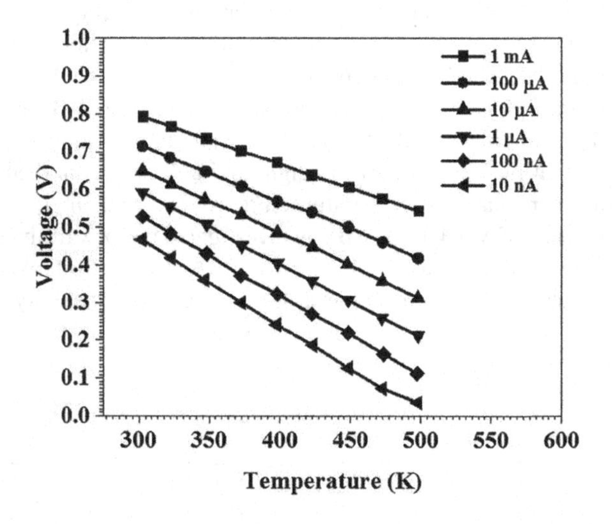

FIGURE 7.10 Voltage vs. temperature characteristics of Mo/4H-SiC SBD for various values of current I_D [77].

The temperature-sensing performance of tungsten/4H-SiC SBD was investigated by analyzing the voltage–temperature characteristics by varying I_D [78]. At $I_D = 5.97\,nA$, a maximum sensitivity of 2.33 mV/K was reported.

An overview of electrical parameters, namely, ideality factor and barrier height of different 4H-SiC-based SBDs, is presented in Table 7.2.

7.4 ZnO-BASED SBDs

ZnO semiconductor material has a direct and wide energy gap, good thermal and chemical stability, and high mobility of conducting electrons. These properties make

TABLE 7.2
Ideality Factor and Barrier Height of Various 4H-SiC-Based SBDs

Ref.	SBD	Ideality Factor (n)	Barrier Height (ϕ) (eV)
[79]	Ti/4H-SiC (750°C)	1.04	1.33
[80]	Ni/4H-SiC(0 0 0 1)	1.6	1.25
[81]	Ti/4H-SiC (500°C)	1.17	1.21
[74]	Ti/Al 4H SiC SBD(60 minutes/873 K)	1.86	0.85
[82]	Ti/4H-SiC SBD (700°C)	1.23	1.19
[83]	Ni/4H-SiC SBD(500°C)	1.66	1.3
[84]	4H SiC SBD with Ni_2Si Contact	2.72	1.358
[85]	4H SiC JBS (150°C)	2.06	0.78
[86]	Ni/Si/Ni/4H SiC SBD	1.118	1.693
[87]	Ni/4H-nSiC SBD (473 K)	1.225	1.625
[88]	Ti/4H SiC SBD (sputter-deposited)	1.04	1.33

ZnO semiconductor material suitable for use in the fabrication of high-performance Schottky ultraviolet photodetectors, laser diodes, gas sensors, nanogenerators, solar cells, and light-emitting diodes [89–94]. Temperature characteristics of ZnO SBD with iridium contact electrode were studied by varying the temperature from 25°C to 150°C [95]. Both Cheung's method and Norde's model are used to investigate the variation in barrier height with variation in temperature. With an increase in temperature, increase in leakage current is observed in ZnO SBD (Figure 7.11). At room temperature, barrier height greater than 0.8 eV was reported by measurement from

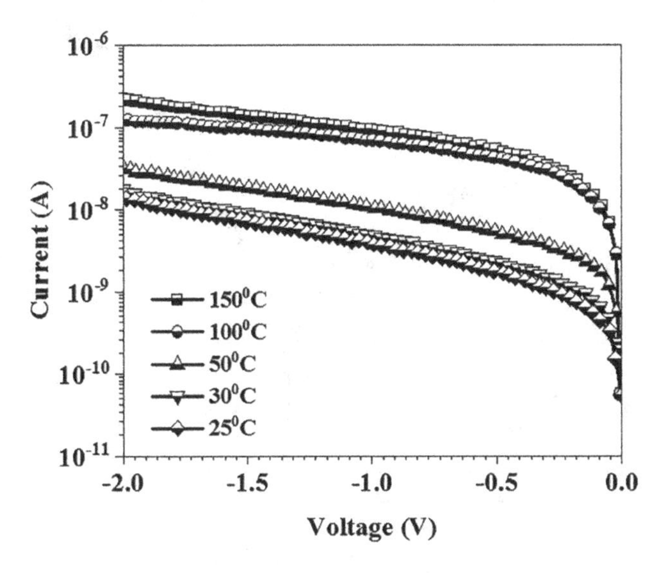

FIGURE 7.11 Reverse voltage–current plot of ZnO SBD for different temperatures: 25°C, 30°C, 50°C, 100°C, and 150°C [95].

both Cheung's method and Norde model, which makes iridium contact electrode an effective material in ZnO SBD photodetectors and metal–semiconductor–metal (MSM) photodetectors.

In 2013, Zhang et al. [96] fabricated flexible Ag/ZnO SBD using a pulsed laser deposition method. Barrier height of 0.54 eV and ideality factor of 2.8 were reported from the voltage–current characteristics. The electrical properties of SBD were investigated with and without bending. Even under bending condition, no significant variation is observed in the electrical performance of Ag/ZnO SBD, which makes them suitable for applications involving flexible electronics. Figure 7.12 shows the voltage–current plot of flexible Ag/ZnO SBD before and after bending.

The effect of various sputtering conditions in indium gallium zinc oxide SBD, which finds application in high-speed electronics, was investigated [97]. The structure of SBD used was Pd/IGZO/Ti/Au. At room temperature, without any thermal treatment, RF magnetron sputtering was used in the fabrication of InGaZnO SBD. The electrical properties of SBD were analyzed by varying RF sputtering power and oxygen partial pressure (Figures 7.13 and 7.14).An increase in barrier height was reported with an increase in oxygen partial pressure. However, degradation of interface quality, reduction in rectification ratio, and increase in ideality factor were observed for high oxygen partial pressure and high RF sputtering power. At 2.5% oxygen partial pressure and 50 W RF sputtering power, ideality factor of 1.14 and barrier height of 0.73 eV were reported.

An overview of electrical parameters, namely, ideality factor and barrier height of different ZnO-based SBDs, has been presented in Table 7.3.

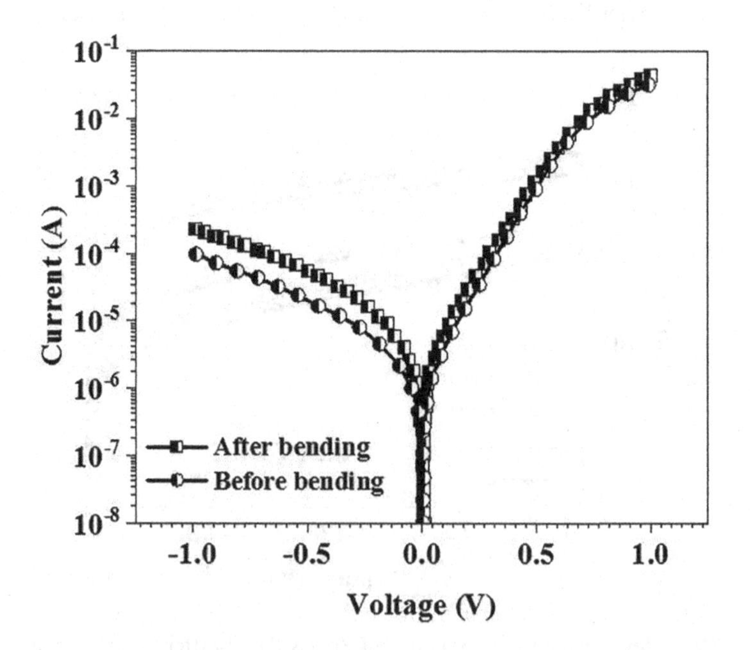

FIGURE 7.12 Voltage–current plot of flexible Ag/ZnO SBD before and after bending [96].

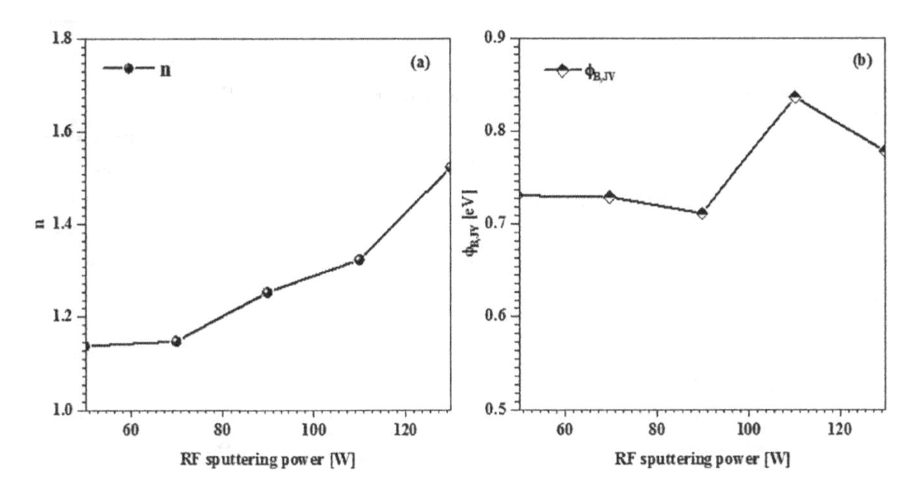

FIGURE 7.13 (a) RF sputtering power vs. ideality factor characteristics of InGaZnO SBD [97] and (b) RF sputtering power vs. barrier height characteristics of InGaZnO SBD [97].

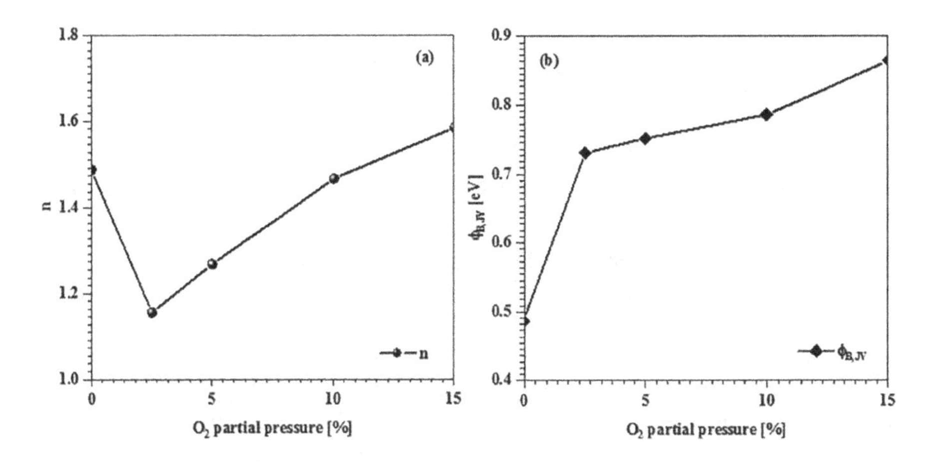

FIGURE 7.14 (a) O_2 partial pressure vs. ideality factor characteristics of InGaZnO SBD [97] and (b) O_2 partial pressure vs. barrier height characteristics of InGaZnO SBD [97].

7.5 OTHER EMERGING SBDs

In recent years, various types of SBDs incorporating different semiconductor materials have been extensively investigated for different types of applications. Light weight, easy processing technique, and low cost have made conducting polymers as well as their composites find wide applications in the fabrication of various optoelectronic and electronic devices including organic Schottky diodes. Low-temperature, processing, and good flexibility are the main advantages of organic semiconductor thin films. The main drawback of organic SBD is its poor environmental

TABLE 7.3
Ideality Factor and Barrier Height of Various ZnO-Based SBDs

Ref.	SBD	Ideality Factor (n)	Barrier Height (ϕ) (eV)
[98]	Graphene/ZnO nanowire SBD	1.7	0.28
[99]	Indium gallium ZnO SBD	2.1	0.51
[100]	Ag/ZnO SBD (10 μW/cm^2 illumination intensity)	1.71	0.69
[101]	Ni/Cu-doped ZnO SBD	3.16	0.7245
[102]	PEDOT:PSS/a-IGZO (160nm)/Mo SBD	1.63	0.76
[103]	ZnO/single-layer graphene SBD	1.82	0.684
[104]	Au/ZnO SBD	2.27	0.54
[105]	4.0 at% Ce-doped ZnO nanorods SBD	3.0	0.764
[106]	In-Ga-ZnO SBD	0.88	0.76
[107]	Pd/Cu-doped p-ZnO/n-Si SBD	2.2	0.79
[108]	Pd/ZnO/n$^+$-Si/AuSb SBD	2.84	0.66
[109]	PEDOT:PSS/ZnO SBD (bending radius 4mm)	3.7	0.83
[110]	AgO_x/IGZO SBD	1.7	1.14
[111]	Pd/ZnO/Sn-Si/AuSb SBD	2.40	0.77
[112]	Ag/ZnO/Al SBD	3.75	0.67
[113]	SBD based on 0.4 at% (Ce and Sm) co-doped ZnO nanorods	5.11	0.704
[114]	Ag_xO (8% R[O_2])/ IGZO SBD	1.03	0.96
[115]	Cu/ZnO SBD	2.78	0.54
[119]	FTO/CP/a-IGZO/Mo SBD	2.41	0.56

stability [117–120]. Organic SBD finds application in the fabrication of organic strain sensors [121]. Srivastava et al. [122] fabricated a novel polycarbazole-based organic SBD (a-si/PCz/ITO), which exhibited very good rectifying behavior. Barrier height of 0.96 eV, ideality factor of 1.3, and reverse saturation current density of 8.78×10^{-12} A were reported. Diamond SBDs were used in the fabrication of high-frequency and high-power electronic devices [123–129]. The major drawback of diamond SBD is its low breakdown voltage [130,131]. The breakdown voltage in diamond SBD can be improved using the floating metal ring (FMR) edge termination technique [132]. R_N (ring number), R_S (ring spacing), and R_W (ring width) are the three parameters of the FMR method. The breakdown voltage variation by changing one ring parameter and keeping the other two parameters fixed has been studied. Figure 7.15 shows the plot of variation in breakdown voltage as a function of ring spacing. The ring spacing is varied to 7 from 3 μm by keeping the ring width constant at 5, 10, and 15 μm. At 5 μm, a ring spacing decrease in breakdown voltage is reported.

Figure 7.16 shows the plot of variation in breakdown voltage as a function of ring width. The ring spacing is kept constant at 5, 10, and 15 μm. With an increase in ring width, increase in breakdown voltage is observed in diamond SBD.

Figure 7.17 shows the plot of variation in breakdown voltage as a function of ring number. At $R_N = 3$, a decrease in breakdown voltage is reported. It was observed

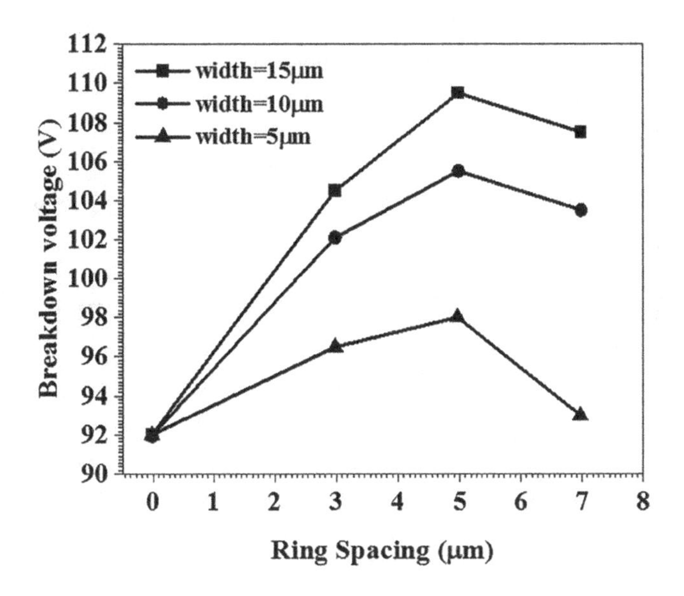

FIGURE 7.15 Ring spacing vs. breakdown voltage plot of diamond SBD [132].

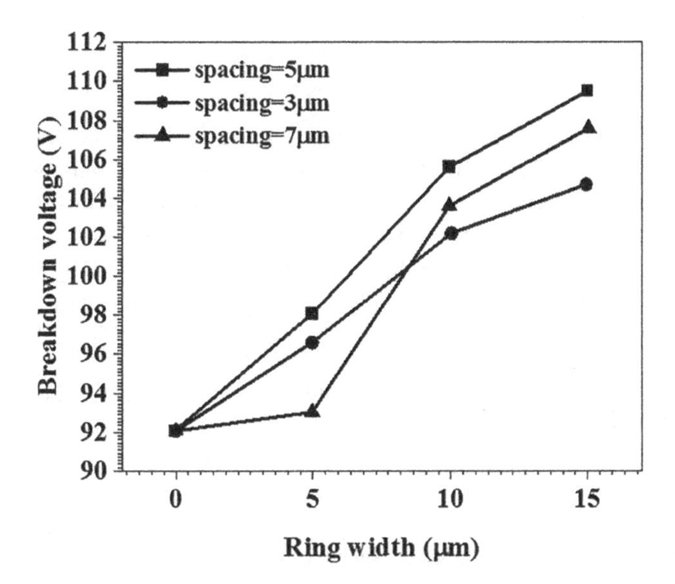

FIGURE 7.16 Ring width vs. breakdown voltage plot of diamond SBD [132].

that these ring parameters have a significant impact on the breakdown voltage of diamond SBD.

Feng-Renn Juang et al. [133] fabricated Au/SnO_2/n-type low-temperature polysilicon MOS SBD, which is used as an effective carbon monoxide sensor. Further, 546% relative response ratio was observed in Au/SnO_2/n-type low-temperature polysilicon

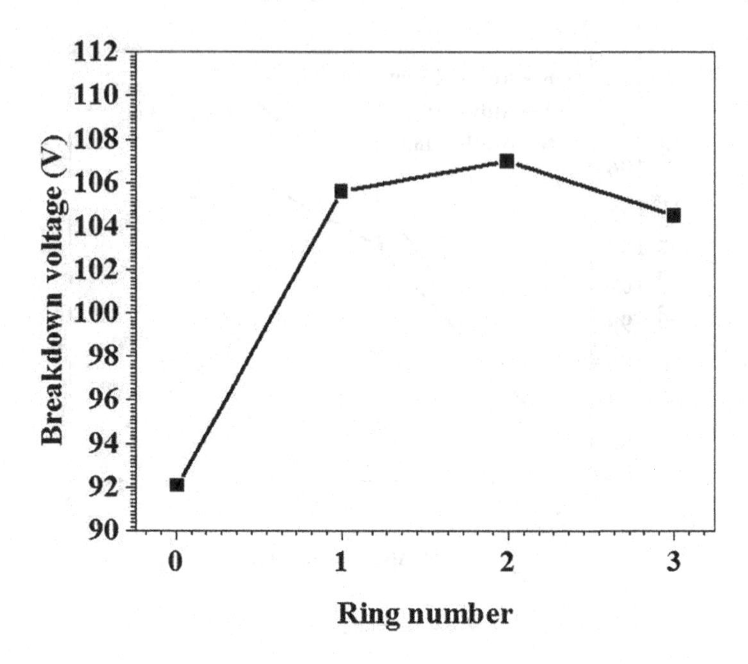

FIGURE 7.17 Ring number vs. breakdown voltage plot of diamond SBD [132].

MOS SBD to 100 ppm carbon monoxide exposure. Au/SnO_2/n-type low-temperature polysilicon MOS SBD was reported to be a high-performance and low-cost carbon monoxide sensor. In betavoltaic batteries, Au-Si SBD was used as an effective energy conversion device [134]. When compared to the conventional p–n junction, which was used as an energy conversion device in betavoltaic batteries, Au-Si SBD exhibited higher short circuit current and higher radiation resistance. An overview of electrical parameters, namely, ideality factor and barrier height of various types of SBDs, is presented in Table 7.4.

The use of advanced semiconductor materials further improves the performance of SBDs [166–174].

7.6 CONCLUSIONS

SBDs are suitable to be employed in various high-current, low-voltage applications. They have high switching speed, which makes them suitable for high-frequency and microwave applications. SBDs also find applications in detectors, linearizers, multipliers, and in the mixer as a nonlinear device. It is also widely used as efficient hydrogen and CO_2 gas sensors. High-performance temperature sensors have been fabricated with the help of SBDs. In this chapter, we have reviewed different emerging SBDs such as GaN SBD, 4H-SiC SBD, ZnO SBD, diamond SBD, and organic SBD, as well as their applications. It was observed that Cu/diamond SBD is ideal as a high-power device operating at several hundred °C.

TABLE 7.4
Ideality Factor and Barrier Height of Various Types of SBDs

Ref.	SBD	Ideality Factor (n)	Barrier Height (ϕ) (eV)
[135]	InGaP/GaAs SBD	1.20	1.01
[136]	Erbium Silicide SBD	1.23	0.76
[137]	Pd/InP(p) pseudo-SBD	0.80–0.83	1.14–1.28
[138]	Au/graphene (500°C)/n-type silicon SBD	1.82 ± 0.04	0.78 ± 0.01
[139]	Au/PVDF/n-InP SBD	1.14	0.73
[140]	Ag/MgPc/n-GaAs/Au-Ge SBD	3.64	0.53
[141]	Pt/p-NiO/n-PANI/n-Si SBD	7.425	0.7021
[142]	Pt/n-Ge SBD (183 K)	1.65	0.44
[143]	Ni/p-Si SBD	8.88	0.205
[4]	MgPc/n-Si SBD	1.1	0.98
[144]	Au/n-InP MS SBD	1.94	0.74
[145]	n-Si/BLG	2.13	0.54
[146]	Au/ZnO nanostructures/ITO SBD	1.96	0.81
[147]	Al/n-type si SBD with 20% Au and 80% CuPC nanocomposite interlayer	1.96	0.79
[148]	Ag/TiO_2 nanotube array SBD	2.39	0.92
[149]	Au/n-Si SBD (290 K)	1.11	0.798
[150]	Au-n-GaAs SBD	1.05	0.783
[151]	Ti/Au/n-GaAs planar SBD	1.2	0.840
[152]	Al/Cu_5FeS_4/FTO SBD	2.54	0.22
[153]	MoS_2–graphene composite-based SBD	1.31	0.63
[154]	Fe@MEA metallohydrogel-based SBD	1.33	0.72
[155]	Cu-Al-Mn SMA-based SBD	3.52	0.58
[156]	Cu/MoO_3-ZrO_2/p-Si SBD (15wt% Zr)	2.98	0.664
[157]	Graphene/HfO_2/Si SBD	1.66	0.51
[158]	FeGa/n-Si SBD	2.57	0.76
[159]	Al/HfO_2/n-Si SBD (600°C)	3.4	0.776
[160]	Cu/Ce-V_2O_5/n-Si SBD (6wt% Ce)	1.95	0.79
[161]	Au/Ga Se:Ce SBD (100 K)	3.97	0.31
[162]	Al/ 8wt% Zr:SnO_2/p-Si SBD	2.78	0.96
[163]	Au/Ni/β-Ga_2O_3 SBD (100 K)	2.76	0.40
[164]	Al/Ce:ZrO_2/p-Si SBD	2.71	0.99
[165]	3D-Gr Si SBD (laser-patterned)	6.67	0.61

REFERENCES

[1] T.T. Mnatsakanov, M.E. Levinshtein, A.G. Tandoev, S.N. Yurkov, J.W. Palmour, Minority carrier injection and current–voltage characteristics of Schottky diodes at high injection level, *Solid-State Electronics*, 121 (2016) 41–46.

[2] M.E. Aydin, F. Yakuphanoglu, G. Öztürk, Modification of electrical properties of the Au/1,1′dimethyl ferrocenecarboxylate/n-Si Schottky diode, *Synthetic Metals*, 160 (2010) 2186–2190.

[3] K. Driche, S. Rugen, N. Kaminski, H. Umezawa, H. Okumura, E. Gheeraert, Electric field distribution using floating metal guard rings edge-termination for Schottky diodes, *Diamond & Related Materials*, 82 (2018) 160–164.
[4] I. Missoum, Y.S. Ocak, M. Benhaliliba, C.E. Benouis, A. Chaker, Microelectronic properties of organic Schottky diodes based on MgPc for solar cell applications, *Synthetic Metals*, 214 (2016) 76–81.
[5] M.B. Askari, M. Shahryari, S. Nanekarani, S.B. Dehaghi, Effect of layer thickness on electrical characterization of Ag/Si schottky diode fabricated by thermal evaporation technique, *Optik*, 127 (2016) 11151–11155.
[6] Ö. Aksoy, İ. Uzun, G. Topal, Y.S. Ocak, Ö. Çelik, D. Batibay, Synthesis, characterization, and Schottky diode applications of low-cost new chitin derivatives, *Polymer Bulletin*, 75 (2018) 2265–2283.
[7] E. Giovine, R. Casini, D. Dominijanni, A. Notargiacomo, M. Ortolani, V. Foglietti, Fabrication of Schottky diodes for terahertz imaging, *Microelectronic Engineering*, 88 (2011) 2544–2546.
[8] R.O. Ocaya, A current–voltage–temperature method for fast extraction of Schottky diode static parameters, *Measurement*, 49 (2014) 246–255.
[9] R.K. Mamedov, A.R. Aslanova, Features of current transport in Schottky diodes with additional electric field, *Superlattices and Microstructures*, 136 (2019) 106297.
[10] F. Gity, L. Ansari, C. König, G.A. Verni, J.D. Holmes, B. Long, M. Lanius, P. Schüffelgen, G. Mussler, D. Grützmacher, J.C. Greer, Metal-semimetal Schottky diode relying on quantum confinement, *Microelectronic Engineering*, 195 (2018) 21–25.
[11] W. Schottky, Halbleitertheorie der Sperrschicht, *Naturwissenschaften*, 26 (1938) 843.
[12] N.F. Mott, Note on the contact between a metal and an insulator or semi-conductor, *Mathematical Proceedings of the Cambridge Philosophical Society*, 34 (1938) 568–572.
[13] B.S. Sannakashappanavar, A.B. Yadav, C.R. Byrareddy, N.V.L. Narasimha Murty, Fabrication and characterization of Schottky diode on ultra thin ZnO film and its application for UV detection, *Material Research Express*, 6 (2019) 116445.
[14] S. Chand, J. Kumar, Effects of barrier height distribution on the behavior of a Schottky diode, *Journal of Applied Physics*, 82 (1997) 5005.
[15] W. Jung, M. Guziewicz, Schottky diode parameters extraction using Lambert W function, *Materials Science and Engineering B*, 165 (2009) 57–59.
[16] S. Chand, P. Kaushal, J. Osvald, Numerical simulation study of current–voltage characteristics of a Schottky diode with inverse doped surface layer, *Materials Science in Semiconductor Processing* 16 (2013) 454–460.
[17] J.O. Bodunrin, D.A. Oeba, S.J. Moloi, Current-voltage characteristics of iron-implanted silicon based Schottky diodes, *Materials Science in Semiconductor Processing*, 123 (2021) 105524.
[18] E. Bernuchon, F. Aniel, N. Zerounian, A.S. Grimault-Jacquin, Monte Carlo modelling of Schottky diode for rectenna simulation, *Solid-State Electronics*, 135 (2017) 71–77.
[19] O. Pakma, Ş. Çavdar, H. Koralay, N. Tuğluoğlu, Ö.F. Yüksel, Improvement of diode parameters in Al/n-Si Schottky diodes with Coronene interlayer using variation of the illumination intensity, *Physica B*, 527 (2017) 1–6.
[20] A.M. Cowley, Surface States and Barrier height of metal-semiconductor systems, *Journal of Applied Physics*, 36 (1965) 3212.
[21] I.S. Yahia, H.Y. Zahran, F.H. Alamri, M.A. Manthrammel, S. AlFaify, A.M. Ali, Microelectronic properties of the organic Schottky diode with pyronin-Y: Admittance spectroscopy, and negative capacitance, *Physica B: Condensed Matter*, 543 (2018) 46–53.
[22] H. Norde, A modified forward I-V plot for Schottky diodes with high series resistance, *Journal of Applied Physics*, 50 (1979) 5052–5053.

[23] C.-D. Lien, F.C.T. So, M.-A. Nicolet, An improved forward I-V method for nonideal Schottky diodes with high series resistance, *IEEE Transactions on Electron Devices*, 31 (1984) 1502–1503.

[24] A. Rabehi, M. Amrani, Z. Benamara, B. Akkal, A. Ziane, M. Guermoui, A. Hatem-Kacha, G. Monier, B. Gruzza, L. Bideux, C. Robert-Goumet, Simulation and experimental studies of illumination effects on the current transport of nitridated GaAs Schottky diode, *Semiconductors*, 52 (2018) 1998–2006.

[25] A. Rabehi, B. Nail, H. Helal, A. Douara, A. Ziane, M. Amrani, B. Akkal, Z. Benamara, Optimal estimation of Schottky diode parameters using a novel optimization algorithm: Equilibrium optimizer, *Superlattices and Microstructures*, 146 (2020) 106665.

[26] X. Sun, C. Zhang, L. Gao, Y. Li, Z. Wang, Modeling of a Schottky diode in CMOS process with a flexible "open-through" On-Chip De-embedding Method, *Tsinghua Science and Technology*, 16 (2011) 175–180.

[27] T.H. Tsai, H.I. Chen, K.W. Lin, Y.W. Kuo, C.F. Chang, C.W. Hung, L.Y. Chen, T.P. Chen, Y.C. Liu, W.C. Liu, SiO_2 passivation effect on the hydrogen adsorption performance of a Pd/AlGaN-based Schottky diode, *Sensors and Actuators B*, 136 (2009) 338–343.

[28] Y. Zhang, M. Sun, D. Piedra, M. Azize, X. Zhang, T. Fujishima, T. Palacios, GaN-on-Si vertical Schottky and p-n diodes, *IEEE Electron Device Letters*, 35 (2014) 618–620.

[29] Z. Jiang, W. Zhang, A. Luo, M.R.M. Atalla, G. You, X. Li, L. Wang, J. Liu, A.M. Elahi, L. Wei, Y. Zhang, J. Xu, Bias-enhanced visible-rejection of GaN Schottky barrier ultraviolet photodetectors, *IEEE Photonics Technology Letters*, 27 (2015) 994–997.

[30] H.C. Chiu, J.F. Chi, H.L. Kao, C.Y. Chu, K.L. Cho, F.T. Chien, The ESD protection characteristic and low-frequency noise analysis of GaN Schottky barrier diode with fluorine-based plasma treatment, *Microelectronics Reliability*, 59 (2016) 44–48.

[31] P. Ferrandis, M. Charles, C. Gillot, R. Escoffier, E. Morvan, A. Torres, G. Reimbold, Effects of negative bias stress on trapping properties of AlGaN/GaN Schottky barrier diodes, *Microelectronic Engineering*, 178 (2017) 158–163.

[32] A. Colón, E.A. Douglas, A.J. Pope, B.A. Klein, C.A. Stephenson, M.S. Van Heukelom, A. Tauke-Pedretti, A.G. Baca, Demonstration of a 9 kV reverse breakdown and 59 mΩ-cm^2 specific on-resistance AlGaN/GaN Schottky barrier diode, *Solid-State Electronics*, 151 (2019) 47–51.

[33] L. Li, J. Chen, X. Gu, X. Li, T. Pu, J.-P. Ao, Temperature sensor using thermally stable TiN anode GaN Schottky barrier diode for high power device application, *Superlattices and Microstructures*, 123 (2018) 274–279.

[34] J. Osvald, T. Lalinský, G. Vanko, High temperature current transport in gate oxides based (GaN)/AlGaN/GaN Schottky diodes, *Applied Surface Science*, 461 (2018) 206–211.

[35] A. Elhaji, J.H. Evans-Freeman, M.M. El-Nahass, M.J. Kappers, C.J. Humphries, Electrical characterization and DLTS analysis of a gold/n-type gallium nitride Schottky diode, *Materials Science in Semiconductor Processing*, 17 (2014) 94–99.

[36] I.P. Liu, C.H. Chang, B.Y. Ke, K.W. Lin, Study of a GaN Schottky diode based hydrogen sensor with a hydrogen peroxide oxidation approach and platinum catalytic metal, *International Journal of Hydrogen Energy*, 44 (2019) 32351–32361.

[37] A. Zhong, T. Sasaki, K. Hane, Platinum/porous GaN nanonetwork metal-semiconductor Schottkydiode for room temperature hydrogen sensor, *Sensors and Actuators A*, 209 (2014) 52–56.

[38] S. Jang, S. Jung, K.H. Baik, Hydrogen sensing characteristics of Pt Schottky diode on nonpolar m-plane (1100) GaN single crystals, *Thin Solid Films*, 660 (2018) 646–650.

[39] K.H. Baik, S. Jung, C.-Y. Cho, K.-H. Park, F. Ren, S.J. Pearton, S. Jang, AlGaN/GaN heterostructure based Pt nanonetwork Schottky diode with water-blocking layer, *Sensors & Actuators: B. Chemical*, 317 (2020) 128234.

[40] W. Li, K. Nomoto, M. Pilla, M. Pan, X. Gao, D. Jena, H.G. Xing, Design and realization of GaN trench junction-barrier-Schottky-diodes, *IEEE Transactions on Electron Devices*, 64 (2017) 1635–1641.
[41] K.R. Peta, M.D. Kim, Leakage current transport mechanism under reverse bias in Au/Ni/GaN Schottky barrier diode, *Superlattices and Microstructures*, 113 (2018) 678–683.
[42] Z.-K. Bian, H. Zhou, S.-R. Xu, T. Zhang, K. Dang, J.-B. Chen, J.-C. Zhang, Y. Hao, High-performance quasi-vertical GaN Schottky diode with low turn-on voltage, *Superlattices and Microstructures*, 125 (2019) 295–301.
[43] K. Kim, J. Jang, Improving Ni/GaN Schottky diode performance through interfacial passivation layer formed via ultraviolet/ozone treatment, *Current Applied Physics*, 20 (2020) 293–297.
[44] W. Wang, X.-X. Li, Z.-Q. Xiao, W. Huang, Z.-G. Ji, T.-K. Chiang, D.W. Zhang, H.-L. Lu, An analytical model for merged GaN heterojunction barrier Schottky diodes with inserting p-Si technology, *Superlattices and Microstructures*, 150 (2021) 106744.
[45] S. Besendörfer, E. Meissner, F. Medjdoub, J. Derluyn, J. Friedrich, T. Erlbacher, The impact of dislocations on AlGaN/GaN Schottky diodes and on gate failure of high electron mobility transistors, *Scientific Reports*, 10 (2020) 17252.
[46] R.-S. Ki, J.-G. Lee, H.-Y. Cha, K.-S. Seo, The effect of edge- terminated structure for lateral AlGaN/GaN Schottky barrier diodes with gated ohmic anode, *Solid State Electronics*, 166 (2020) 107768.
[47] X. Li, T. Hoshi, L. Li, T. Pu, T. Zhang, T. Xie, X. Li, J.-P. Ao, GaN Schottky barrier diode with thermally stable nickel nitride electrode deposited by reactive sputtering, *Materials Science in Semiconductor Processing*, 93 (2019) 1–5.
[48] W.B. Bouiadjra, A. Saidane, A. Mostefa, M. Henini, M. Shafi, Effect of nitrogen incorporation on electrical properties of Ti/Au/GaAsN Schottky diodes, *Superlattices and Microstructures*, 71 (2014) 225–237.
[49] E. Erdoğan, M. Kundakçı, Room temperature current-voltage (I-V) characteristics of Ag/InGaN/n-Si Schottky barrier diode, *Physica B: Condensed Matter*, 506 (2017) 105–108.
[50] S. Amor, A. Ahaitouf, A. Ahaitouf, J.P. Salvestrini, A. Ougazzaden, Evidence of minority carrier traps contribution in deep level transient spectroscopy measurement in n–GaN Schottky diode, *Superlattices and Microstructures*, 101 (2017) 529–536.
[51] S. Sanjay, K. Baskar, Fabrication of Schottky barrier diodes on clump of gallium nitride nanowires grown by chemical vapour deposition, *Applied Surface Science*, 456 (2018) 526–531.
[52] L. Yang, B. Zhang, Y. Li, D. Chen, Improved Schottky barrier characteristics for AlInN/GaN diodes by oxygen plasma treatment, *Materials Science in Semiconductor Processing*, 74 (2018) 42–45.
[53] P. Kruszewski, P. Prystawko, M. Grabowski, T. Sochacki, A. Sidor, M. Bockowski, J. Jasinski, L. Lukasiak, R. Kisiel, M. Leszczynski, Electrical properties of vertical GaN Schottky diodes on Ammono-GaN substrate, *Materials Science in Semiconductor Processing*, 96 (2019) 132–136.
[54] X. Li, T. Pu, H. Taiki, T. Zhang, T. Xie, S.J.L. Fujiwara, H. Kitahata, L. Li, S. Kobayashi, M. Ito, X. Li, J.-P. Ao, GaN Schottky barrier diodes with nickel nitride anodes sputtered at different nitrogen partial pressure, *Vacuum*, 162 (2019) 72–77.
[55] M. Garg, A. Kumar, H. Sun, C.-H. Liao, X. Li, R. Singh, Temperature dependent electrical studies on Cu/AlGaN/GaN Schottky barrier diodes with its microstructural characterization, *Journal of Alloys and Compounds*, 806 (2019) 852–857.
[56] H. Gu, C. Hu, J. Wang, Y. Lu, J.-P. Ao, F. Tian, Y. Zhang, M. Wang, X. Liu, K. Xu, Vertical GaN Schottky barrier diodes on Ge-doped free-standing GaN substrates, *Journal of Alloys and Compounds*, 780 (2019) 476–481.

[57] H. Helal, Z. Benamara, A.H. Kacha, M. Amrani, A. Rabehi, B. Akkal, G. Monier, C. Robert- Goumet, Comparative study of ionic bombardment and heat treatment on the electrical behavior of Au/GaN/n-GaAs Schottky diodes, *Superlattices and Microstructures*, 135 (2019) 106276.

[58] V. Janardhanam, I. Jyothi, S.-N. Lee, V.R. Reddy, C.-J. Choi, Rectifying and breakdown voltage enhancement of Au/n-GaN Schottky diode with Al-doped ZnO films and its structural characterization, *Thin Solid Films*, 676 (2019) 125–132.

[59] Y.-J. Lin, Electronic transport and Schottky barrier heights of Pt/ n -type GaN Schottky diodes in the extrinsic region, *Journal of Applied Physics*, 106 (2009) 013702.

[60] E.I. Shabunina, M.E. Levinshtein, N.M. Shmidt, P.A. Ivanov, J.W. Palmour, 1/f noise in forward biased high voltage 4H-SiC Schottky diodes, *Solid-State Electronics*, 96 (2014) 44–47.

[61] S. Rao, G. Pangallo, F.G.D. Corte, 4H-SiC p-i-n diode ashighly linear temperature sensor, *IEEE Transactions on Electron Devices*, 63 (2016) 414–418.

[62] V. Kumar, A.S. Maan, J. Akhtar, Barrier height inhomogeneities induced anomaly in thermal sensitivity of Ni/4H-SiC Schottky diode temperature sensor, *Journal of Vacuum Science & Technology B*, 32 (2014) 41203.

[63] G. Brezeanu, F. Draghici, F. Craciunioiu, C. Boianceanu, F. Bernea, D. Puscasu, I. Rusu, 4H-SiC Schottky diodes for temperature sensing applications in harsh environments, *Materials Science Forum*, 679–680 (2011) 575–578.

[64] N. Zhang, Chih-Ming Lin, D.G. Senesky, A.P. Pisano, Temperature sensor based on 4H- silicon carbide pn diode operational from 20°C to 600°C, *Applied Physics Letters*, 104 (2014) 073504.

[65] S. Rao, G. Pangallo, F. Pezzimenti, F.G.D. Corte, High-performance temperature sensor based on 4H-SiC Schottky diodes, *IEEE Electron Device Letters*, 36 (2015) 720–722.

[66] T. Nakamura, T. Miyanagi, I. Kamata, T. Jikimoto, H. Tsuchida, A 4.15 kV 9.07-mΩ. cm^2 4H–SiC Schottky-barrier diode using Mo contact annealed at high temperature, *IEEE Electron Device Letters*, 26 (2005) 99–101.

[67] G. Pristavu, G. Brezeanu, R. Pascu, F. Drăghici, M. Bădilă, Characterization of non-uniform Ni/4H-SiC Schottky diodes for improved responsivity in high-temperature sensing, *Materials Science in Semiconductor Processing*, 94 (2019) 64–69.

[68] R. Aiba, K. Matsui, M. Baba, S. Harada, H. Yano, N. Iwamuro, Demonstration of superior electrical characteristics for 1.2 kV SiC Schottky barrier diode-wall integrated trench MOSFET with higher Schottky barrier height metal, *IEEE Electron Device Letters*, 41 (2020) 1810–1813.

[69] S. Rao, G. Pangallo, L.D. Benedetto, A. Rubino, G.D. Licciardo, F.G.D. Corte, A V_2O_5/4H-SiC Schottky diode-based PTAT sensor operating in a wide range of bias currents, *Sensors and Actuators A*, 269 (2018) 171–174.

[70] A. Kestle, S.P. Wilks, P.R. Dunstan, M. Prilcliard, P.A. Mawby, Improved Ni/SiC Schottky diode formation, *Electronics Letters*, 36 (2000) 267–268.

[71] T.N. Oder, T.L. Sung, M. Barlow, J.R. Williams, A.C. Ahyi, T. Isaacs-Smith, Improved Ni Schottky contacts on n-type 4H-SiC using thermal processing, *Journal of Electronic Materials*, 38 (2009) 772–777.

[72] M. Sochacki, J. Szmidt, M. Bakowski, A. Werbowy, Influence of annealing on reverse current of 4H-SiC Schottky diodes, *Diamond and Related Materials*, 11 (2002) 1263–1267.

[73] S. Kyoung, E.-S. Jung, M.Y. Sung, Post-annealing processes to improve inhomogeneity of Schottky barrier height in Ti/Al 4H-SiC Schottky barrier diode, *Microelectronic Engineering*, 154 (2016) 69–73.

[74] D.J. Morrison, N.G. Wright, A.B. Horsfall, C.M. Johnson, A.G. O'Neill, A.P. Knights, K.P. Hilton, M.J. Uren, Effect of post-implantation anneal on the electrical characteristics of Ni 4H-SiC Schottky barrier diodes terminated using self-aligned argon ion implantation, *Solid-State Electronics*, 44 (2000) 1879–1885.

[75] M. Huang, Z. Yang, S. Wang, J. Liu, M. Gong, Y. Ma, J. Liu, P. Zhai, Y. Sun, Y. Li, Recrystallization effects in GeV Bi ion implanted 4H-SiC Schottky barrier diode investigated by cross-sectional Micro-Raman spectroscopy, *Nuclear Instruments and Methods in Physics Research Section B: Beam Interactions with Materials and Atoms*, 478 (2020) 5–10.

[76] V.E. Gora, F.D. Auret, H.T. Danga, S.M. Tunhuma, C. Nyamhere, E. Igumbor, A Chawanda, Barrier height inhomogeneities on Pd/n-4H-SiC Schottky diodes in a wide temperature range, *Materials Science & Engineering B*, 247 (2019) 114370.

[77] K. Zeghdar, L. Dehimi, F. Pezzimenti, M.L. Megherbi, F.G.D. Corte, Analysis of the electrical characteristics of Mo/4H-SiC Schottky barrier diodes for temperature- sensing applications, *Journal of Electronic Materials*, 49 (2020) 1322–1329.

[78] K. Zeghdar, H. Bencherif, L. Dehimi, F. Pezzimenti, F.G. DellaCorte, Simulation and analysis of the forward bias current–voltage–temperature characteristics of W/4H-SiC Schottky barrier diodes for temperature-sensing applications, *Solid State Electronics Letters*, 2 (2020) 49–54.

[79] D.H. Kim, J.H. Lee, J.H. Moon, M.S. Oh, H.K. Song, J.H. Yim, J.B. Lee, H.J. Kim, Improvement of the reverse characteristics of Ti/4H-SiC Schottky barrier diodes by thermal treatments, *Solid State Phenomena*, 124–126 (2007) 105–108.

[80] S.K. Gupta, A. Azam, J. Akhtar, Improved electrical parameters of vacuum annealed Ni/4H-SiC(0001) Schottky barrier diode, *Physica B*, 406 (2011) 3030–3035.

[81] S.B. Yun, J.H. Kim, Y.H. Kang, J.H. Lee, K.-H. Kim, S.-S. Kim, E.S. Jung, I.-H. Kang, H.K. Shin, C.H. Yang, Optimized annealing temperature of Ti/4H-SiC Schottky barrier diode, *Journal of Nanoscience and Nanotechnology*, 17 (2017) 3406–3408.

[82] H. Linchao, H. Shen, K. Liu, Y. Wang, Y. Tang, Y. Bai, X. Hengyu, Y. Wu, X. Liu, Annealing temperature influence on the degree of inhomogeneity of the Schottky barrier in Ti/4H—SiC contacts, *Chinese Physics B*, 23 (2014) 127302.

[83] P.V. Raja, N.V.L.N. Murty, Thermal annealing studies in epitaxial 4H-SiC Schottky barrier diodes over wide temperature range, *Microelectronics Reliability*, 87 (2018) 213–221.

[84] G. Lioliou, N.R. Gemmell, M. Mazzillo, A. Sciuto, A.M. Barnett, 4H-SiC Schottky diodes with Ni2Si contacts for X-ray detection, *Nuclear Instruments and Methods in Physics Research Section A: Accelerators, Spectrometers, Detectors and Associated Equipment*, 940 (2019) 328–336.

[85] Y. Tang, L. Ge, H. Gu, Y. Bai, Y. Luo, C. Li, X. Liu, Degradation in electrothermal characteristics of 4H-SiC junction barrier Schottky diodes under high temperature power cycling stress, *Microelectronics Reliability*, 102 (2019) 113451.

[86] M.-M. Gao, L.-Y. Fan, Z.-Z. Chen, Ideal Ni-based 4H–SiC Schottky barrier diodes with Si intercalation, *Materials Science in Semiconductor Processing*, 107 (2020) 104866.

[87] V. Kumar, J. Verma, A.S. Maan, J. Akhtar, Epitaxial 4H–SiC based Schottky diode temperature sensors in ultra-low current range, *Vacuum*, 182 (2020) 109590.

[88] L. Stöber, M. Schneider, U. Schmid, Impact of contact material deposition technique on the properties of Ti/4H-SiC Schottky structures, *Materials Science Forum*, 858 (2016) 569–572.

[89] B. Lee, C. Kim, Y. Lee, S. Lee, D.Y. Kim, Dependence of photocurrent on UV wavelength in ZnO/Pt bottom-contact Schottky diode, *Current Applied Physics*, 15 (2015) 29–33.

[90] Q. Feng, J. Liu, J. Lu, Y. Mei, Z. Song, P. Tao, D. Pan, Y. Yang, M. Li, Fabrication and characterization of single ZnO microwire Schottky light emitting diodes, *Materials Science in Semiconductor Processing*, 40 (2015) 436–438.

[91] L. Rajan, C. Periasamy, V. Sahula, Electrical characterization of Au/ZnO thinfilm Schottky diode on silicon substrate, *Perspectives in Science*, 8 (2016) 66–68.

[92] Ş. Aydoğan, M.L. Grilli, M. Yilmaz, Z. Çaldiran, H. Kaçuş, A Facile growth of spray based ZnO films and device performance investigation for Schottky diodes: Determination of interface state density distribution, *Journal of Alloys and Compounds*, 708 (2017) 55–66.
[93] Y. Caglar, M. Caglar, S. Ilican, XRD, SEM, XPS studies of Sb doped ZnO films and electrical properties of its based Schottky diodes, *Optik*, 164 (2018) 424–432.
[94] M. Singh, M. Rajoriya, M. Sahni, P. Gupta, Effect of Aluminum doping on potential barrier of gold-ZnO-Si Schottky barrier diode, *Materials Today: Proceedings*, 34 (2021) 588–592.
[95] S.J. Young, S.J. Chang, L.W. Ji, T.H. Meen, C.H. Hsiao, K.W. Liu, K.J. Chen, Z.S. Hu, Thermally stable Ir/n-ZnO Schottky diodes, *Microelectronic Engineering*, 88 (2011) 113–116.
[96] X. Zhang, J. Zhai, X. Yu, L. Ding, W. Zhang, Fabrication and characterization of flexible Ag/ZnO Schottky diodes on polyimide substrates, *Thin Solid Films*, 548 (2013) 623–626.
[97] Q. Xin, L. Yan, L. Du, J. Zhang, Y. Luo, Q. Wang, A. Song, Influence of sputtering conditions on room-temperature fabricated InGaZnO-based Schottky diodes, *Thin Solid Films*, 616 (2016) 569–572.
[98] R. Liu, X.-C. You, X.-W. Fu, F. Lin, J. Meng, D.-P. Yu, Z.-M. Liao, Gate modulation of graphene-ZnO nanowire Schottky diode, *Scientific Reports*, 5 (2015) 10125.
[99] J. Zhang, Y. Li, B. Zhang, H. Wang, Q. Xin, A. Song, Flexible indium–gallium–zinc–oxide Schottky diode operating beyond 2.45 GHz, *Nature Communications*, 6 (2015) 7561.
[100] R. Zhu, X. Zhang, J. Zhao, R. Li, W. Zhang, Influence of illumination intensity on the electrical characteristics and photoresponsivity of the Ag/ZnO Schottky diodes, *Journal of Alloys and Compounds*, 631 (2015) 125–128.
[101] L. Agarwal, B.K. Singh, S. Tripathi, P. Chakrabarti, Fabrication and characterization of Pd/Cu doped ZnO/Si and Ni/Cu doped ZnO/Si Schottky diodes, *Thin Solid Films*, 612 (2016) 259–266.
[102] C.-H. Chang, C.-J. Hsu, C.-C. Wu, Rectified Schottky diodes based on PEDOT:PSS/ InGaZnO junctions, *Organic Electronics*, 48 (2017) 35–40.
[103] H. Lee, N. An, S. Jeong, S. Kang, S. Kwon, J. Lee, Y. Lee, D.Y. Kim, S. Lee, Strong dependence of photocurrent on illumination-light colors for ZnO/graphene Schottky diode, *Current Applied Physics*, 17 (2017) 552–556.
[104] T. Varma, C. Periasamy, D. Boolchandani, Performance analyses of Schottky diodes with Au/ Pd contacts on n-ZnO thin films as UV detectors, *Superlattices and Microstructures*, 112 (2017) 151–163.
[105] M.A.M. Ahmed, W.E. Meyer, J.M. Nel, Structural, optical and electrical properties of a Schottky diode fabricated on Ce doped ZnO nanorods grown using a two step chemical bath deposition, *Materials Science in Semiconductor Processing*, 87 (2018) 187–194.
[106] J.-W. Kim, T.-J. Jung, S.-M. Yoon, Device characteristics of Schottky barrier diodes using In-Ga-Zn-O semiconductor thin films with different atomic ratios, *Journal of Alloys and Compounds*, 771 (2019) 658–663.
[107] B.K. Singh, S. Tripathi, Performance analysis of Schottky diodes based on Bi doped p- ZnO thin films, *Superlattices and Microstructures*, 120 (2018) 288–297.
[108] M.A.M. Ahmed, W.E. Meyer, J.M. Nel, Structural, optical and electrical properties of the fabricated Schottky diodes based on ZnO, Ce and Sm doped ZnO films prepared via wet chemical technique, *Materials Research Bulletin*, 115 (2019) 12–18.
[109] N. Hernandez-Como, M. Lopez-Castillo, F.J. Hernandez-Cuevas, H. Baez-Medina, R. Baca-Arroyo, M. Aleman, Flexible PEDOT:PSS/ZnO Schottky diodes on polyimide substrates, *Microelectronic Engineering*, 216 (2019) 111060.

[110] L.A. Santana, L.M. Reséndiz, A.I. Díaz, F.J. Hernandez-Cuevas, M. Aleman, N. Hernandez-Como, Schottky barrier diodes fabricated with metal oxides AgO_x/IGZO, *Microelectronic Engineering*, 220 (2020) 111182.

[111] M.A.M. Ahmed, W.E. Meyer, J.M. Nel, Effect of (Ce, Al) co-doped ZnO thin films on the Schottky diode properties fabricated using the sol-gel spin coating, *Materials Science in Semiconductor Processing*, 103 (2019) 104612.

[112] F. Gül, Addressing the sneak-path problem in crossbar RRAM devices using memristor-based one Schottky diode-one resistor array, *Results in Physics*, 12 (2019) 1091–1096.

[113] M.A. Ahmed, L. Coetsee, W.E. Meyer, J.M. Nel, Influence (Ce and Sm) co-doping ZnO nanorods on the structural, optical and electrical properties of the fabricated Schottky diode using chemical bath deposition, *Journal of Alloys and Compounds*, 810 (2019) 151929.

[114] Y. Magari, H. Makino, S. Hashimoto, M. Furuta, Origin of work function engineering of silver oxide for an In–Ga–Zn–O Schottky diode, *Applied Surface Science*, 512 (2020) 144519.

[115] V.S. Rana, J.K. Rajput, T.K. Pathak, L.P. Purohit, Cu sputtered Cu/ZnO Schottky diodes on fluorine doped tin oxide substrate for optoelectronic applications, *Thin Solid Films*, 679 (2019) 79–85.

[116] C.-Y. Huang, P.-T. Lin, H.-C. Cheng, F.-C. Lo, P.-S. Lee, Y.-W. Huang, Q.-Y. Huang, Y.-C. Kuo, S.-W. Lin, Y.-R. Liu, Rectified Schottky diodes that use low-cost carbon paste/InGaZnO junctions, *Organic Electronics*, 68 (2019) 212–217.

[117] V. Chaudhary, N. Kumar, A.K. Singh, Solubility dependent trap density in poly (3-hexylthiophene) organic Schottky diodes at room temperature, *Synthetic Metals*, 250 (2019) 88–93.

[118] K.S. Kang, K.J. Han, J. Kim, Polymer-based flexible schottky diode made with pentacene–PEDOT:PSS, *IEEE Transactions on Nanotechnology*, 8 (2009) 627–630.

[119] B. Gupta, A.K. Singh, A.A. Melvin, R. Prakash, Influence of monomer concentration on polycarbazoleepolyindole (PCz-PIn) copolymer properties: Application in Schottky diode, *Solid State Sciences*, 35 (2014) 56–61.

[120] I.S. Yahia, A.A.M. Farag, F. Yakuphanoglu, W.A. Farooq, Temperature dependence of electronic parameters of organic Schottky diode based on fluorescein sodium salt, *Synthetic Metals*, 161 (2011) 881–887.

[121] Y. Cho, P.J. Jeon, J.S. Kim, S. Im, Organic strain sensor comprised of heptazole-based thin film transistor and Schottky diode, *Organic Electronics*, 40 (2017) 24–29.

[122] A. Srivastava, P. Chakrabarti, An organic Schottky diode (OSD) based on a- silicon/polycarbazole contact, *Synthetic Metals*, 207 (2015) 96–101.

[123] K. Ueda, K. Kawamoto, H. Asano, High-temperature and high-voltage characteristics of Cu/diamond Schottky diodes, *Diamond and Related Materials*, 57 (2015) 28–31.

[124] D. Zhao, Z. Liu, J. Wang, W. Yi, R. Wang, W. Wang, K. Wang, H.-X. Wang, Schottky barrier diode fabricated on oxygen-terminated diamond using a selective growth approach, *Diamond & Related Materials*, 99 (2019) 107529.

[125] D. Zhao, C. Hu, Z. Liu, H.-X. Wang, W. Wang, J. Zhang, Diamond MIP structure Schottky diode with different drift layer thickness, *Diamond & Related Materials*, 73 (2017) 15–18.

[126] N. Ozawa, T. Makino, H. Kato, M. Ogura, Y. Kato, D. Takeuchi, H. Okushi, S. Yamasaki, Temperature dependence of electrical characteristics for diamond Schottky-pn diode in forward bias, *Diamond & Related Materials*, 85 (2018) 49–52.

[127] T. Matsumoto, T. Mukose, T. Makino, D. Takeuchi, S. Yamasaki, T. Inokuma, N. Tokuda, Diamond Schottky-pn diode using lightly nitrogen-doped layer, *Diamond & Related Materials*, 75 (2017) 152–154.

[128] V.A. Kukushkin, Simulation of a perfect CVD diamond Schottky diode steep forward current–voltage characteristic, *Physica B*, 498 (2016) 1–6.

[129] N. Basman, N. Aslan, O. Uzun, G. Cankaya, U. Kolemen, Electrical characterization of metal/diamond-like carbon/inorganic semiconductor MIS Schottky barrier diodes, *Microelectronic Engineering*, 140 (2015) 18–22.
[130] X. Yu, J. Zhou, Y. Wang, F. Qiu, Y. Kong, H. Wang, T. Chen, Breakdown enhancement of diamond Schottky barrier diodes using boron implanted edge terminations, *Diamond & Related Materials*, 92 (2019) 146–149.
[131] C. Hitchcock, T.P. Chow, Degradation of forward current density with increasing blocking voltage in diamond Schottky-pn diodes, *Diamond & Related Materials*, 104 (2020) 107736.
[132] J. Wang, D. Zhao, W. Wang, X. Zhang, Y. Wang, X. Chang, Z. Liu, J. Fu, K. Wang, H.-X. Wang, Diamond Schottky barrier diodes with floating metal rings for high breakdown voltage, *Materials Science in Semiconductor Processing*, 97 (2019) 101–105.
[133] F.-R. Juang, Y.-K. Fang, Y.-T. Chiang, T.-H. Chou, C.-I. Lin, C.-W. Lin, The low temperature polysilicon (LTPS) thin film MOS Schottky diode on glass substrate for low cost and high performance CO sensing applications, *Sensors and Actuators B*, 156 (2011) 338–342.
[134] Y. Liu, R. Hu, Y. Yang, G. Wang, S. Luo, N. Liu, Investigation on a radiation tolerant betavoltaic battery based on Schottky barrier diode, *Applied Radiation and Isotopes*, 70 (2012) 438–441.
[135] S. Sassen, B. Witzigmann, C. Wölk, H. Brugger, Barrier height engineering on GaAs THz Schottky diodes by means of high–low doping, InGaAs- and InGaP-layers, *IEEE Transactions on Electron Devices*, 47 (2000) 24–32.
[136] M. Jang, Y. Kim, J. Shin, S. Lee, Characterization of erbium-silicided Schottky diode junction, *IEEE Electron Device Letters*, 26 (2005) 354–356.
[137] C. Varenne, J. Brunet, A. Pauly, B. Lauron, A comparative study of Schottky barrier height enhancement by realized pseudo-Schottky diodes on p-InP, *Physica B*, 404 (2009) 1082–1086.
[138] D.-J. Kim, G.-S. Kim, N.-W. Park, W.-Y. Lee, Y. Sim, K.-S. Kim, M.-J. Seong, J.-H. Koh, S.-K. Lee, Effect of annealing of graphene layer on electrical transport and degradation of Au/graphene/n-type silicon Schottky diodes, *Journal of Alloys and Compounds*, 612 (2014) 265–272.
[139] V.R. Reddy, Electrical properties of Au/polyvinylidene fluoride/n-InP Schottky diode with polymer interlayer, *Thin Solid Films*, 556 (2014) 300–306.
[140] I. Missoum, M. Benhaliliba, A. Chaker, Y.S. Ocak, C.E. Benouis, A novel device behavior of Ag/MgPc/n-GaAs/Au-Ge organic based Schottky diode, *Synthetic Metals*, 207 (2015) 42–45.
[141] S. Ameen, M.S. Akhtar, H.S. Shin, Manipulating the structure of polyaniline by exploiting redox chemistry:Novel p-NiO/n-polyaniline/n-Si Schottky diode based chemosensor for the electrochemical detection of hydrazinobenzene, *Electrochimica Acta*, 215 (2016) 200–211.
[142] E. Guo, Z. Zeng, Y. Zhang, X. Long, H. Zhou, X. Wang, The effect of annealing temperature on the electronic parameters and carrier transport mechanism of Pt/n-type Ge Schottky diode, *Microelectronics Reliability*, 62 (2016) 63–69.
[143] R. Kumar, S. Chand, Fabrication and electrical characterization of nickel/p-Si Schottky diode at low temperature, *Solid State Sciences*, 58 (2016) 115–121.
[144] P.P. Thapaswini, R. Padma, N. Balaram, B. Bindu, V.R. Reddy, Modification of electrical properties of Au/n-type InP Schottky diode with a high-k $Ba_{0.6}Sr_{0.4}TiO_3$ interlayer, *Superlattices and Microstructures*, 93 (2016) 82–91.
[145] H. Aydin, C. Bacaksiz, N. Yagmurcukardes, C. Karakaya, O. Mermer, M. Can, R.T. Senger, H. Sahin, Y. Selamet, Experimental and computational investigation of graphene/SAMs/n-Si Schottky diodes, *Applied Surface Science*, 428 (2018) 1010–1017.

[146] B.S. Mwankemwa, S. Akinkuade, K. Maabong, J.M. Nel, M. Diale, Effects of surface morphology on the optical and electrical properties of Schottky diodes of CBD deposited ZnO nanostructures, *Physica B: Condensed Matter*, 535 (2018) 175–180.

[147] P.R.S. Reddy, V. Janardhanam, I. Jyothi, H.S. Chang, S.N. Lee, M.S. Lee, V.R. Reddy, C.J. Choi, Microstructural and electrical properties of Al/n-type Si Schottky diodes with Au-CuPc nanocomposite films as interlayer, *Superlattices and Microstructures*, 111 (2017) 506–517.

[148] M. Yilmaz, B.B. Cirak, S. Aydogan, M.L. Grilli, M. Biber, Facile electrochemical-assisted synthesis of TiO_2 nanotubes and their role in Schottky barrier diode applications, *Superlattices and Microstructures*, 113 (2018) 310–318.

[149] S. Mahato, J. Puigdollers, Temperature dependent current-voltage characteristics of Au/n-Si Schottky barrier diodes and the effect of transition metal oxides as an interface layer, *Physica B: Physics of Condensed Matter*, 530 (2018) 327–335.

[150] R.K. Mamedov, A.R. Aslanova, Features of current-voltage characteristic of nonequilibrium trench MOS barrier Schottky diode, *Superlattices and Microstructures*, 118 (2018) 298–307.

[151] A. Shurakov, P.Mikhalev, D. Mikhailov, V. Mityashkin, I. Tretyakov, A. Kardakova, I. Belikov, N. Kaurova, B. Voronov, I. Vasil'evskii, G. Gol'tsman, Ti/Au/n-GaAs planar Schottky diode with a moderately Si-doped matching sublayer, *Microelectronic Engineering*, 195 (2018) 26–31.

[152] S. Sil, A. Dey, J. Datta, M. Das, R. Jana, S. Halder, J. Dhar, D. Sanyal, P.P. Ray, Analysis of interfaces in Bornite (Cu_5FeS_4) fabricated Schottky diode using impedance spectroscopy method and its photosensitive behavior, *Materials Research Bulletin*, 106 (2018) 337–345.

[153] S. Halder, B. Pal, A. Dey, S.Sil, P. Das, A. Biswas, P.P. Ray, Effect of graphene on improved photosensitivity of MoS_2-graphene composite based Schottky diode, *Materials Research Bulletin*, 118 (2019) 110507.

[154] S. Dhibar, R. Jana, P.P. Ray, B. Dey, Monoethanolamine and Fe(III) based metallohydrogel: An efficient Schottky barrier diode, *Journal of Molecular Liquids*, 289 (2019) 111126.

[155] E. Aldırmaz, A. Tataroğlu, A. Dere, M. Güler, E. Güler, A. Karabulut, F. Yakuphanoglu, Cu-Al-Mn shape memory alloy based Schottky diode formed on Si, *Physica B: Condensed Matter*, 560 (2019) 261–266.

[156] P. Vivek, J. Chandrasekaran, R. Marnadu, S. Maruthamuthu, V. Balasubramani, P. Balraju, Zirconia modified nanostructured MoO3 thin films deposited by spray pyrolysis technique for Cu/MoO_3-ZrO_2/p-Si structured Schottky barrier diode application, *Optik - International Journal for Light and Electron Optics*, 199 (2019) 163351.

[157] Y. Xu, J. Bi, Y. Li, K. Xi, L. Fan, M. Liu, M. Sandip, L. Luo, The total ionizing dose effects of X-ray irradiation on graphene/Si Schottky diodes with a HfO_2 insertion layer, *Microelectronics Reliability*, 100–101 (2019) 113355.

[158] G. Bhattacharya, N.V.P. Chaudhary, S.S. Takri, R.R. Kumar, A. Venimadhav, Investigation of current transport in Galfenol based Schottky diodes, *Materials Today: Proceedings*, 33 (2020) 5116–5122.

[159] P. Harishsenthil, J. Chandrasekaran, R. Marnadu, P. Balraju, C. Mahendarn, Influence of high dielectric HfO_2 thin films on the electrical properties of Al/HfO_2/n-Si (MIS) structured Schottky barrier diodes, *Physica B: Physics of Condensed Matter*, 594 (2020) 412336.

[160] V. Balasubramani, J. Chandrasekaran, T.D. Nguyen, S. Maruthamuthu, R. Marnadu, P. Vivek, S. Sugarthi, Colossal photosensitive boost in Schottky diode behaviour with Ce-V_2O_5 interfaced layer of MIS structure, *Sensors and Actuators A*, 315 (2020) 112333.

[161] H. Ertap, H. Kacus, S. Aydogan, M. Karabulut, Analysis of temperature dependent electrical characteristics of Au/GaSe Schottky barrier diode improved by Ce-doping, *Sensors and Actuators A*, 315 (2020) 112264.
[162] K. Ravikumar, S. Agilan, M. Raja, R. Marnadu, T. Alshahrani, M. Shkir, M. Balaji, R. Ganesh, Investigation on microstructural and opto-electrical properties of Zr-doped SnO_2 thin films for Al/Zr:SnO_2/p-Si Schottky barrier diode application, *Physica B: Condensed Matter*, 599 (2020) 412452.
[163] P.R.S. Reddy, V. Janardhanam, K.-H. Shim, V.R. Reddy, S.-N. Lee, S.-J. Park, C.-J. Choi, Temperature-dependent Schottky barrier parameters of Ni/Au on n-type (001) β-Ga_2O_3 Schottky barrier diode, *Vacuum*, 171 (2020) 109012.
[164] K. Sasikumar, R. Bharathikannan, M. Raja, B. Mohanbabu, Fabrication and characterization of rare earth (Ce, Gd, and Y) doped ZrO_2 based metal-insulator- semiconductor (MIS) type Schottky barrier diodes, *Superlattices and Microstructures*, 139 (2020) 106424.
[165] E.O. Orhan, E. Efil, O. Bayram, N. Kaymak, H. Berberoglu, O. Candemir, I. Pavlov, S.B. Ocak, 3D-graphene-laser patterned p-type silicon Schottky diode, *Materials Science in Semiconductor Processing*, 121 (2021) 105454.
[166] J. Ajayan, D. Nirmal, R. Ramesh, S. Bhattacharya, S. Tayal, L.L. Joseph, L.R. Thoutam, D. Ajitha, A critical review of AlGaN/GaN- heterostructure based Schottky diode/ HEMT hydrogen (H_2) sensors for aerospace and industrial applications, *Measurement*, 186 (2021) 110100.
[167] L. Arivazhagan, D. Nirmal, D. Godfrey, J. Ajayan, P. Prajoon, A.S. Augustine Fletcher, A.A.A. Jone, J.R. Kumar, Improved RF and DC performance in AlGaN/GaN HEMT by P-type doping in GaN buffer for millimetre-wave applications, *AEU - International Journal of Electronics and Communications*, 108 (2019) 189–194.
[168] J. Ajayan, D. Nirmal, A review of InP/InAlAs/InGaAs based transistors for high frequency applications, *Superlattices and Microstructures*, 86 (2015) 1–19.
[169] J. Ajayan, D. Nirmal, T. Ravichandran, P. Mohankumar, P. Prajoon, L. Arivazhagan, C.K. Sarkar, InP high electron mobility transistors for submillimetre wave and terahertz frequency applications: A review, *AEU-International Journal of Electronics and Communications*, 94 (2018) 199–214.
[170] A.S. Augustine Fletcher, D. Nirmal, J. Ajayan, L. Arivazhagan, Analysis of AlGaN/ GaN HEMT using discrete field plate technique for high power and high frequency applications, *AEU-International Journal of Electronics and Communications*, 99 (2019) 325–330.
[171] S. Tayal, S. Jadav, Power-delay trade-offs in complementary metal-oxide semiconductor circuits using self and optimum bulk control, *Sensor Letters*, 18 (2020) 210–215.
[172] S. Tayal, A. Nandi, Study of temperature effect on junctionless Si nanotube FET concerning analog/RF performance, *Cryogenics*, 92 (2018) 71–75.
[173] S. Tayal, A. Nandi, Optimization of gate-stack in junctionless Si- nanotube FET for analog/RF applications, *Materials Science in Semiconductor Processing*, 80 (2018) 63–67.
[174] S. Tayal, A. Nandi, Analog/RF performance analysis of inner gate engineered junctionless Si nanotube, *Superlattices and Microstructures*, 111 (2017) 862–871.

8 Numerical Study of a Symmetric Underlap S/D High-κ Spacer on JAM-GAA FinFET for Low-Power Applications

B. Kumar and Rishu Chaujar
Delhi Technological University

CONTENTS

8.1 INTRODUCTION

Transistors need to have features like less power dissipation, lower leakage current, higher current-driving capability, enhanced operational frequency, etc., to meet the present-day demands of ULSI industries [1]. Consequently, the quantity of transistors has been increased, whereas the transistor size has been gradually reduced to the sub-10 nm regime to accomplish these demands [2]. The continuous reduction in transistor size results in unwanted short-channel effects (SCEs) such as threshold voltage roll-off, mobility degradation, subthreshold swing (SS), and drain-induced barrier lowering (DIBL) [3–6]. The I–V characteristics of the device deteriorate considering the minimized gate control area over the channel in short-channel devices. Numerous device structures have been proposed to suppress these effects, such as multigate MOSFET [7], cylindrical gate MOSFET [8,9], recessed channel MOSFET [10–12], TFETs [13], and FinFETs [14,15]. FinFET, due to its characteristics, like enhanced drive-current capability, amplified electrostatic control over the channel, reduced leakage current, etc.,

DOI: 10.1201/9781003240778-8

has gained acceptance as a capable device to overcome the SCEs effectively [16]. The gate all around (GAA) structure was put forward to escalate the subthreshold FinFET performance [17,18]. Due to the high gate control area over the channel, the GAA FinFET is more electrostatically stable than the conventional FinFET. Subsequently, the device size can be further reduced without compensating the performance.

The sharp and ultra-shallow p-n junction fabrications at the sub-10 nm regime are a difficult piece of work. Inspired by Lilienfeld's work, Colinge proposed junctionless (JL) transistors to elude problems [19,20]. JL transistors can be easily fabricated as they do not comprise any p-n junction and are economical. The JL devices have better immunity to SCEs with the productively enhanced channel length [21]. Still, they cannot deplete the channel entirely in the OFF state to achieve superlative turn-off characteristics [22]. The rapid depletion of the charge carriers in the OFF state can be done by providing a lower channel region doping concentration, although it will increase the drain/source series resistance and reduce the drain current. Consequently, junctionless accumulation mode (JAM) transistors with amplified source/drain doping are taken into consideration to overcome these problems [23–25].

The device performance can be enhanced, and SCEs can be suppressed using underlap high-κ spacer engineering. With underlap introduction into the device, the effect of drain bias in the channel region is reduced, reducing the SCEs [26]. The high-κ spacer enhances the low I_{on} reported by the underlap MOSFETs [27]. Studies have confirmed that enhanced I_{on}, reduced I_{off}, and excellent control over the channel are attainable using high-κ spacers in the underlap region [28–30]. This chapter considers S/D underlap JAM-GAA FinFET with high-κ spacers and studies its influence on parameters like intrinsic gain, early voltage, quality factor, device efficiency, transconductance, drain current, TFP, GBP, cut-off frequency, etc. The purpose of this study is to quantitatively describe the impact of high-κ spacers on device performance and optimize its value and spacer length in terms of analog/RF parameters.

The chapter is organized as follows. Section 8.2 delineates the structure details of the device. Section 8.3 focuses on the simulation framework along with the calibration of experimental and simulation data. Section 8.4 discusses the impact of symmetric underlap S/D high-κ spacer on analog/RF performance parameters. Further, the analog/RF performance of JAM-GAA FinFET is investigated for different values of the high-κ spacer and spacer lengths. Section 8.5 authenticates the paper's originality with concluding remarks.

8.2 DEVICE STRUCTURE

The three-dimensional and two-dimensional structures of the proposed JAM-GAA FinFET with the high-κ spacer are portrayed in Figure 8.1a and b. The whole fin region contains silicon (Si) material. The uniform doping profiles are assumed, and the entire fin region from source to drain is doped with n-type doping during the analysis. As explained in the introduction section, the channel region doping concentration (N_{Ch}) is lower than the S/D region doping concentration ($N_{S/D}$) to make the device work in JAM mode. The fin width (W_{Fin}) is a multiple of fin height (H_{Fin}) throughout the simulations to obey the width quantization property [31]. Polysilicon gates exhibit a polydepletion effect that reduces the device's effective oxide thickness

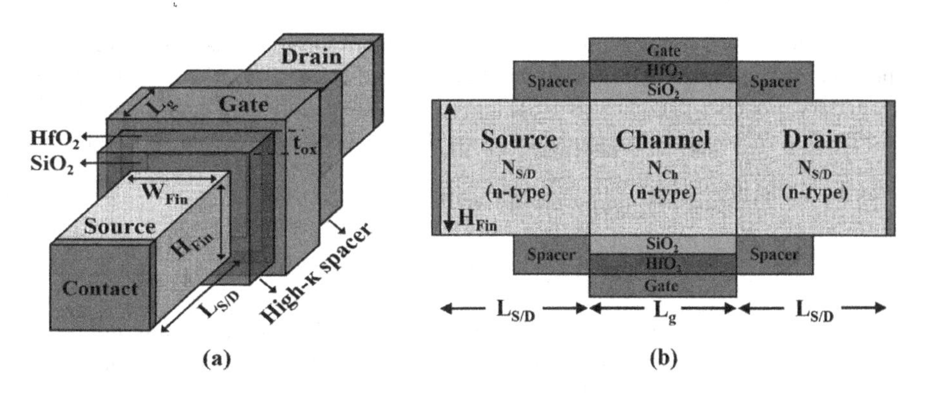

FIGURE 8.1 (a) JAM-GAA FinFET 3-D structure with a symmetric underlap S/D high-κ spacer and (b) a horizontally cut 2D structure of the proposed device.

TABLE 8.1
Distinct Device Parameter Dimensions Considered in Device Simulations

Parameters	CONVL FinFET	GAA FinFET	GAAwS FinFET
S/D doping concentration ($N_{S/D}$, cm^{-3})	5×10^{18}	5×10^{18}	5×10^{18}
Channel doping concentration (N_{Ch}, cm^{-3})	1×10^{16}	1×10^{16}	1×10^{16}
Fin width (W_{Fin}, nm)	5	5	5
Fin height (H_{Fin}, nm)	10	10	10
Work function (ϕ_m, eV)	4.65 (TiN)	4.65 (TiN)	4.65 (TiN)
Oxide thickness (t_{ox}, nm)	1	1	1
S/D region length ($L_{S/D}$, nm)	15	15	15
Gate length (L_g, nm)	10	10	10
High-κ spacer (κ)	-	-	3.9 (SiO_2)
Underlap high-κ spacer length (L_{sp}, nm)	-	-	5
Temperature (T, Kelvin)	300	300	300

(EOT), thereby reducing the device's performance. In addition, polysilicon gates are not chemically stable with high-κ dielectrics, due to which metal gates are considered over the polysilicon gate at the moment [32]. In metal gates, titanium nitride (TiN) is chosen due to its standout features such as compatibility during CMOS processing, low resistivity, high purity, thermal stability, etc. [33,34]. The high-κ spacer ranges from $\kappa=3.9$ (SiO_2) to $\kappa=40$ (TiO_2) with a 5 nm default spacer length (L_{sp}). All of the device parameters used in the analysis are mentioned in Table 8.1.

8.3 SIMULATION METHODOLOGY AND EXPERIMENTAL CALIBRATION

The Silvaco ATLAS 3D simulator has been used to run numerical simulations of different devices [35]. In numerical simulations, the accuracy of the device performance depends on the selection of the appropriate models. The numerical models comprise a set of integral equations and must be implemented properly. For device simulations, the general framework is provided by Poisson's and continuity equations. However, secondary models and equations are required to acquire more convincing and precise results. The various physical models used for device simulations are mentioned below.

(i) *Bohm Quantum Potential Model (BQP) [36]*

$$Q = 0.5\hbar^2 \times \gamma[\nabla\{M^{-1}\nabla(n^{\alpha})\}/n^{\alpha}] \tag{8.1}$$

Quantum confinement effects cannot be neglected in the sub-10 nm regime devices. The Bohm interpretation of quantum mechanics is used to derive the BQP model. Here, M^{-1} represents the inverse effective mass tensor, $\hbar$ is Planck constant, γ and α are the adjustable parameters with 1.4 and 0.3 as the default values for Si, respectively, and n denotes electron density.

(ii) *Bandgap Narrowing Model (BGN) [37]*

$$\Delta E_g = BGN.E\left[ln(N/BGN.N) + \{(ln\ N/BGN.N)^2 + BGN.C\}^{1/2}\right] \tag{8.2}$$

The bandgap separation decreases when the doping is higher than 10^{18}cm^{-3}. The conduction band is lowered almost as much as the valence band is increased. Thus, the BGN model is introduced to implement the bandgap narrowing effects. Here, ΔE_g signifies the bandgap variation; N denotes doping concentration; and BGN.C, BGN.N, and BGN.E are user-definable parameters.

(iii) *Crowell-Sze Impact Ionization Model (Crowell) [38]*

$$\alpha_{n,p} = (1/\lambda) \times exp\{C_0(r) + C_1(r)x + C_2(r)x^2\} \tag{8.3}$$

The impact ionization effects are introduced with the help of this model. Here, $C_0(r)$, $C_1(r)$, $C_2(r)$ are the ionization coefficients, and the carrier mean free path for optical phonon generation is represented by λ.

(iv) *Klaassen Band-to-Band Tunneling Model (BBT.KL) [39]*

$$G_{BBT} = D \times BB.A \times E^{BB.GAMMA} \times exp(-BB.B/E) \tag{8.4}$$

It is used to incorporate the tunneling of electrons between conduction and valence band. Here, D denotes the statistical factor; E represents the electric field magnitude; and BB.A, BB.B, and BB.GAMMA are user-definable parameters.

(v) *Shockley–Read–Hall Recombination Model (CONSRH) [40,41]*

$$R_{SRH} = (pn - n_{ie}^2)/\left[\tau_p\{n + n_{ie}exp(E_{Trap}/kT_L)\} + \tau_n\{p + n_{ie}exp(-E_{Trap}/kT_L)\}\right] \tag{8.5}$$

SRH model is invoked to consider the generation and recombination effects. Here, E_{TRAP} is the difference between trap energy and intrinsic

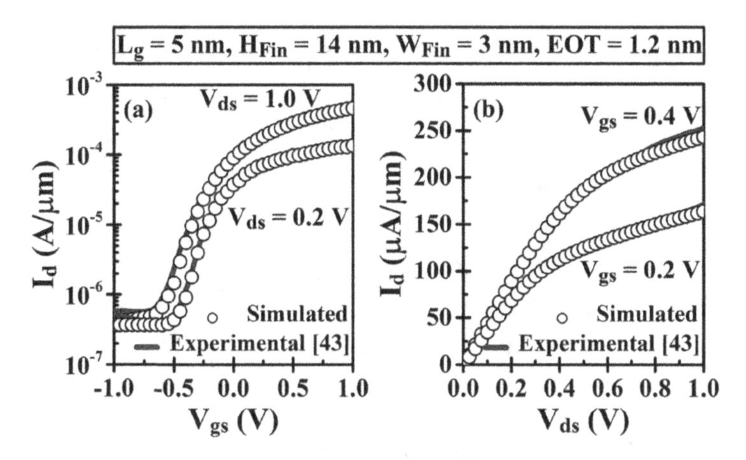

FIGURE 8.2 Simulated and experimental (a) I_d–V_{gs} (b) I_d–V_{ds} characteristics of all-around-gate (AAG) Si FinFET.

Fermi levels, n_{ie} is the intrinsic concentration, and τ_n and τ_p signify the electron and hole lifetimes, respectively.

(vi) *Fermi–Dirac Statistics Model (Fermi) [42]*

$$f(\varepsilon) = 1/\left[1 + exp\{(\varepsilon - E_F)/kT_L\}\right] \tag{8.6}$$

It is included to enhance the result accuracy. Here, ε signifies energy, T_L is the lattice temperature, k indicates the Boltzmann's constant, and E_F indicates the Fermi level.

(vii) *Concentration-Dependent Low-Field Mobility Model (CONMOB) [35]*

CONMOB model associates the low-field carrier mobility with the impurity concentration.

Numerical methods are used to obtain solutions to semiconductor device problems. In this chapter, NEWTON and BLOCK methods are incorporated to solve the equations. It is necessary to calibrate the simulated data with the experimental data to validate the physical models used in the simulations. Thus, we precisely extracted Hyunjin's experimental data and compared it with the simulated device structure [43]. The same device dimensions and physical models are used to confirm the simulations. The I_d–V_{gs} and I_d–V_{ds} curves of all-around-gate (AAG) Si FinFET are portrayed in Figure 8.2a and b, which shows that experimental results are close to simulated results, thereby confirming the models selected for analysis.

8.4 RESULTS AND DISCUSSION

This section discusses the efficiency of a symmetric underlap S/D high-κ spacer on analog and RF performance parameters by comparing gate all around with spacer (GAAwS) FinFET with the conventional (CONVL) FinFET and GAA FinFET. Figure 8.3 represents the I_d–V_{gs} curve in log and linear for all three simulated devices.

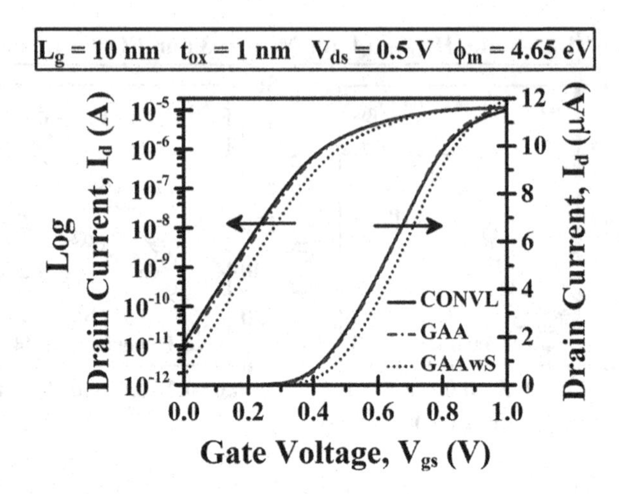

FIGURE 8.3 Variation of transfer characteristics in linear and log scales for all three simulated devices.

The drain current (I_d) increases with the increase in the gate voltage (V_{gs}). The maximum on current (I_{on}) is observed for GAAwS FinFET due to the fringing field effects in the underlap regions. Moreover, the off current (I_{off}) observed for GAAwS FinFET is ~10^{-12} A compared to ~10^{-11} A for the other two device structures. Thus, high-κ spacer introduction improves both the I_{on} and I_{off} of the device. Figure 8.4a and b exhibits the switching ratio (I_{on}/I_{off}) and threshold voltage (V_{th}) of the simulated devices. As explained earlier, due to increased I_{on} and reduced I_{off}, the I_{on}/I_{off} ratio is enhanced significantly for GAAwS FinFET. Compared to the CONVL FinFET, the threshold voltage increases from 0.308 to 0.358 V for GAAwS FinFET. The *SS* and transconductance (g_m) of the three devices are portrayed in Figure 8.4c and d. *SS* is a vital parameter, and it reduces from 76.51 to 75.60 mV/decade for the GAA FinFET structure due to the enhanced electrostatic control. Further, it reduces to 72.18 mV/decade with the use of a high-κ spacer indicating better immunity to SCEs. At a fixed drain–source voltage (V_{ds}), deviation in the I_d–V_{gs} is measured by transconductance, and $g_m = \partial I_d / \partial V_{gs}$ [44]. The g_m increases with the increase in V_{gs} and is recorded the highest at $V_{gs} = 0.7$ V for GAAwS (29.38 μS) compared to the CONVL (27.12 μS) and GAA (27.93 μS) device structures.

The switching behavior of the device is evaluated by an essential parameter known as quality factor ($QF = g_m/SS$) [45,46]. The higher the *QF*, the better the switching behavior. Figure 8.5a represents the *QF* for the three device structures considered. QF is evaluated for the peak value of g_m ($V_{gs} = 0.7$ V). Due to the enhanced g_m and reduced *SS*, *QF* increases by 14.32% for GAAwS FinFET compared to CONVL FinFET. Transconductance generation factor (TGF) is the fraction of generated gain per unit power loss, i.e., $TGF = g_m/I_d$ [47]. The higher the value of *TGF*, the more efficient the device operation at a lower supply voltage. TGF is plotted against V_{gs} for all simulated devices in Figure 8.5b. Compared to the other two device structures, the maximum value of TGF is obtained for GAAwS FinFET due to improved g_m. The device-driving ability is defined by the output conductance (g_d) [48]. The variation

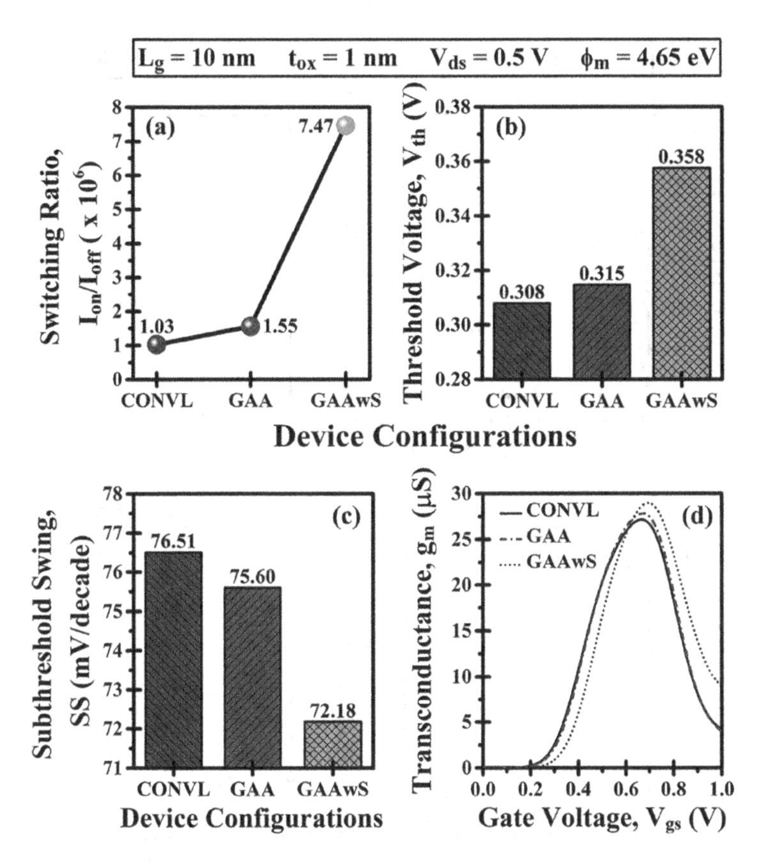

FIGURE 8.4 Plot of (a) I_{on}/I_{off} ratio, (b) V_{th}, (c) SS, and (d) g_m for different device configurations.

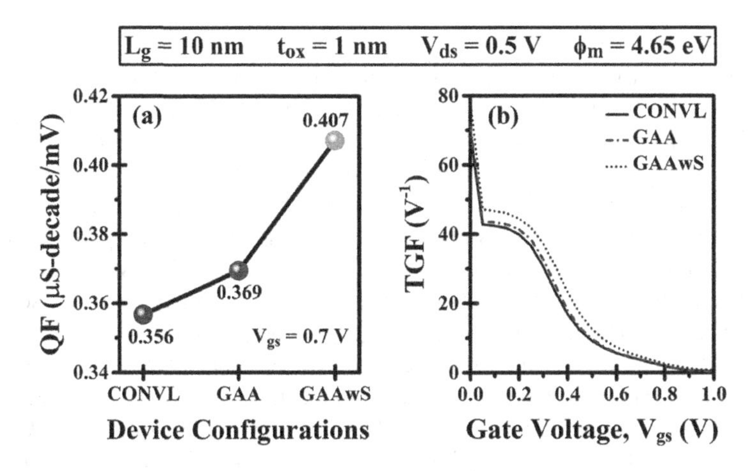

FIGURE 8.5 Change in (a) QF and (b) TGF plot for all simulated devices.

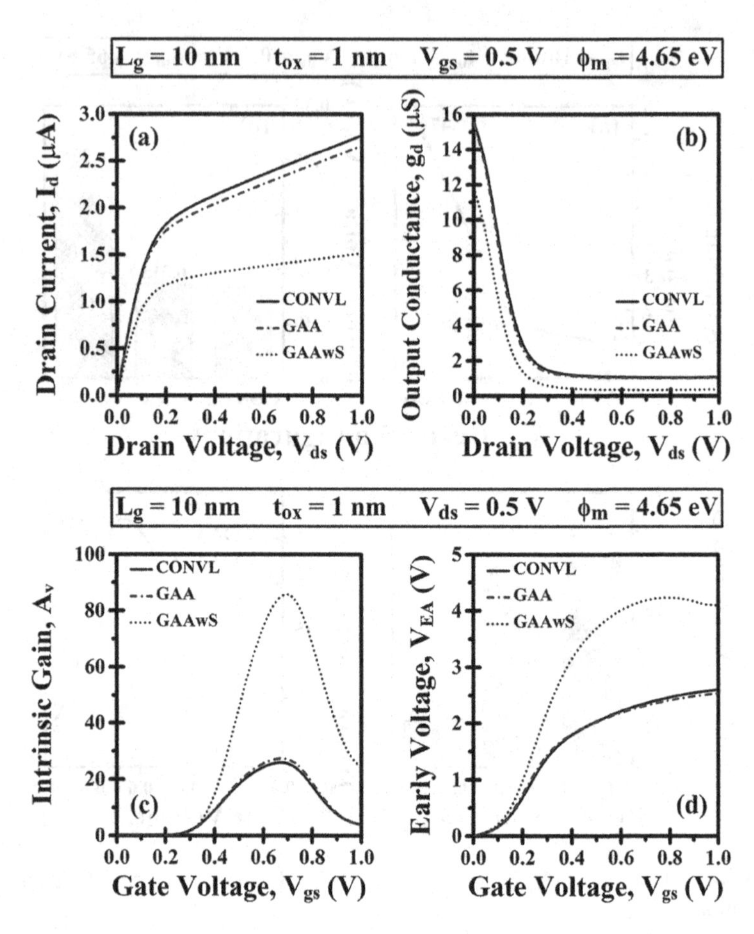

FIGURE 8.6 Variation of (a) I_d and (b) g_d against drain–source voltage. Alteration of (c) A_v and (d) V_{EA} against gate–source voltage for different device structures.

of I_d and g_d against V_{ds} at fixed V_{gs} is displayed in Figure 8.6a and b. First, the I_d increases sharply and then linearly with the increase in the V_{ds}. Compared to CONVL FinFET, the drain current is reduced for GAA FinFET and significantly reduced further for the GAAwS FinFET structure. The reason for this decrease is the suppressed SCEs and enhanced gate controllability. The output conductance decreases before maintaining a constant value with the increase in V_{ds}. The g_d is significantly lower for GAAwS FinFET than for the other two structures because the smaller the drain current, the higher the output resistance, and consequently, the lower the output conductance ($R_o = 1/g_d$). Intrinsic gain ($A_v = g_m/g_d$) and early voltage ($V_{EA} = I_d/g_d$) are crucial analog parameters and should be as high as possible [48,49]. Figure 8.6c and d depicts the variation of A_v and V_{EA} against V_{gs} for the three devices considered. The A_v is enhanced more than three times, while V_{EA} is increased by 62.55% for GAAwS FinFET compared with CONVL FinFET. This considerable increase in both the parameters is because of the suppressed g_d and improved g_m and I_d.

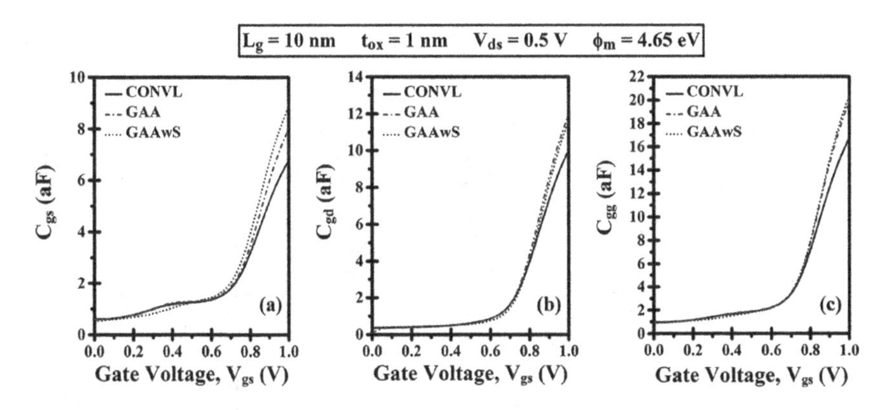

FIGURE 8.7 Parasitic capacitances (a) C_{gs}, (b) C_{gd}, and (c) C_{gg} comparison against gate–source voltage for different structural designs.

The change in gate–source capacitance (C_{gs}), gate–drain capacitance (C_{gd}), and gate–gate capacitance (C_{gg}) against V_{gs} for each considered structure is presented in Figure 8.7a–c. These intrinsic gate capacitances are extracted using AC small-signal analysis performed at 1 MHz operating frequency. In the subthreshold region, the intrinsic gate capacitances of all devices remain almost the same. However, a significant difference in the values of C_{gs}, C_{gd}, and C_{gg} beyond V_{gs} = 0.7 V is detected. The maximum value of C_{gs} and C_{gg} is obtained for the GAAwS FinFET structure compared to the other two structures. This is due to the direct dependence of gate capacitance on the dielectric permittivity ($C \propto \kappa$) and enhanced charge carrier's movement from source to drain side, thereby increasing the lateral field. As expressed in equations (8.7) and (8.8), cut-off frequency (f_T) and maximum oscillation frequency (f_{max}) are the frequencies at which current gain and maximum unilateral power gain become unity (0 dB), respectively [50,51]. Figure 8.8a and b exhibits the peak value plot of f_T and f_{max} obtained at V_{gs} = 0.55 V. A slight decrease in f_T (from 1.917 to 1.889 THz) is observed for the GAAwS FinFET compared to the CONVL FinFET because of the increased intrinsic gate capacitances. f_{max} also deteriorates from 21.21 to 15.91 GHz with the use of the high-κ spacer. The reason for this significant decrease in f_{max} is the enhanced intrinsic gate capacitances and drain–source output conductance (g_{ds}) and reduced cut-off frequency.

$$f_T = g_m / \left\{2\pi\left(C_{gs} + C_{gd}\right)\right\} \tag{8.7}$$

$$f_{max} = f_T / \sqrt{\left\{4R_g\left(g_{ds} + 2\pi f_T C_{gd}\right)\right\}} \tag{8.8}$$

$$TFP = \left(g_m / I_d\right) \times f_T \tag{8.9}$$

$$GBP = g_m / \left(20\pi \times C_{gd}\right) \tag{8.10}$$

Transconductance frequency product (TFP) and gain bandwidth product (GBP) are crucial for high-frequency applications. Both the parameters need to have a high value for enhanced performance. An agreement between bandwidth and power is

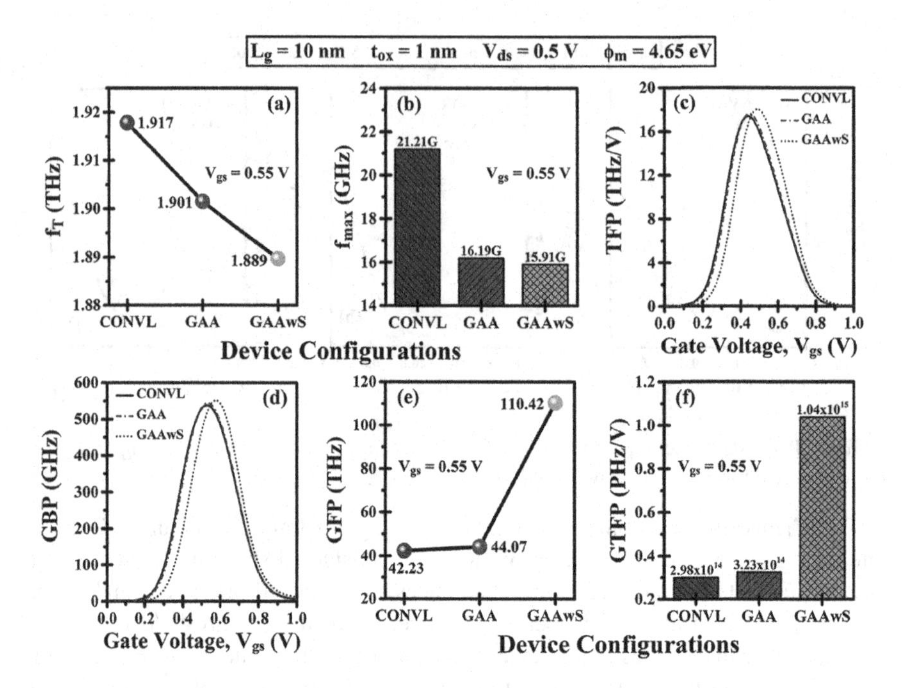

FIGURE 8.8 Peak value plot of (a) f_T and (b) f_{max} for each device configuration. Variation of (c) TFP and (GBP) against V_{gs}. (e) GFP and (f) GTFP maximum value curve for all simulated devices.

exhibited through TFP and GBP [52,53]. Variation of TFP and GBP against V_{gs} for all simulated devices is portrayed in Figure 8.8c and d. Both parameters increase with V_{gs}, attain a peak value, and then decrease with further V_{gs}. The maximum value of *TFP* and *GBP* is obtained for the GAAwS FinFET because as shown in equation (8.9), *TFP* is the product of *TGF* and f_T, and the increase in the *TGF* for the GAAwS FinFET is sufficient to overcome the decrease in f_T. Likewise, GBP contains transconductance and drain–source capacitance in the form of variable parameters (as depicted in equation (8.10)), and the overall product is higher for the GAAwS FinFET than the other two structures. Other important parameters for high-frequency applications are gain frequency product (*GFP*) and gain transconductance frequency product (*GTFP*) expressed in equations (8.11 and 8.12), respectively [54]. Figure 8.8e and f presents the peak value plot of *GFP* and *GTFP* obtained at $V_{gs}=0.55$ V. A significant improvement is observed in *GFP* (from 42.23 to 110.42 THz) and *GTFP* (248.99% increase) for the GAAwS FinFET compared to the CONVL FinFET. This substantial increase in both the parameters is mainly because of the increased intrinsic gain and *TFP*.

$$GFP = \left(g_m / g_d\right) \times f_T \tag{8.11}$$

$$GTFP = \left(g_m / I_d\right) \times f_T \times \left(g_m / g_d\right) = TFP \times A_v \tag{8.12}$$

8.4.1 High-κ Spacer Influence on Analog/RF Performance

In the previous section, we have seen that the analog/RF performance of JAM-GAA FinFET improves prominently with the implication of a symmetric underlap high-κ spacer. Thus, in this section, various essential analog/RF parameters are inspected for the different values of high-κ spacer varying from κ = 3.9 (SiO_2) to κ = 40 (TiO_2) with a 5 nm default spacer length (L_{sp}) on both sides of the drain and the source. Figure 8.9 represents the I_d–V_{gs} curve in linear and log scales for each high-κ spacer considered. Due to increased fringing field effects in the underlap region, the I_{on} current of the device increases with the increase in the high-κ value. Moreover, the I_{off} current observed for TiO_2 is lower than the I_{off} for SiO_2. Consequently, I_{on}/I_{off} ratio improves by more than five times for TiO_2 than SiO_2 portrayed in Figure 8.10a. Thus, the increase in the high-κ values improves the device's I_{on}, I_{off}, and I_{on}/I_{off} ratio. The V_{th} also increases with the increase in the high-κ value, as portrayed in Figure 8.10b. Figure 8.10c and d exhibits the SS and transconductance for the different materials used as a high-κ spacer. *SS* is reduced from 72.18 mV/decade (SiO_2) to 66.70 mV/decade (TiO_2) owing to the reduced I_{off} current and fringing field effects, whereas g_m increases with the increase in the underlap dielectric value. Transconductance depends mainly on the *S/D* charge difference (Q_S–Q_D) [55,56]. The feedback field through the *S/D* spacer increase this charge difference, which increase the device's transconductance.

Figure 8.11a and b represents the QF and TGF for all of the different high-κ spacers considered. The increase in the underlap dielectric value from 3.9 to 40 increase the QF from 0.407 to 0.539 μS-decade/mV. This increase is obvious because of the enhanced g_m and reduced *SS*. Likewise, TGF also increases significantly with the underlap dielectric value. Due to the improved g_m and I_d, TGF increases by 17.46% for TiO_2 than the SiO_2 dielectric spacer. The output characteristics and output conductance variation for different high-κ spacers are presented in Figure 8.12a and b.

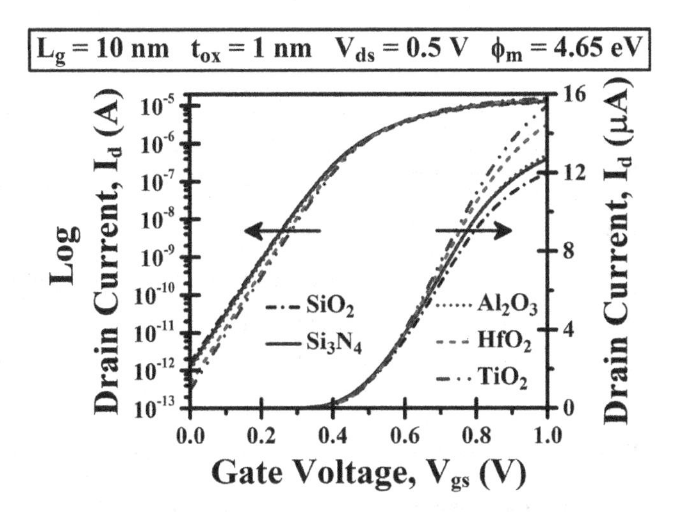

FIGURE 8.9 I_d–V_{gs} characteristics in linear and log scales for each high-κ spacer considered.

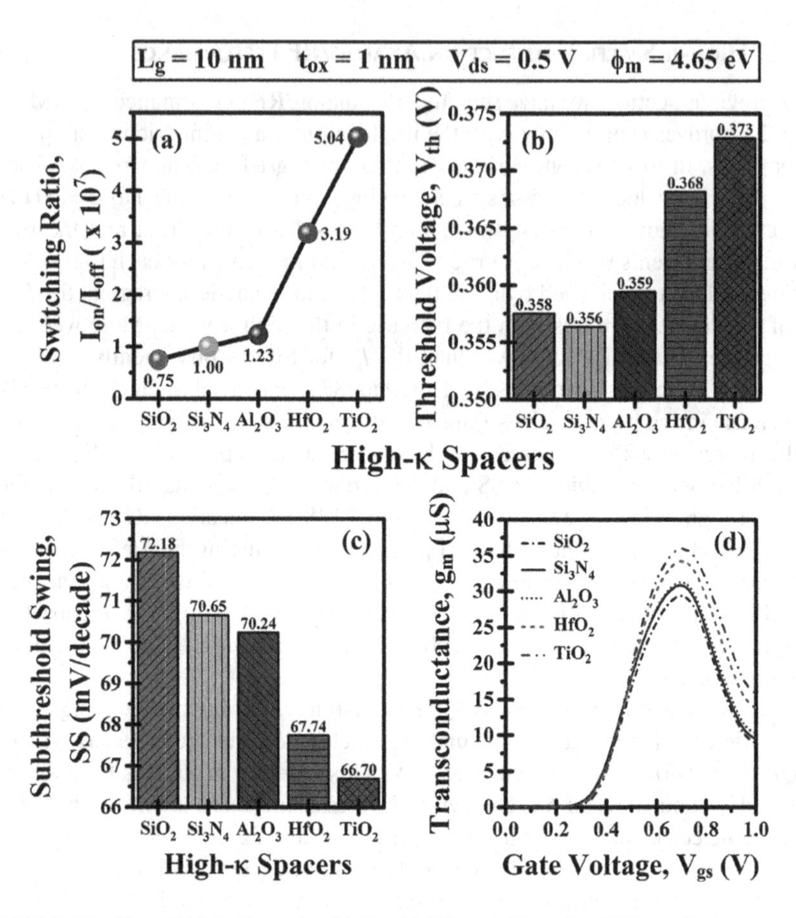

FIGURE 8.10 Plot of (a) I_{on}/I_{off} ratio, (b) V_{th}, (c) SS, and (d) g_m for different high-κ spacers.

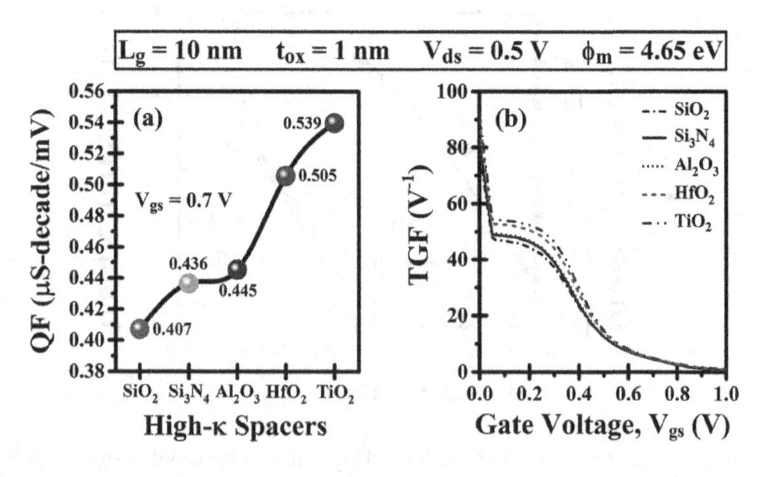

FIGURE 8.11 Impact of high-κ spacer value on (a) QF and (b) TGF plot.

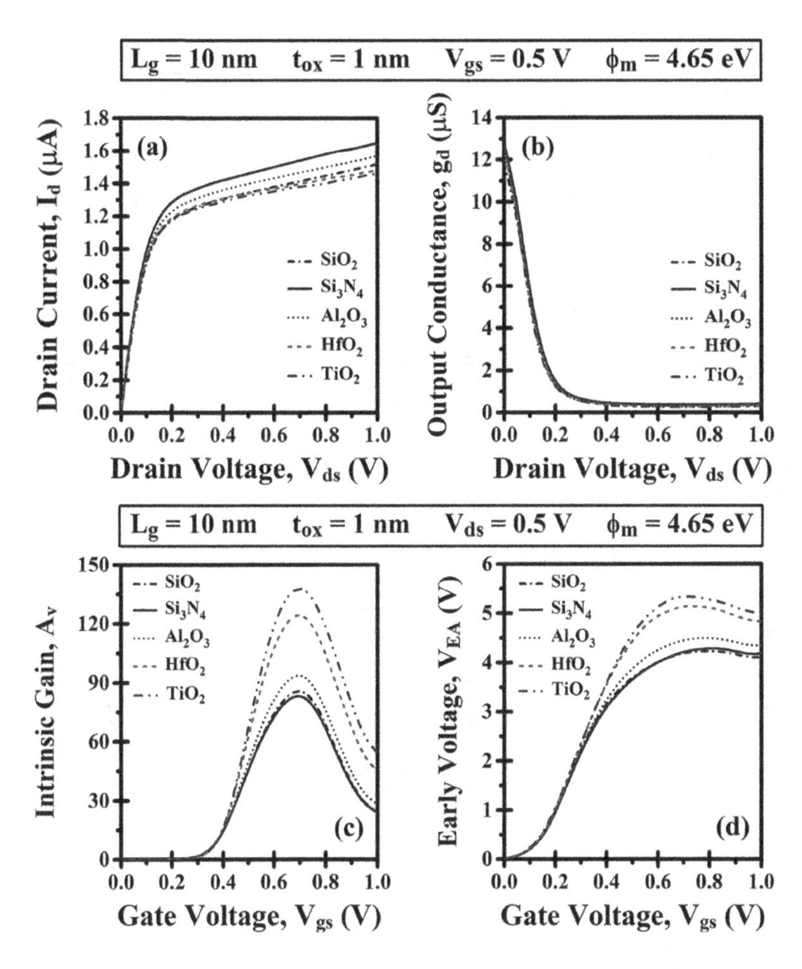

FIGURE 8.12 Variation of (a) I_d and (b) g_d against V_{ds} and (c) A_v and (d) V_{EA} against V_{gs} for all high-κ spacers considered.

The decrease in both parameters with the increase in the high-κ spacer value signifies the reduction in the SCEs. This reduction in g_d and I_d reflects in the curve of intrinsic gain and early voltage depicted in Figure 8.12c and d. The A_v is enhanced by 60.10%, while V_{EA} is increase by 27.59% when SiO_2 is replaced with TiO_2 in the underlap spacer region. This considerable increase in both the parameters is because of the reduced g_d and enhanced g_m and I_d. Thus, with the change in the underlap spacer value from 3.9 to 40, the JAM-GAA FinFET analog performance improves considerably.

The change in C_{gs}, C_{gd}, and C_{gg} against V_{gs} for each considered high-κ spacer is presented in Figure 8.13a–c. In the subthreshold region, the intrinsic gate capacitances of all devices remain almost the same. However, a significant difference is observed in intrinsic gate capacitance beyond the subthreshold regions, especially in TiO_2. This is due to the direct dependence of gate capacitance on the dielectric permittivity ($C \propto \kappa$) and enhanced charge carrier's movement from source to drain side

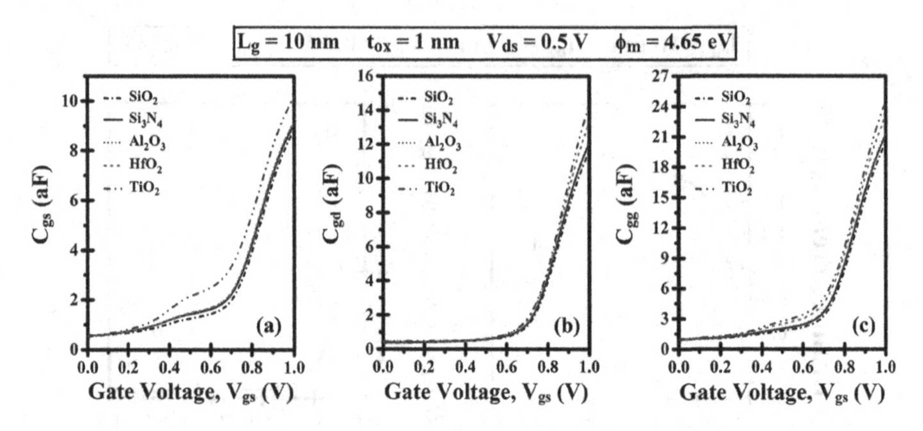

FIGURE 8.13 Influence of high-κ spacer on parasitic capacitances (a) C_{gs}, (b) C_{gd}, and (c) C_{gg} against gate–source voltage.

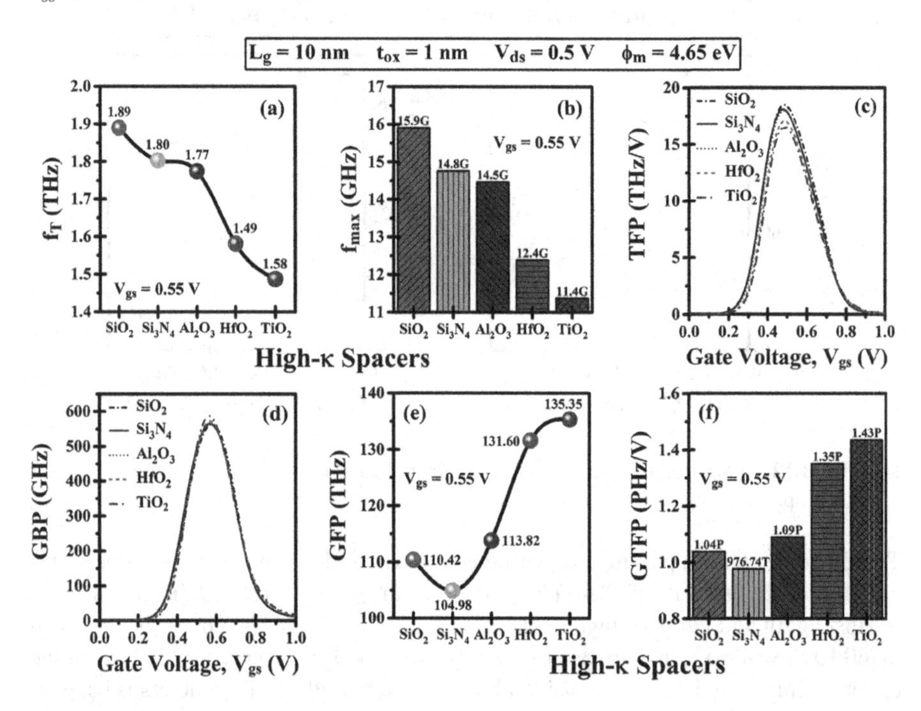

FIGURE 8.14 Peak value plot of (a) f_T and (b) f_{max} for different high-κ spacers. Variation of (c) TFP and (GBP) against the different high-κ spacers. Impact of high-κ spacer value on (e) GFP and (f) GTFP.

due to the gate-induced fringing field. Figure 8.14a and b exhibits the deterioration of the device's f_T and f_{max} because of a rise in the intrinsic gate capacitances. When the value of the underlap spacer changes from 3.9 to 40, the peak value of f_T and f_{max} obtained at $V_{gs} = 0.55$ V is reduced by 16.40% and 28.30%, respectively. Variation of *TFP* and *GBP* against V_{gs} for different high-κ spacers materials is portrayed in

Figure 8.14c and d. Both parameters increase with V_{gs}, attain a peak value, and then decrease with further V_{gs}. The maximum value of TFP decreases when the underlap spacer material is replaced from SiO_2 to TiO_2 because of the reduced cut-off frequency. On the other hand, GBP is increased with the increase in the underlap spacer value because the surge in the transconductance is sufficient to overcome the increase in the drain–source capacitance. Figure 8.14e and f presents the peak value plot of *GFP* and *GTFP* obtained at V_{gs}=0.55 V. A significant improvement is observed in *GFP* (from 110.42 to 135.35 THz), whereas *GTFP* is increased by 37.5% for the TiO_2 spacer compared to the SiO_2 spacer. This substantial increase in both the parameters is mainly because of the increased intrinsic gain (increased g_m and reduced g_d).

8.4.2 Spacer Length Influence on Analog/RF Performance

It has been observed in the previous section that the analog parameters improve significantly, whereas RF parameters exhibit mixed reactions for the different values of the high-κ spacer. Therefore, in this section, we have considered TiO_2 as the default high-κ spacer and inspected various analog/RF parameters for the different *S/D* underlap spacer lengths (L_{sp}) to detect the action of spacer engineering. The L_{sp} is varied from 2.5 to 12.5 nm with a step size of 2.5 nm. Figure 8.15 represents the I_d–V_{gs} curve in log and linear scales for different spacer lengths. Due to increased accumulation of charge in the underlap region, the I_{on} current of the device increases up to 7.5 nm spacer length and decreases with a further increase in the L_{sp}. Moreover, the I_{off} current follows the same trend and is lowest for the 7.5 nm L_{sp}. Consequently, the I_{on}/I_{off} ratio and the threshold voltage are recorded highest at the 7.5 nm spacer length as portrayed in Fig. 8.16a and b. Figure 8.16c and d exhibits the SS and transconductance for the different L_{sp} values. *SS* reduces with the increase in the underlap spacer length, thereby indicating the improved SCEs. Since g_m is a derivative of drain current, it follows the same pattern. First, it increases with L_{sp}, attains a peak value at 7.5 nm, and decreases with further L_{sp}. Transconductance inversely depends on

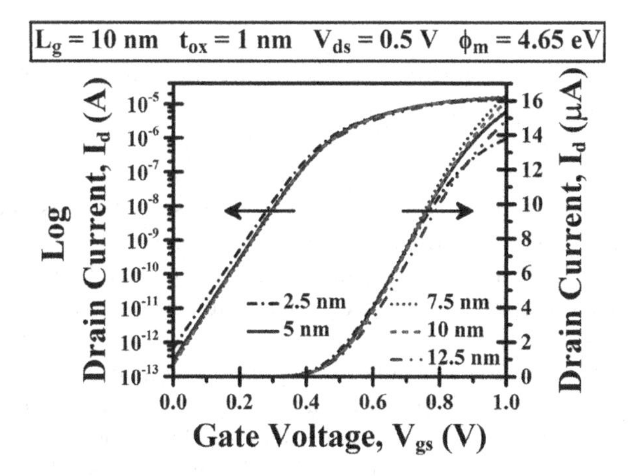

FIGURE 8.15 I_d–V_{gs} characteristics in linear and log scales for each spacer length considered.

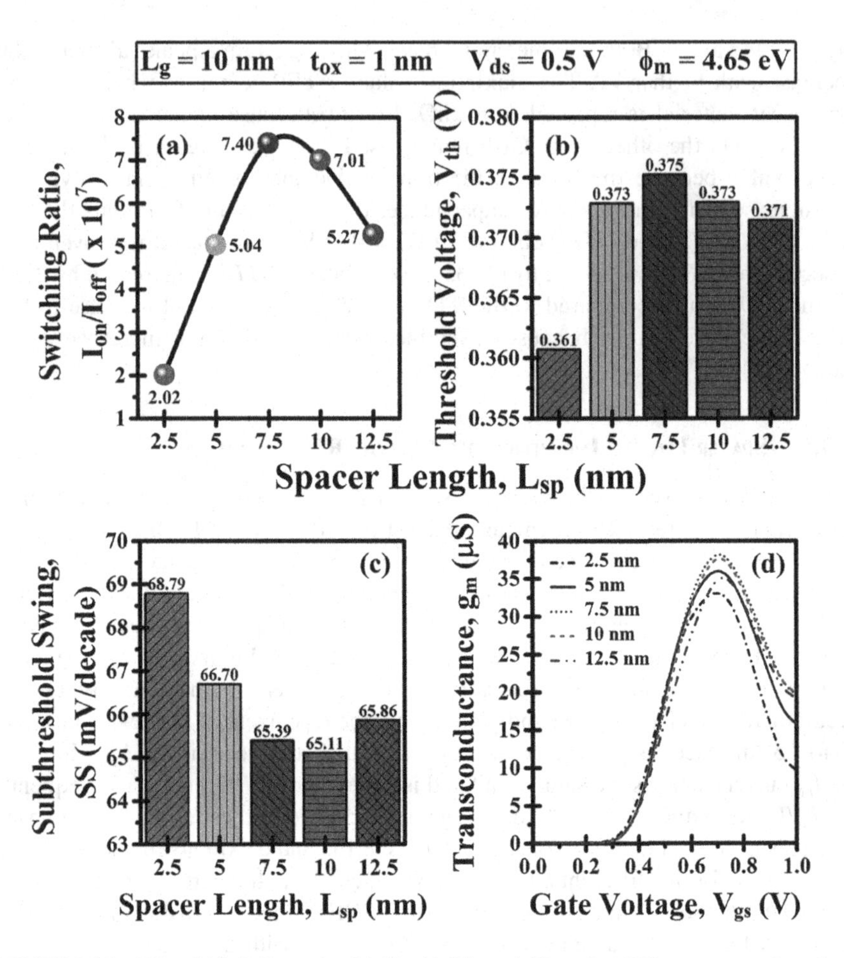

FIGURE 8.16 Plot of (a) I_{on}/I_{off} ratio, (b) V_{th}, (c) SS, and (d) g_m for different spacer lengths.

the channel length, and the increase in spacer length increases the effective channel length, thereby reducing the g_m after the optimum value.

The variation of *QF* and *TGF* against spacer lengths is represented in Figure 8.17a and b. The QF increases initially with spacer length, attains a maximum value at 7.5 nm, and then decreases afterward. This increase is obvious because of the enhanced g_m and reduced *SS*. Likewise, TGF is also recorded at a maximum value for 7.5 nm L_{sp} due to the improved g_m and I_d. Figure 8.18a and b displays the output characteristics and output conductance variation for different spacer lengths. The decrease in both parameters with the increase in the spacer length value signifies the reduction in the SCEs. The curve of intrinsic gain and early voltage against spacer lengths is depicted in Figure 8.18c and d. Again, the maximum value of both the parameters is obtained at 7.5 nm spacer length. Thus, JAM-GAA FinFET with TiO_2 as spacer exhibits the most improved analog performance for 7.5 nm spacer length.

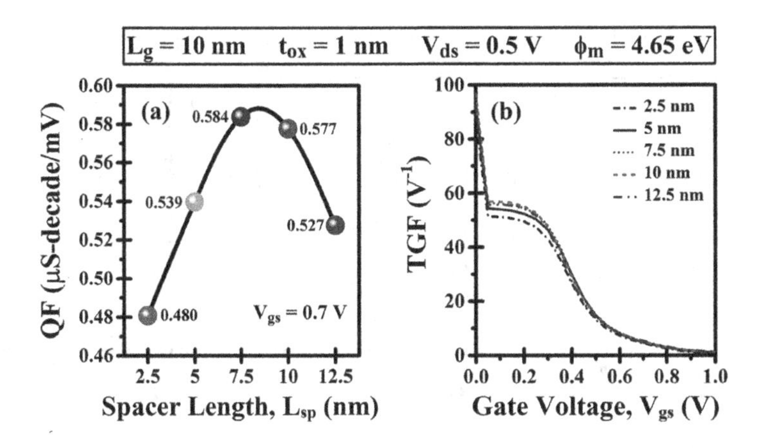

FIGURE 8.17 Impact of different spacer lengths on (a) QF and (b) TGF plot.

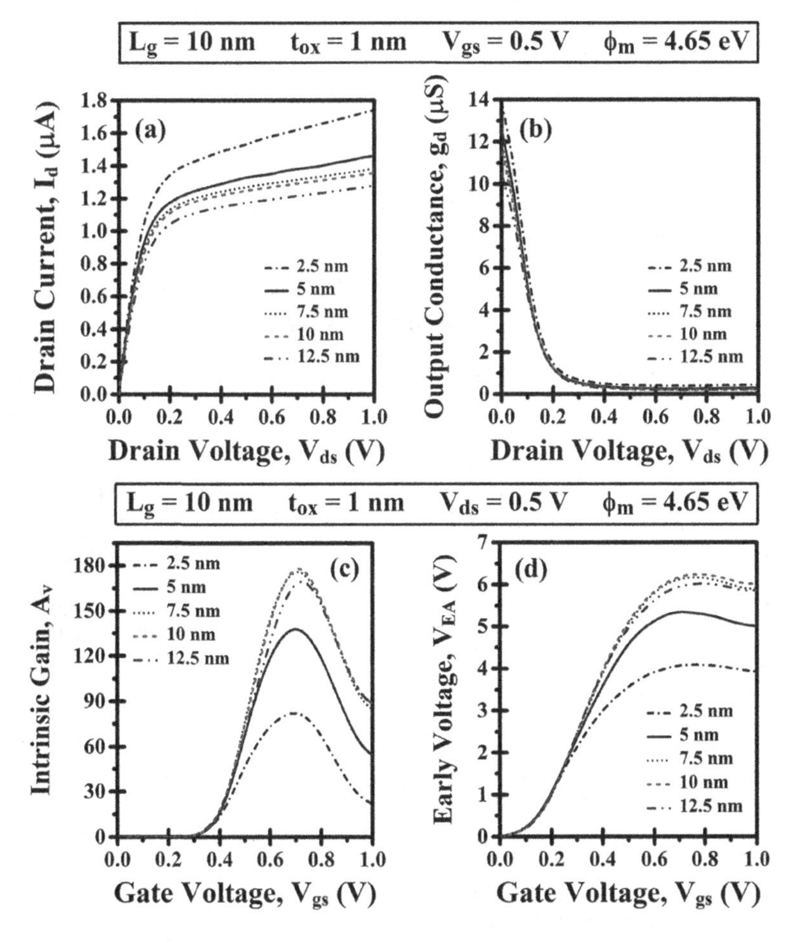

FIGURE 8.18 Variation of (a) I_d and (b) g_d against V_{ds} and (c) A_v and (d) V_{EA} against V_{gs} for all spacer lengths considered.

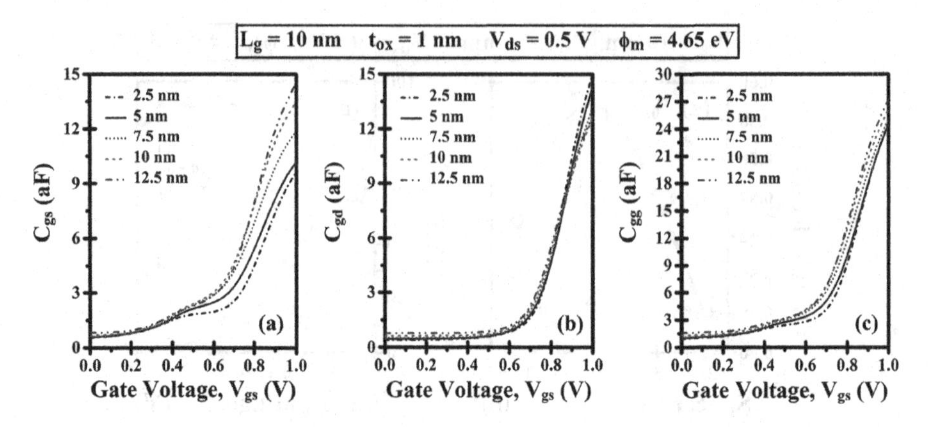

FIGURE 8.19 Influence of spacer lengths on parasitic capacitances (a) C_{gs}, (b) C_{gd}, and (c) C_{gg} against gate–source voltage.

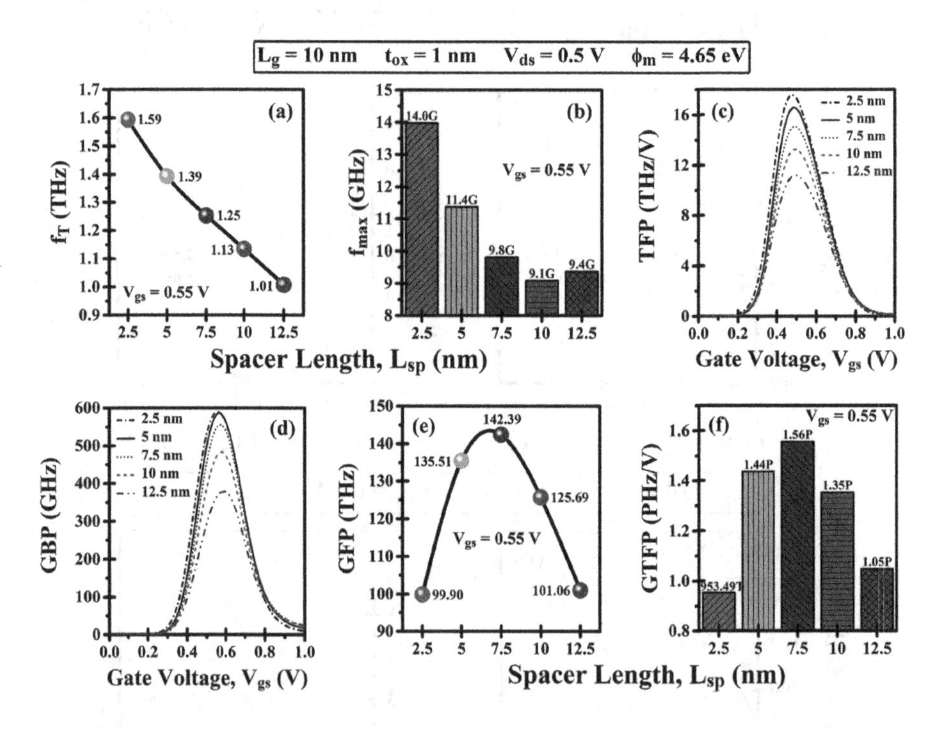

FIGURE 8.20 Peak value plot of (a) f_T and (b) f_{max} for different spacer lengths. Variation of (c) TFP and GBP against the different spacer lengths. Impact of spacer length value on (e) GFP and (f) GTFP.

Figure 8.19a–c presents the change in C_{gs}, C_{gd}, and C_{gg} against V_{gs} for each considered spacer length. In the subthreshold region, the increase in the intrinsic gate capacitances with the spacer length is not that much. However, a significant difference is observed in intrinsic gate capacitance beyond the subthreshold regions. C_{gs} increases with the increase in the spacer length, whereas the reverse effect is observed for the C_{gd}. The overall combination of C_{gs} and C_{gd}, that is, gate–gate capacitance (C_{gg}), increases with spacer length. The intrinsic gate capacitance increase reflects in the deterioration of the device's f_T and f_{max}, as depicted in Figure 8.20a and b. When the value of the underlap spacer length is increased from 2.5 to 12.5 nm, the peak value of f_T and f_{max} obtained at V_{gs} = 0.55 V is reduced by 36.48% and 32.86%, respectively. Variation of TFP and GBP against V_{gs} for different spacer lengths is portrayed in Figure 8.20c and d. The maximum value of TFP decreases when the underlap spacer length is enhanced from 2.5 to 12.5 nm because of the reduced cut-off frequency. Similarly, GBP is also decreased with the increase in the underlap spacer length value. Figure 8.20e and f presents the peak value plot of GFP and GTFP obtained at V_{gs} = 0.55 V. Because of the increased intrinsic gain (increased g_m and reduced g_d) for 7.5 nm L_{sp}, the peak value of both the parameters is attained at the same spacer length.

8.5 CONCLUSIONS

For the first time, this work has examined the symmetric underlap S/D high-κ spacer effect on the analog/RF performance of JAM-GAA FinFET. The results obtained using TCAD simulations reveal that GAA FinFET with high-κ spacer exhibits enhanced device performance owing to the fringing field effects compared to the GAA FinFET and conventional trigate FinFET. The implication of a high-κ spacer improves the switching ratio by about fivefold, increases the intrinsic gain by 216.33%, nearly doubles the early voltage, and substantially improves the RF parameters such as GFP (↑ 150.55%) and GTFP (↑ 221.98%). When the high-κ spacer value increases from κ = 3.9 (SiO_2) to κ = 40 (TiO_2), the channel region conducts more efficiently owing to the gate fringing field, thereby improving the electrical parameters of the proposed device. It is discovered that using TiO_2 as a high-κ spacer significantly increases the A_v (60.10%) and GFP (22.58%) and improves the leakage current and SS by 80.97% and 7.59%, respectively. Further, the S/D underlap spacer length (L_{sp}) is optimized with TiO_2 as the default high-κ spacer to inspect the action of spacer engineering on various analog/RF parameters. Improved analog and RF performance parameters are observed for 7.5 nm spacer length. Consequently, the proposed symmetric S/D underlap JAM-GAA FinFET with κ = 40 and L_{sp} = 7.5 nm can be looked upon as a tempting option for low-power applications.

ACKNOWLEDGMENTS

The authors are obliged to the Microelectronics Research Laboratory, Department of Applied Physics, Delhi Technological University, for assisting with this research work.

REFERENCES

1. Arora ND (1993) *MOSFET Models for VLSI Circuit Simulation: Theory and Practice.* Vienna: Springer
2. Moore GE (1 998) Cramming more components onto integrated circuits. *Proc IEEE* 86:82–85 https://doi.org/10.1109/JPROC.1998.658762.

3. Chaudhary A, Kumar MJ (2004) Controlling short-channel effects in deep-submicron SOI MOSFETs for improved reliability: A review. *IEEE Trans Device Mater Reliab* 4:99–109 https://doi.org/10.1109/TDMR.2004.824359.
4. Iwai H (2009) Roadmap for 22 nm and beyond (invited paper). *Microelectron Eng* 86:1520–1528 https://doi.org/10.1016/j.mee.2009.03.129.
5. Kumar A, Gupta N, Chaujar R (2016) TCAD RF performance investigation of transparent gate recessed channel MOSFET. *Microelectron J* 49:36–42 https://doi.org/10.1016/j.mejo.2015.12.007.
6. Kumar B, Chaujar R (2021) Analog and RF performance evaluation of Junctionless Accumulation Mode (JAM) Gate Stack Gate All Around (GS-GAA) FinFET. *Silicon* 13:919–927 https://doi.org/10.1007/s12633-020-00910-7.
7. Barsan RM (1981) Analysis and modeling of dual-gate MOSFET's. *IEEE Trans Electron Devices* 28:523–534 https://doi.org/10.1007/978-3-540-79076-1_3.
8. Auth CP, Plummer JD (1997) Scaling theory for cylindrical, fully-depleted, surrounding-gate MOSFET's. *IEEE Electron Device Lett* 18:74–76 https://doi.org/10.1109/55.553049.
9. Sarkar A, De S, Dey A, Sarkar CK (2012) Analog and RF performance investigation of cylindrical surrounding-gate MOSFET with an analytical pseudo-2D model. *J Comput Electron* 11:182–195 https://doi.org/10.1007/s10825-012-0396-9.
10. Chaujar R, Kaur R, Saxena M, Gupta M, Gupta RS (2008) TCAD assessment of gate electrode workfunction engineered recessed channel (GEWE-RC) MOSFET and its multi-layered gate architecture, part I: Hot-carrier-reliability evaluation. *IEEE Trans Electron Devices* 55:2602–2613 https://doi.org/10.1016/j.spmi.2009.07.027.
11. Kumar A, Tripathi MM, Chaujar R (2018) Reliability issues of In_2O_5Sn gate electrode recessed channel MOSFET: Impact of Interface Trap Charges and Temperature. *IEEE Trans Electron Devices* 65:860–866 https://doi.org/10.1109/TED.2018.2793853.
12. Kumar A, Tripathi MM, Chaujar R (2018) Comprehensive analysis of sub-20 nm black phosphorus based junctionless-recessed channel MOSFET for analog/RF applications. *Superlattices Microstruct* 116:171–180 https://doi.org/10.1016/j.spmi.2018.02.018.
13. Rahi SB, Asthana P, Gupta S (2017) Heterogate junctionless tunnel field-effect transistor: Future of low-power devices. *J Comput Electron* 16:30–38 https://doi.org/10.1007/s10825-016-0936-9.
14. Chang CY, Chang CH, Hou CH, Lin KL, Lee KY, Yu XF, Chui CO (2019) Semiconductor devices, Finfet devices and methods of forming the same. US Patent App 15/876,223.
15. Sreenivasulu VB, Narendar V (2021) A comprehensive analysis of junctionless Tri-Gate (TG) FinFET towards low-power and high-frequency applications at 5-nm gate length. *Silicon*. https://doi.org/10.1007/s12633-021-00987-8.
16. Samal A, Pradhan KP, Mohapatra SK (2021) Improvising the switching ratio through low-k/High-k spacer and dielectric gate stack in 3D FinFET - a simulation perspective. *Silicon* 13:2655–2660 https://doi.org/10.1007/s12633-020-00618-8.
17. Huang YC, Chiang MH, Wang SJ, Fossum JG (2017) GAAFET versus pragmatic FinFET at the 5nm Si-based CMOS Technology node. *IEEE J Electron Devices Soc* 5:164–169 https://doi.org/10.1109/JEDS.2017.2689738.
18. Kumar B, Kumar A, Chaujar R (2020) The effect of gate stack and high-κ spacer on device performance of a junctionless GAA FinFET. IEEE VLSI Device, Circuit Systems Conference 159–163. https://doi.org/10.1109/VLSIDCS47293..2020.9179855.
19. Ansari L, Feldman B, Fagas G, Colinge JP, Greer JC (2010) Simulation of junctionless Si nanowire transistors with 3 nm gate length. *Appl Phys Lett* 97:062105. https://doi.org/10.1063/1.3478012.
20. Colinge JP, Lee CW, Afzalian A, Akhavan ND, Yan R, Ferain I, Razavi P, O'Neill B, Blake A, White M, Kelleher AM, McCarthy B, Murphy R (2010) Nanowire transistors without junctions. *Nat Nanotechnol* 5:225–229 https://doi.org/10.1038/nnano.2010.15.

21. Gupta N, Kumar A (2021) Numerical assessment of high-k spacer on symmetric S/D underlap GAA junctionless accumulation mode silicon nanowire MOSFET for RFIC design. *Appl Phys A Mater Sci Process* 127:76. https://doi.org/10..1007/s00339-020-04234-6.
22. Biswas K, Sarkar A, Sarkar CK (2018) Fin shape influence on analog and RF performance of junctionless accumulation-mode bulk FinFETs. *Microsyst Technol* 24:2317–2324 https://doi.org/10.1007/s00542-018-3729-1.
23. Kim TK, Kim DH, Yoon YG, Moon JM, Hwang BW, Moon DI, Lee GS, Lee DW, Yoo DE, Hwang HC, Kim JS, Choi YK, Cho BJ, Lee SH (2013) First demonstration of junctionless accumulation mode bulk FinFETs with robust junction isolation. *IEEE Electron Device Lett* 34:1479–1481 https://doi.org/10.1109/LED.2013.2283291.
24. Lee CW, Ferain I, Afzalian A, Yan R, Akhavan ND, Razavi P, Colinge JP (2010) Performance estimation of junctionless multigate transistors. *Solid State Electron* 54:97–103 https://doi.org/10.1016/j.sse.2009.12.003.
25. Kumar B, Chaujar R (2021) Numerical study of JAM-GS-GAA FinFET: A fin aspect ratio optimization for upgraded analog and intermodulation distortion performance. *Silicon*. https://doi.org/10.1007/s12633-021-01395-8.
26. Xu JP, Ji F, Lai PT, Guan JG (2008) Influence of sidewall spacer on threshold voltage of MOSFET with high-k gate dielectric. *Microelectron Reliab* 48:181–186 https://doi.org/10.1016/j.microrel.2007.03.001.
27. Koley K, Dutta A, Syamal B, Saha SK, Sarkar CK (2013) Subthreshold analog/RF performance enhancement of underlap DG FETs with high-k spacer for low power applications. *IEEE Trans Electron Devices* 60:63–69 https://doi.org/10.1109/TED.2012.2226724.
28. Pal PK, Kaushik BK, Dasgupta S (2014) Investigation of symmetric dual-k spacer trigate FinFETs from delay perspective. *IEEE Trans Electron Devices* 61:3579–3585 https://doi.org/10.1109/TED.2014.2351616.
29. Shan C, Wang Y, Luo X, Bao M, Yu C, Cao F (2017) A high-performance channel engineered charge-plasma-based MOSFET with high-κ spacer. *Superlattices Microstruct* 112:499–506 https://doi.org/10.1016/j.spmi.2017.10.002.
30. Gracia D, Nirmal D, Moni DJ (2018) Impact of leakage current in germanium channel based DMDG TFET using drain-gate underlap technique. *AEU - Int J Electron Commun* 96:164–169 https://doi.org/10.1016/j.aeue.2018.09.024.
31. Bhattacharya D, Jha NK (2014) FinFETs: from devices to architectures. *Adv Electron* 2014:21–55 https://doi.org/10.1155/2014/365689.
32. Sjöblom G (2006) Metal gate technology for advanced CMOS devices, Ph.D. dissertation Dept. Engg. Sci., Uppsala Univ., Sweden.
33. Liu Y, Kijima S, Sugimata E, Masahara M, Endo K, Matasukawa T, Ishii K, Sakamoto K, Sekigawa T, Yamauchi H, Takanashi Y, Suzuki E (2006) Investigation of the TiN gate electrode with tunable work function and its application for FinFET fabrication. *IEEE Trans Nanotechnol* 5:723–728 https://doi.org/10.1109/TNANO.2006.885035.
34. Vitale SA, Kedzierski J, Healey P, Wyatt PW, Keast CL (2011) Work-function-tuned TiN metal gate FDSOI transistors for subthreshold operation. *IEEE Trans. Electron Devices* 58:419–426 https://doi.org/10.1109/TED.2010.2092779.
35. ATLAS (2016) *User's Manual*. SILVACO International, Santa Clara, CA.
36. Iannaccone G, Curatola G, Fiori G (2004) Effective Bohm quantum potential for device simulators based on drift-diffusion and energy transport. *Simul Semicond Process Devices* 275–278. https://doi.org/10..1007/978-3-7091-0624-2_64.
37. Slotboom JW, de Graaff HC (1976) Measurements of bandgap narrowing in Si bipolar transistors. *Solid State Electron* 19:857–862 https://doi.org/10.1016/0038-1101(76)90043-5.
38. Crowell CR, Sze SM (1966) Temperature dependence of avalanche multiplication in semiconductors. *Appl Phys Lett* 9:242–244 https://doi.org/10.1063/1.1754731.

39. Hurkx GAM, Klaassen DBM, Knuvers MPG (1992) A new recombination model for device simulation including tunneling. *IEEE Trans Electron Devices* 39:331–338 https://doi.org/10.1109/16.121690.
40. Shockley W, Read WT (1952) Statistics of the recombinations of holes and electrons. *Phys Rev* 87:835–842 https://doi.org/10.1103/PhysRev.87.835.
41. Hall RN (1952) Electron-hole recombination in Germanium. *Phys Rev* 87:387. https://doi.org/10.1103/PhysRev.87.387.
42. Dirac PAM (1926) On the theory of quantum mechanics. *Proc R Soc London Ser A.. Contain Pap a Math Phys Character* 112:661–677 https://doi.org/10.1098/rspa.1926.0133.
43. Lee H, Yu LE, Ryu SW, Han JW, Jeon K, Jang DY, Kim KH, Lee J, Kim JH, Jeon SC, Lee GS, Oh JS, Park YC, Bae WH, Lee HM, Yang JM, Yoo JJ, Kim SI, Choi YK (2006) Sub-5nm all-around gate FinFET for ultimate scaling. Dig Tech Pap - Symp VLSI Technol 58–59. https://doi.org/10.1109/vlsit..2006.1705215.
44. Kumar B, Chaujar R (2021) TCAD temperature analysis of gate stack gate all around (GS-GAA) FinFET for improved RF and wireless performance. *Silicon* 13:3741–3753 https://doi.org/10.1007/s12633-021-01040-4.
45. Doornbos G, Passlack M (2010) Benchmarking of III-V nMOSFET maturity and feasibility for future CMOS. *IEEE Electron Device Lett* 31:1110–1112 https://doi.org/10.1109/LED.2010.2063012.
46. Gupta N, Jain A, Kumar A (2021) 20nm GAA-GaN/Al_2O_3 nanowire MOSFET for improved analog/linearity performance metrics and suppressed distortion. *Appl Phys A Mater Sci Process* 127:1–9 https://doi.org/10.1007/s00339-021-04673-9.
47. Kumar B, Chaujar R (2021) Fin aspect ratio optimization of novel junctionless gate stack gate all around (GS-GAA) FinFET for analog/RF applications. *Microelectronics, Circuits and Systems.. Lecture Notes in Electrical Engineering* 755:59–67 https://doi.org/10.1007/978-981-16-1570-2_6.
48. Pradhan KP, Mohapatra SK, Sahu PK, Behera DK (2014) Impact of high-k gate dielectric on analog and RF performance of nanoscale DG-MOSFET. *Microelectronics J* 45:144–151 https://doi.org/10.1016/j.mejo.2013.11.016.
49. Sharma M, Chaujar R (2021) Design and investigation of recessed-T-gate double channel HEMT with InGaN back barrier for enhanced performance. *Arab J Sci Eng* 47:1109–1116 https://doi.org/10..1007/s13369-021-06157-7.
50. Mohapatra SK, Pradhan KP, Artola L, Sahu PK (2015) Estimation of analog/RF figures-of-merit using device design engineering in gate stack double gate MOSFET. *Mater Sci Semicond Process* 31:455–462. https://doi.org/10.1016/j.mssp.2014.12.026.
51. Naima G, Rahi SB, Boussahla G (2021) Impact of dielectric engineering on analog/RF and linearity performance of double gate tunnel FET. *Int J Nanoelectron Mater* 14:281–303.
52. Kumar A, Tripathi MM, Chaujar R (2017) Investigation of parasitic capacitances of In2O5Sn gate electrode recessed channel MOSFET for ULSI switching applications. *Microsyst Technol* 23:5867–5874 https://doi.org/10.1007/s00542-017-3348-2.
53. Gupta N, Kumar A, Chaujar R, Kumar B, Tripathi MM (2020) Gate engineered GAA silicon-nanowire MOSFET for high switching performance. IEEE VLSI Device, Circuit Systems Conference 258–262. https://doi.org/10.1109/VLSIDCS47293.2020.9179932.
54. Tayal S, Nandi A (2017) Analog/RF performance analysis of channel engineered high-K gate-stack based junctionless Trigate-FinFET. *Superlattices Microstruct* 112:287–295 https://doi.org/10.1016/j.spmi.2017.09.031.
55. Moldovan O, Jiménez D, Guitart JR, Chaves FA, Iñiguez B (2007) Explicit analytical charge and capacitance models of undoped double-gate MOSFETs. *IEEE Trans Electron Devices* 54:1718–1724 https://doi.org/10.1109/TED.2007.899402.
56. Biswas K, Sarkar A, Sarkar CK (2016) Spacer engineering for performance enhancement of junctionless accumulation-mode bulk FinFETs. *IET Circuits, Devices Syst* 11:80–88 https://doi.org/10.1049/iet-cds.2016.0151.

9 Potential Prospects of Negative Capacitance Field Effect Transistors for Low-Power Applications

Shalini Chaudhary, Nawaz Shafi, Basudha Dewan, Chitrakant Sahu, and Menka
Malaviya National Institute of Technology Jaipur

CONTENTS

DOI: 10.1201/9781003240778-9

9.1 INTRODUCTION

At present, human beings are in a way buried in electronic devices. Human beings have not only become dependent on such devices for ease and application but are also obsessed with them. To understand this, we can rely on the simple observation that in last few decades, the number of electronic devices an individual possesses and uses in everyday life has gone up manifold. We are surrounded by microelectronic devices. However, all of these visible devices are only a part of things; for the performance of these visible devices, we have the internet of things in the background like virtual space, massive data centers, cloud computing, etc., which create a huge power demand, as shown in Figure 9.1. All of the developed industries that affect the everyday life of human beings owe their achievements to the advancement of microelectronics. Silicon technology has advanced with an exponential rate in terms of performance and productivity during the past few decades, which was anticipated by Moore in 1965.

Moore's law arises from the fact that the cost decreases by increasing the number of transistors, but it also increases the chance of defects. Moore stated that the number of transistors on an integrated circuit would be doubled every 2 years. This law became a self-fulfilling prophecy for the semiconductor industry and is valid to date, which is evidenced by the International Roadmap for Semiconductors (ITRS) [2]. The scaling in microelectronics has been driven by the demand for a higher-density integration in order to minimize the fabrication cost and also increase the speed. The scaling based on Moore's law did not confront any sever obstacles until these past years, where the transistor dimensions reached the sub-100 nm regime [3]. By increasing the number density of transistors, the power consumption increases exponentially.

The problem started with the advancement of technology and the quest for faster calculations, which translates to higher frequencies. The consumed power was mostly dissipated during the switching of transistors. Hence, increasing the number density of transistors results in an exponential rise in power consumption. It is also anticipated that aggressive scaling will finally stop due to the inability to remove the heat generated by transistors [4].

Therefore, we can see that power is the critical factor for the growth of microelectronics at the microlevel (from the point of view of scaling) and the macrolevel

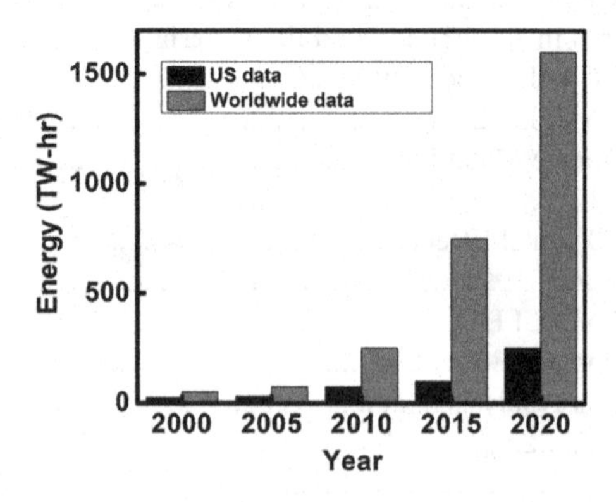

FIGURE 9.1 Energy consumption by data centers [1].

(from the point of view of increasing demand) as shown in Figure 9.2. The only solution to both these problems is the fabrication of low-power electronic devices. For obtaining power-efficient devices, the subthreshold swing (*SS*) of the devices should be reduced. But it is stuck at 60 mv/decade due to the Boltzmann Tyranny limit [6]. According to this tyranny, *SS* cannot be below 60 mV/decade when the transport mechanism is the same as that of the conventional transistors. *SS* is defined as

$$SS = \left[\frac{d \log I_d}{dV_g} \right]^{-1} \quad (9.1)$$

if it is written in terms of surface potential, it becomes

$$SS = \left[\frac{d\psi_s}{dV_g} \right] \left[\frac{d \log I_d}{d\psi_s} \right]^{-1} \quad (9.2)$$

$$= m * n$$

where '*m*' is the voltage amplification, whereas '*n*' depends upon the current flow mechanism (carrier inject phenomenon) of the device. Figure 9.2 describes that static power dissipation is more in comparison to dynamic power dissipation, and this becomes the challenging factor for the CMOS industry. For an ideal switch, I_{on} should be as high as possible and I_{off} should reach zero. Practically, this type of switch is not possible; therefore, we focus on small *SS* switches that have an improved I_{on}/I_{off} ratio over conventional MOSFETs. There are many devices that come under energy-efficient devices such as tunneling field effect transistor (TFET) [7], impact ionization (IMOSFET) [8], and nanoelectromechanical FET (NEMFET), which improves the factor '*n*'. Negative capacitance FET (NCFET) is also a solution for reducing power dissipation and improving the I_{on}/I_{off} ratio [9,10]. It is based on the factor '*m*', which means that it reduced power dissipation by reducing voltage across the device using the concept of negative capacitance.

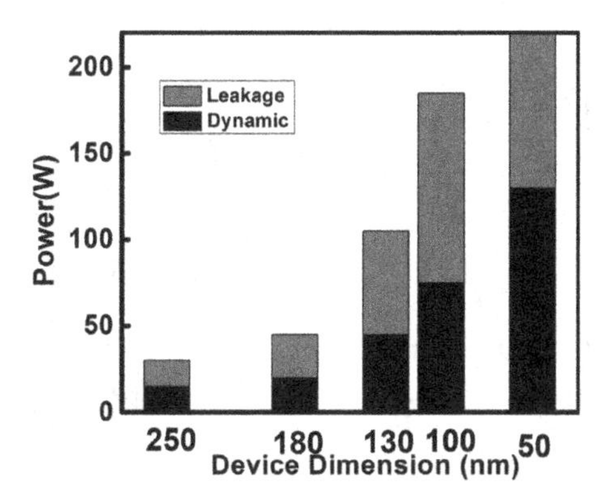

FIGURE 9.2 Static and dynamic power comparison as technology node improves taken from reference [5].

9.2 NEGATIVE CAPACITANCE

To overcome the Boltzman tyranny limit, in 2008, a negative capacitance concept was given by Salahuddin and Datta. According to it, the conventional gate material is replaced with the negative capacitance material. NC is not present in conventional paraelectric material, but it can be obtained using ferroelectric (FE) materials as shown in Figure 9.3. From the energy landscape, we found paraelectric materials having only one energy minima capacitance regions; therefore, they cannot be used to obtain NC effect. But on the other hand FE materials have double well energy diagram, and between two wells there is a region that shows negative capacitance. This NC in FE can be obtained using phenomenological Landau free energy [11]. This phenomenology is based on symmetry and provides a conceptual bridge between the atomic models and the observed megascopic phenomenon. Therefore, it assumes averaging of local fluctuations, and due to this, it is appropriate for long-range interactions as in ferroelectronics.

According to this, the Gibbs free energy density (U) of the FE layer can be expressed in the powers of polarization (P) in the vicinity of a phase transition as:

$$U = \alpha t_f P^2 + \beta t_f P^4 + \gamma t_f P^6 - V_f P \tag{9.3}$$

Here α, β, and γ are constants (Landau parameters), which are material-dependent, and t_f is the thickness of ferroelectric film. β can be both positive and negative depending upon phase transition. For a second phase transition, it is positive while for the first transition, it is negative. γ is always a positive quantity. α is a temperature-dependent term while others are temperature-independent; therefore, α is defined as,

$$\alpha = a_0(T - T_c) \tag{9.4}$$

where a_0 is positive, T is the temperature, and T_c is the Curie temperature. FEs behave in the negative capacitance region for a negative value of α only. For this, the temperature is always less than the Curie temperature so that its energy landscape remains in double well energy form as shown in Figure 9.4.

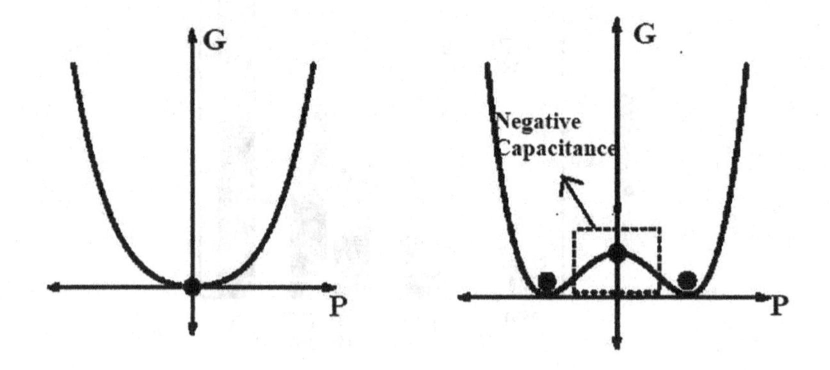

FIGURE 9.3 Energy landscape for paraelectric and FE materials.

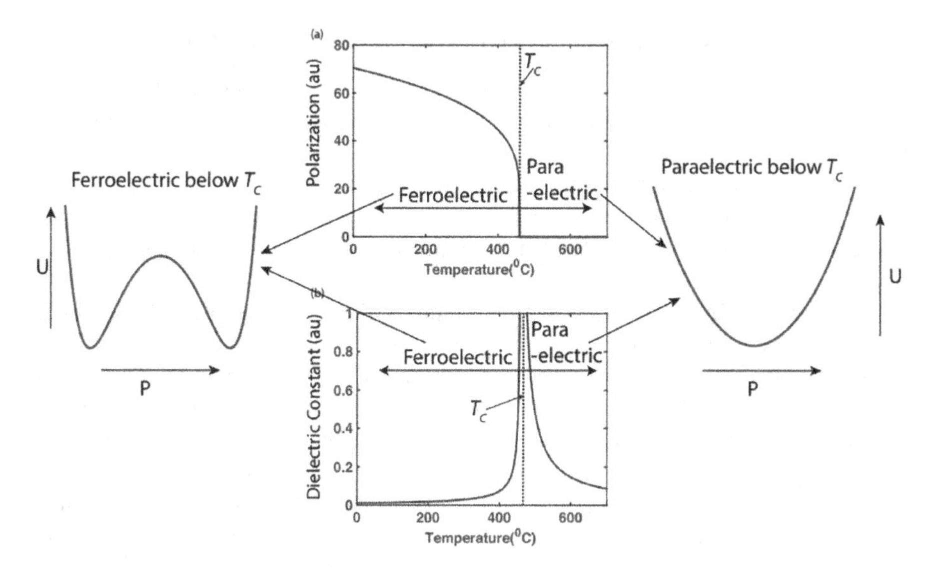

FIGURE 9.4 Evolution of the spontaneous polarization (a), the dielectric constant (b), and the energy landscape (c) as functions of temperature for a FE material with a second-order phase transition.

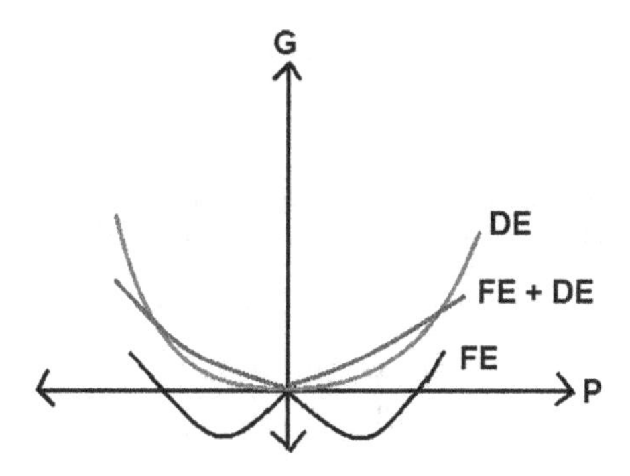

FIGURE 9.5 Energy landscape showing the stable state of a FE capacitor.

This spontaneous polarization (P) for the FE materials can be taken as Q (charge). Voltage across the FE layer can be obtained by taking the derivative of this Gibbs free energy. This Q–V curve is related to the P–V curve of the FE material. By taking the derivative of this Q with the voltage, we can obtain capacitance across the FE layer.

Nevertheless, NC effect is only present in a metastable state of the Landau free energy with negative curvature. But it is not stable itself in FE capacitor. To obtain this NC with a stable state, a paraelectric capacitor is connected in series with a FE capacitor so that the total free energy of the system has a minima in the negative capacitance regime of the FE as shown in Figure 9.5.

9.2.1 Negative Capacitance FETs for Steep Subthreshold Slope

NC reduces the *SS* and makes the device energy-efficient. To understand this, we will consider the transistor as a series combination of the capacitances as in Figure 9.6. One is the gate capacitance

(C_{ox}) and the other semiconductor capacitance (C_s). Both C_{ox} and C_s are positive; hence, the equivalent capacitance of FET will be smaller than both the individual capacitances. On the other hand, when C_{ox} is taken as negative, the equivalent capacitance would be greater than C_{ox} when $C_{ox} > C_s$.

This increase in resultant capacitance reduces the power supply voltage of the device. As the value of the resultant capacitance increases, the voltage will be reduced for generating the same amount of charge Q across C_{ox} and C_s. The device current takes the charge across the C_s. Now the value of the voltage is reduced to produce the same amount of current. Due to the NC effect, the surface potential (ψ_s) will be greater than the gate voltage (V_g), and the derivative of ψ_s with respect to *Vg* will be greater than 1. This *ddVψgs* ratio for conventional transistors is less than 1, and we cannot go beyond the *SS* limit of 60 mv/decade. As ${}^{d}{}_{dV}{}^{\psi}g^{s} > 1$, the minimum value of voltage is reduced to below 60 mV. This can be better explained with the help of *SS*. The region below which current saturates is defined as the subthreshold region, and the *SS* defines the steepness of the curve, which implies how sharply the current increases with voltage. The *SS* is defined as

$$SS = \left[\frac{d \log I_d}{dV_g} \right]^{-1} \tag{9.5}$$

where I_d and V_g are the drain current and gate voltage of the transistor, respectively. This equation can also be written in terms of surface potential of the semiconductor channel (ψ_S).

$$SS = \left[\left(\frac{d\psi_s}{dV_g} \right) \left(\frac{d \log I_d}{d\psi_s} \right) \right]^{-1} \tag{9.6}$$

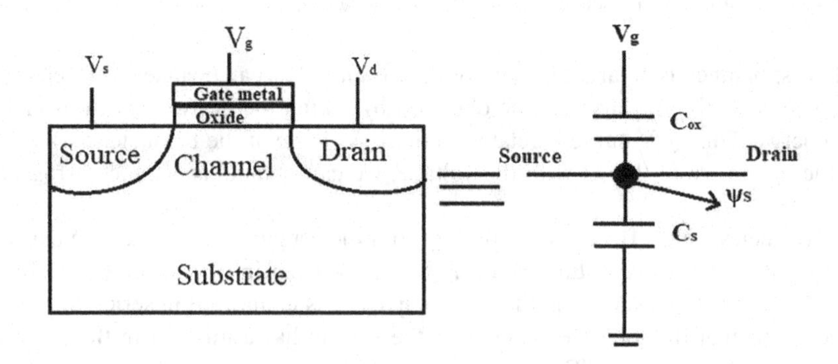

FIGURE 9.6 Schematic diagram of bulk NCFET and its equivalent capacitance model.

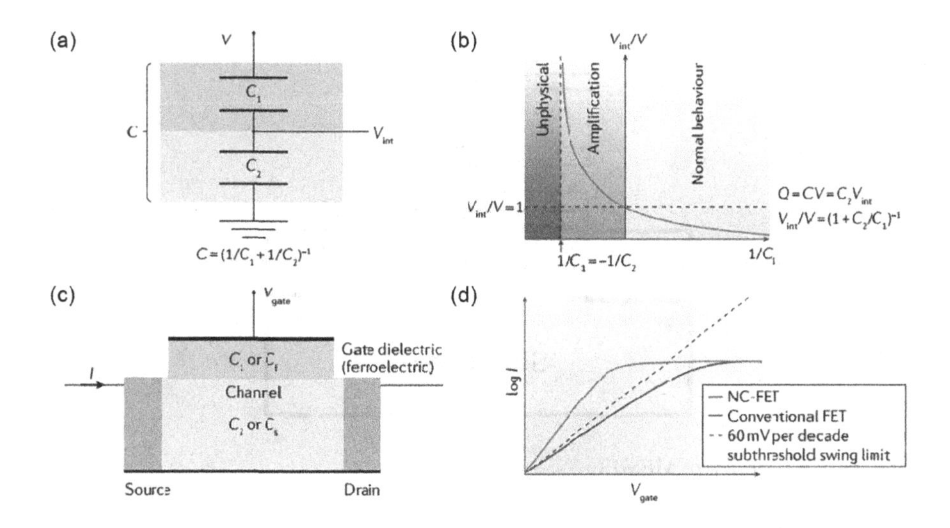

FIGURE 9.7 (a) Schematic of two dielectric capacitors C1 and C2 connected in series represented through internal voltage V_{int}. (b) Describing regime of negative capacitance (NC) that represents voltage enhancement. (c) Representative capacitance model for the NCFET. (d) Steep SS using NC effect taken from reference [12].

There are two terms which define the SS. First is '*m*' and the other is '*n*'. '*n*' is determined by Boltzmann factor, and at room temperature, its value is 60 mV/decade. The factor '*m*' (Body factor) in NC, which is the ratio of change in surface potential and gate voltage, is improved. This body factor can also be written in terms of capacitance as given in equation (9.7).

$$m=\left[1+\frac{C_s}{C_{ox}}\right] \tag{9.7}$$

For conventional transistors, it is greater than 1. But in NC, it is less than 1. Due to this, *SS* is reduced to below 60 mV/decade as explained in Figure 9.7.

9.2.2 Device Integration for NCFET (MFMIS and MFIS)

The NCFET is diagnosed by two structures: MFMIS and MFIS [13,14]. MFIS is metal FE insulator semiconductor and MFMIS is metal ferroelectric metal insulator semiconductor as given in Figure 9.8. In MFMIS, there is an internal metal layer that is present between gate oxide and the ferroelectric layer. This integration is mainly used for the experimental presentation. Earlier, the MFMIS structure was mostly used. Due to this integration, a NCFET is a series combination of two capacitors: one is across MFM and other is MOSFET capacitance. Both can be independently designed and fabricated. In MFMIS, the metallic layer averages out the nonuniformity across the ferroelectric layer, which is due to the domain formation in the FE layer, and we get a constant voltage across it, which is directly measurable for amplification purpose.

But when NCFET goes beyond the sub-20 nm gate length for the advanced CMOS technologies, there is no area where the metal layer can be placed. Therefore, the

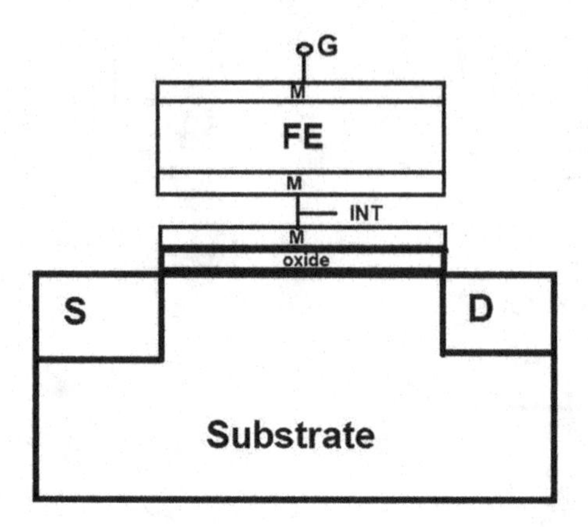

FIGURE 9.8 Design of MFMIS.

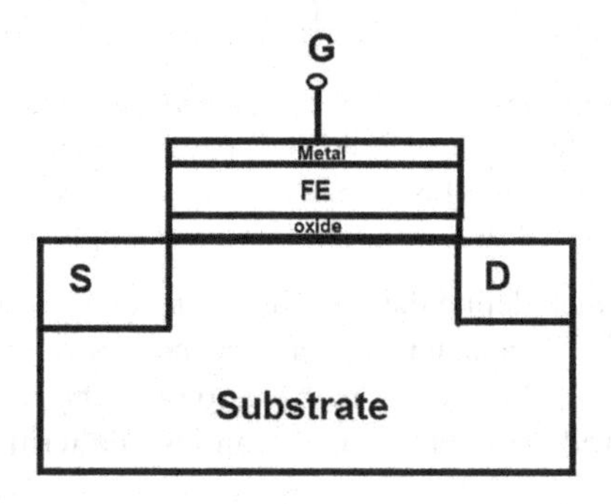

FIGURE 9.9 Design of MFIS .

MFIS structure becomes a feasible choice as in Figure 9.9. Here, the FE material is directly imposed over dielectric material. It overcomes the gate leakage effect over stabilization of FE oxide with gate oxide. At a particular drain-to-source voltage (V_{ds}), there is difference of arrangement of polarization during domain formation along with the channel direction. MFMIS has uniform potential over the channel with the presence of the internal layer, whereas MFIS does not have uniform potential along the FE layer, and surface potential across the source region is amplified. The simulation of both structures are carried out by considering the multidomain array as a one-dimensional array. If there is no gate leakage in the MFMIS structures, then they are better in terms of subthreshold characteristics over MFIS.

9.2.3 Capacitance Matching

NCFET is considered as the conventional MOSFET with an added amplifier as shown in Figure 9.10. For this equivalent model, amplification factor (β) can be written as

$$\beta = \frac{V_{\text{int}}}{V_g} = \frac{C_{FE}}{C_{FE} + C_{\text{int}}}$$

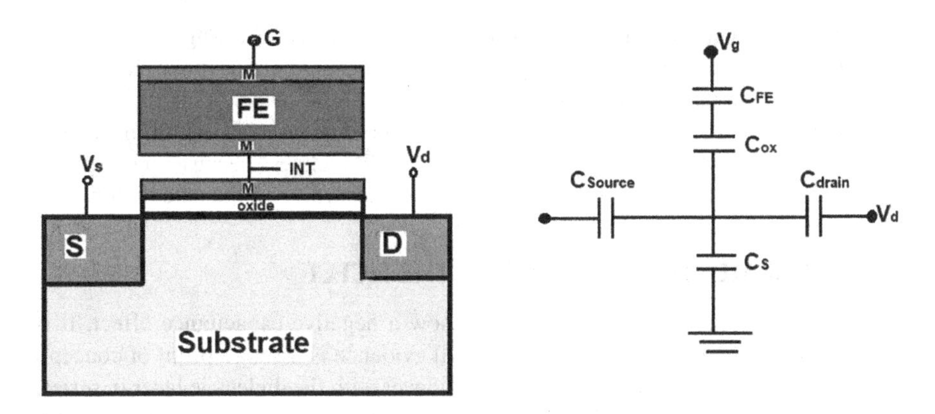

FIGURE 9.10 Simple MOSFET and its equivalent capacitance model.

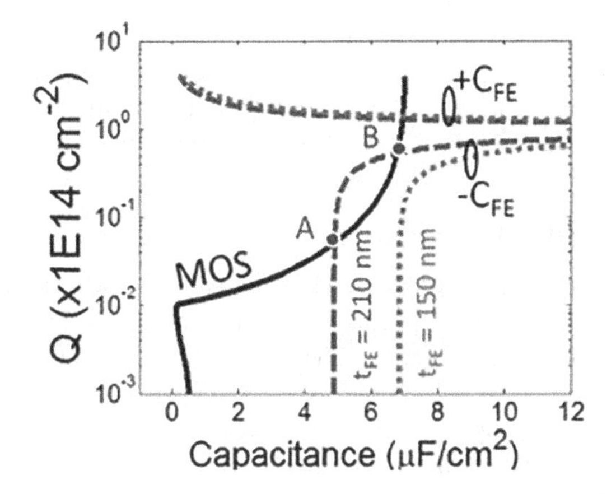

FIGURE 9.11 Charge density versus capacitance that describes capacitance matching of the device [17].

It is considered that β is greater than 1 in the case of NCFET. To obtain a nonhysteresis condition along with $\beta > 1$, there should be capacitance matching between the MOSFET capacitance and ferroelectric capacitance as in Figure 9.11. The following are conditions for operating in the NC region.

1. Value of the ferroelectric capacitance must be greater than MOSFET capacitance (series combination of MOSFET capacitance and oxide capacitance).
2. Ferroelectric capacitance value should be close to that of MOSFET capacitance, and the total capacitance should be positive in the operation.

These two conditions can be obtained by either adjusting the value of MOSFET capacitance or the FE capacitance [15,16]. MOSFET capacitance has a small value of capacitance because of the low density of the space charge in the depletion layer.

It can be improved by using different kinds of channel engineering. Ferroelectric capacitance can be improved by adjusting the value of remnant polarization (P_r) and coercive field (E_c) because for capacitance matching, steeper SS is required. To achieve this, value of P_r should be small and E_c should be large for obtaining a compressed P–V curve. It can also be improved by the FE thickness because in integrated NCFET, the area of ferroelectric capacitor is fixed according to the gate dimension.

9.3 EXPERIMENTAL EVIDENCE OF THE NCFET

There are many experimental works that show a negative capacitance effect. It is divided into two parts. The first experimental evidence is through proof of concept. It is examined by connecting the ferroelectric layer with the dielectric layer in series. Second evidence is obtained through device implementation. In this, reduction in SS is directly obtained using a FE layer in the gate. These above two attempts shows static NC action. Static means when external voltage is applied across the ferroelectric layer there is monotonic increase in P as V increases and the negative capacitance region cannot be accessed. For assessing the NC region, transient analysis is to be done. In this, direct measurement of voltage change across the FE layer is obtained when external voltage is applied. For these types of experiments, the FE capacitor should be isolated, and for this, the MFMIS approach is used.

9.3.1 Proof of Concept

NC was first observed by Bratkovsky and Levanyuk. They used experimental data on ultrathin $SrRuO_3/BaTiO_3/SrRuO_3$ capacitors and replotted the hysteresis curve with voltage across the ferroelectric layer. The curve obtained shows negative slope as shown in Figure 9.12.

The external voltage applied across the ferroelectric capacitor is divided across the ferroelectric layer and the metal FE interface capacitance, which arise due to the finite length at interface.

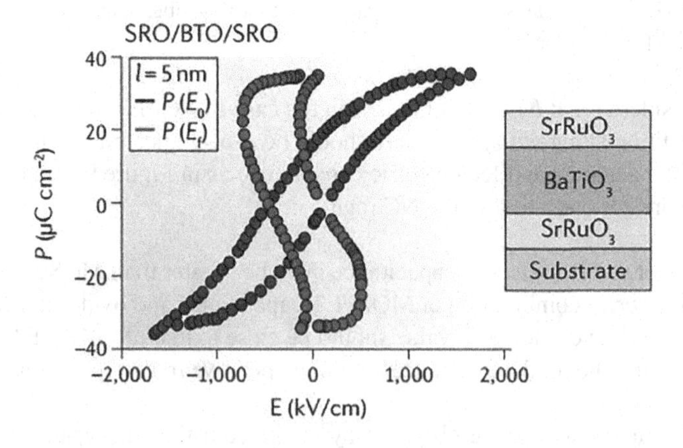

FIGURE 9.12 5 nm-thick epitaxial $BaTiO_3$ (BTO) capacitor with $SrRuO_3$ (SRO) electrodes indicating the NC region with multidomain effects taken from reference [18].

In Figure 9.12, black curves are between polarization and the external electric field (E_0) while the red curve is between polarization and voltage across the FE layer and shows the NC region. A high-quality 5 nm-thick epitaxial $BaTiO_3$ (BTO) capacitor with $SrRuO_3$ (SRO) electrodes is used to obtain NC.

In 2011, Islam Khan, A. et al. compared the capacitance of a $SrTiO_3$ layer with that of a Pb(Zr, Ti)O_3/$SrTiO_3$ bilayer and found that in a certain temperature range, the bilayer had higher capacitance than the $SrTiO_3$ layer [17]. Similar capacitance enhancements were later reported in $BaTiO_3$/$SrTiO_3$ capacitors [19], (Ba, Sr)TiO_3/$LaAlO_3$ superlattices [20], and $BaTiO_3$/Al_2O_3 capacitors [21].

To overcome the domain effects, simple single-domain study has been carried out. Zubko, P. et al. explicitly found that $PbTiO_3$/$SrTiO_3$ and $Pb_{0.5}Sr_{0.5}TiO_3$/$SrTiO_3$ show NC in multidomain structures. These multidomain structures are modeled as capacitors in series and are complex owing to the strong electrostatic interactions between the dielectric and FE layers [22].

9.3.2 Device Implementation

Experiments to prove steep slope operation are performed using two approaches. First is by taking the FE layer in the gate stack and the external connection of the FE capacitor with FET. Organic poly(vinylidenefluoride trifluoroethylene) gate FEs and perovskite PZT gates were explored for direct integrations. Although these oxides can be directly matured on semiconductors, the adverse band adjustment and surface states of metal and FE need a dielectric buffer layer. Hence, in the second approach, this dielectric buffer layer is applied and the structure looks like MFIS and MFMIS. In the last few years, we found FE properties in doped hafnium oxide. These materials have excellent compatibility with the semiconductor processing technique and so become the most favorite choice for NC devices.

Recently better experimental results have been obtained for 14 nm bulk FinFET technology as shown in Figure 9.13. Results show better SS as compared to conventional MOSFET [23]. There is also no reliable degradation present. Although further characterization and optimization will be needed to verify the NC effect, this is a great step toward manufacturable NCFET technology.

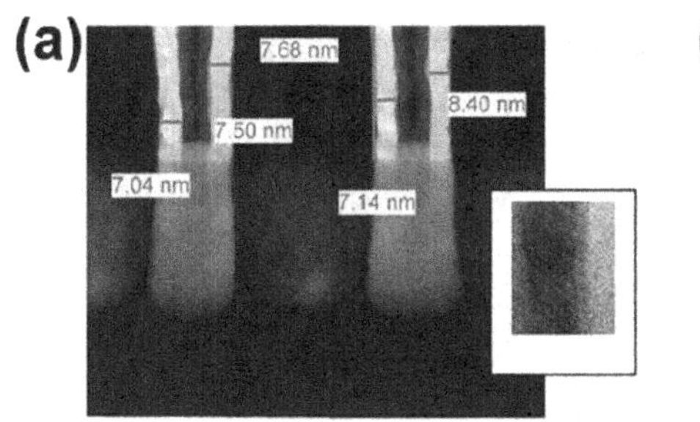

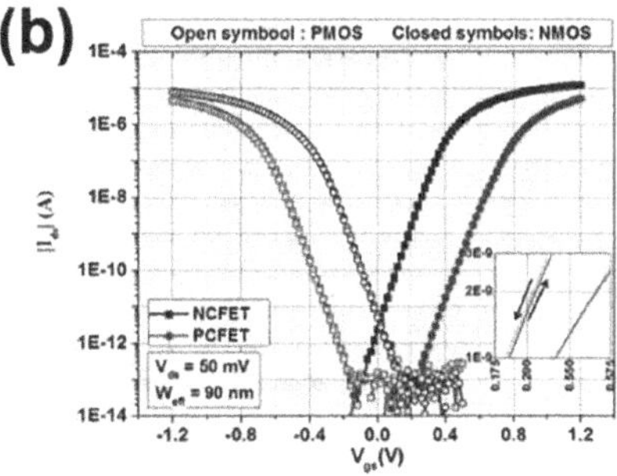

FIGURE 9.13 (a) Cross-sectional TEM image of an NC FinFET based on 14 nm bulk FinFET technology. (b) Measured Id–Vg curves of NCFET and reference MOSFET taken from reference [23].

9.3.3 Transient NC

In transient analysis, voltage across the FE layer is directly measured during the switching process. There are two ways by which we perform transient analysis. First is using the MFMIS type of structure. In this, voltage at the intermediate node is directly monitored during the change in the gate voltage. In this way, the NC region is obtained in the hysteresis curve plotted between polarization and FE voltage.

Another method is to sweep the charge in place of the voltage. For this, a transient system is considered in which a resistance is taken in series with the FE capacitance as shown in Figure 9.14. From this resistance, current and charge can be regulated and the voltage is obtained across FE capacitance. Charge is calculated by integrating the current in the circuit, and then the graph between charge and voltage is plotted as given in Figure 9.15. The S curve is not fully derived from this analysis, but some part can be obtained to prove NC. There are many analyses that use this method, but it remains controversial because a realistic model should include a domain-oriented process to generate NC effects.

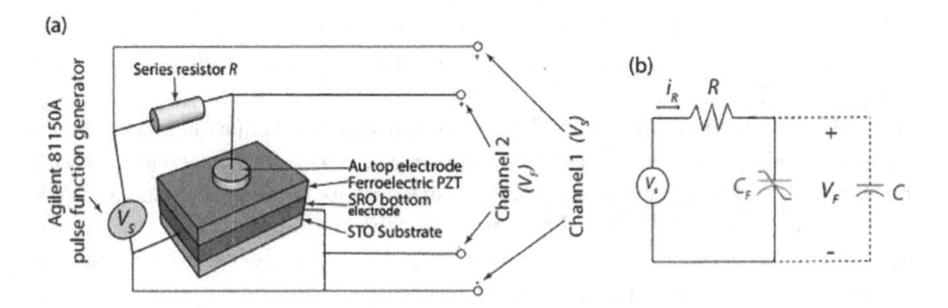

FIGURE 9.14 (a) Experimental setup for transient analysis and (b) its equivalent capacitance model taken from reference [24]

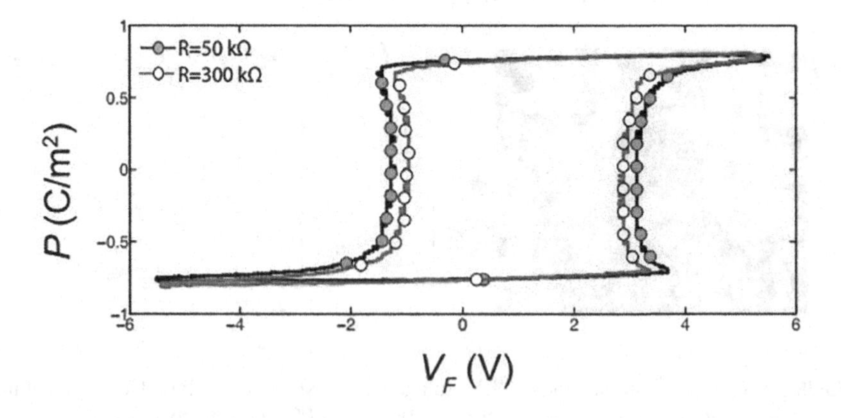

FIGURE 9.15 P–V curve obtained by transient analysis [24].

9.4 ADVANTAGES AND APPLICATIONS

NCFET is a rising steep-SS device. With the discovery of HfO_2-based ferroelectrical materials, it becomes a most promising device. It has many advantages over the conventional devices and other steep-SS devices. These are listed below.

9.4.1 Advantages

Conventional MOSFET uses the drift and diffusion current mechanism for the operation of the device. As scaling increases, many SCEs arise, and to overcome them, other devices, like double-gate FET, junctionless FET (JLFET) [25–27], FinFET, etc., were fabricated. But these were unable to reduce the power issue of the device and therefore new devices like TFET, IMOSFET, NEMFET, etc., were fabricated [28]. The transport phenomenon of these devices are different from that of conventional devices such as TFET, IMOSFET, etc., and due to this, the analysis become somewhat different and complex for these devices. There is a need for such devices that have the same transport mechanism and also solve the power supply issue, and NCFETs solve both of these.

Due to the same transport mechanism, they also have high ON current and improve the I_{on}/I_{off} ratio. There is a symmetric operation between the source and drain for this device. It is also a low-cost device because of its easy fabrication approach and no complicated material requirement. Table 9.1 represents advantages of different devices over the NCFET.

9.4.2 Applications

NCFET is a device that can be combined with any baseline devices such as bulk FET, FDSOI FET, JLFET, DGFET, FinFET, etc. and uses the advantage of these baseline devices in application areas also. It has steep SS, better I_{on}/I_{off} ratio, overcomes SCEs such as DIBL and V_{th} roll-off, has a higher noise margin(NM), and low

TABLE 9.1
Comparison of Different Devices Over NCFET

Device	Tech. (L_g)	SS (mV/dec)	I_{ON} (μA/μm)	I_{OFF} (A/μm)	Advantages	Ref.
Conv. MOSFET	20 nm	96	629	18×10^{-9}	—	[29]
SOI FET	20 nm	86	889	18×10^{-9}	Optimize power/ performance trade-offs	[29]
Multigate MOSFET	20 nm	77	870	18×10^{-9}	Reduce SCEs	[29]
JLFET	1 μm	150	100	50.0×10^{-6}	Easy fabrication	[30]
TFET	13 nm	60	10.0	1.0×10^{-9}	Used as a biosensor	[31]
NCFET	100 nm	57–59	852.0	×10−9	Low-power devices	[15]

TABLE 9.2
Application Area of NCFET

Application Area	Specifications	Ref.
Resistive load inverter	$NM_H = 0.584$, $NM_L = .0551$ Power dissipation up to 24.6% compared to simple MOSFET-based inverters	[32]
Schmitt trigger inverter	$V_hys = 0.0743$, $V_low = 0.272$, $V_high = 0.332$ Only two NCFETs required for Schmitt trigger	[34]
Latch comparator	In the case of NCFET-based circuits, speed increases due to formation of another positive feedback	[36]
Gates using NCFET (NAND, NOR)	Power delay product (PDP) of NCFET-based gates = 29/*micro*Wps PDP for simple MOSFET-based gates = 38/*micro*Wps	[37]
NCFET-based flipflops	Vary drain and body biasing voltage to set the hysteresis region	[38]

power dissipation; therefore, it can be used in digital applications such as resistive load inverter [32], ring oscillators [33], etc.

It overcomes the drain-induced barrier lowering (DIBL) effect and generates drain-induced barrier rising (DIBR) in place of it. This DIBR produces an effect of negative differential resistance, which can be used in analog fields [34]. Here, in Table 9.2, there are some applications that use the concept of benefits and are found to be more advantageous over baseline devices.

9.5 RECENT RESEARCH IN NCFET-RELATED FE MATERIALS

FE materials were first discovered by J. Valasek in 1921 by working on the Rochelle salt 39. There was little or no use of these materials in applications, but in 1943 with the discovery of barium titanate ($BaTiO_3$), their use increased in the field of making capacitors for the electronics need. Before 1960, the main challenge in the FE material was related with its modeling for operation purpose. We needed some new novel FE material that could be integrated with silicon-integrated circuits [35].

After 1960, with the development of the FE thin films, a new way was found for integrating this type of materials in electronic components. The main property of polarization reversal in FEs is used in the electronic industry. FE materials have a hysteresis curve when between polarization (P) and coercive field (Ec) or voltage are plotted. In this curve, after a particular value of Ec, the polarization curve reverses. There are many fields wherever these FE materials are used, such as in memory applications, digital application fields, FE transistors (make them steep slope), substrate film interface, etc. FE are those materials that have a permanent dipole moment and spontaneous polarization (due to the atomic arrangement of ions in the crystal structure). When an electric field is applied in FE materials, then stable states of this material are obtained, as shown in Figure 9.3, having different energy levels due to polarization energy and wells tilted with the applied electric field. Polarization does not change immediately because there are many domains in the FE materials and time is taken to overcome them; in this way, a hysteresis curve is obtained. FE materials are explained with the help of Landau Ginzburg Devonshire theory (L-K) as explained in Section 9.2. The basis of negative capacitance is the presence of FE material. As the demand of device miniaturization is increased, the demand of thin films is inflated. There are various materials and deposition techniques that provide thin FE films such as ceramics, liquid crystals, and polymers. In liquid crystals, the crystal lattice has unit cells that contains = atoms and ions [39]. In polymers, the unit cell contains macro-molecules that form the lower symmetry [40]. Most studied and useful FE material perovskite structure materials Pb(Zr, Ti)O_3 (PZT), $BaTiO_3$ (BTO), and SBT, were obtained after some time. These thin-film materials are used in areas such as FE random access memories (FeRAM).

PZT (lead zirconate titanate) has poor FE properties but is better in terms of high spontaneous polarization, low leakage, and high retention time. But PZT does not have a CMOS-compatible fabrication process due to which the interface between FE and the silicon substrate is incompatible in most of the cases. Another issue related with PZT is the scaling of this material. It cannot be scaled down below the 130nm technology; therefore, there is a need for such types of FE materials that can overcome these issues and are compatible with CMOS technology [41].

9.5.1 NCFET with Doped HfO_2

The discovery of doped high-K oxides such as HfO_2 has created a new path for CMOS-compatible FE devices. There are many advantages of using these, such as obtaining ferroelectricity in thin films up to 10nm without any leakage current [42]. The coercive field in the doped HfO_2 is high; therefore, a large window is available for memory application even for very thin films. It is thermally stable in the 600°C–1,000°C range. NC is found in this also by transient analysis as shown in Figure 9.16a. With the discovery of FE-HfO_2, NCFET becomes a promising candidate in the field of steep-SS devices. It has a high I_{on}/I_{off} ratio, is CMOS-compatible, and has the same current mechanism as that in conventional MOSFET.

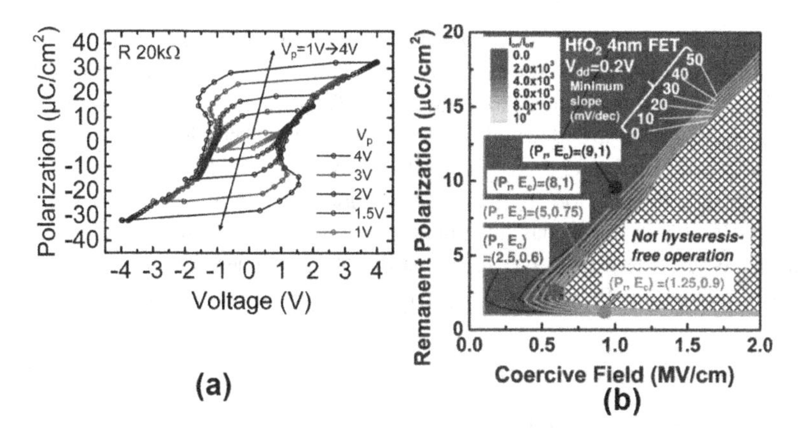

FIGURE 9.16 (a) P–V curve obtained by transient analysis for FE-HfO_2 taken from reference [43]. (b) Simulated NCFET curve using FE-HfO_2 between E_c and P_r taken from reference [44].

There are many analyses that show the superiority of NCFET when using HfO_2 material, as shown in Figure 9.16b, which is a contour plot in terms of remnant polarization and coercive field and shows a higher I_{on}/I_{off} ratio than conventional FETs. Post 2015, many experimental papers presented steeper SS in NCFET with FE-HfO_2.

9.6 CHALLENGES IN NCFET DESIGN

NCFET is a good candidate in the field of steep-SS devices. But there are many issues related to its manufacturing in order to make it comparable with CMOS technologies. These are capacitance matching, reliability, SCEs, irregularity, operation speed, multidomain effect, and transient analysis.

9.6.1 Capacitance Matching

An important feature of NCFET is the presence of negative capacitance using FE materials and the operation depends upon perfect capacitance matching. To maximize the voltage amplification phenomenon, the FE capacitance should be as negative as possible. To achieve this, we have to reduce the temperature at which the multidomain state appears and bring it close to the temperature at which the monostate domain state appears. This depends upon the domain wall mobility and the polarization property of the FE material [45]. Another factor that effects NC is the larger domain wall in FE in comparison to the size of the domain. Therefore, there is the need of controlling the properties and structure of the FE materials to enhance the NC. The interface between the FE layer and the dielectric layer also plays an important role in capacitance matching. When the FE layer is placed over the DE layer there may be chemical effects such as polar instabilities [46]. Effects generated through it can produce other possibilities to tune the NC.

9.6.2 Reliability

There is an inbuilt challenge of reliability in the FE materials itself due to the automatic motion of ions, which degrades the characteristics such as imprint and fatigue [47]. This characteristic degradation does not affect memory application too much, but logic applications are affected due to the high endurance level. Endurance level for memory is defined as approximately 10^4–10^5 when using FE-HfO_2 materials [48]. This endurance level is too high for use in logic applications. The recent material HfO_2 used as FE material has an inherent property of oxygen vacancy, and ferroelectricity is generated from the anions of the oxygen. Therefore, overcoming these effects at the manufacturable level is the main challenge for its reliability.

9.6.3 Short-Channel Effects

When the device is scaled down, SCEs came into play. NC affects the scaled devices because capacitance matching becomes complex due to the presence of parasitic capacitances and other coupling capacitances. But this complex matching is good for NCFET because due to this, we obtained steep SS even at a small gate length. These parasitic capacitances have larger values than the substrate capacitances; therefore, their effect improves capacitance matching. SCEs such as DIBL and V_{th} roll-off in conventional MOSFET become DIBR and V_{th} roll-up in NCFET. In the subthreshold region for the same surface potential, fewer charges are generated across the FE layer. Due to this, voltage across the FE layer decreases as the gate length decreases. In this way, coupling capacitance across the drain increases and we obtain negative DIBL or DIBR. This DIBR supports negative differential resistance (NDR). Hence, SCEs improve; however, if they keep increasing, then the circuit design has to be reconsidered.

9.6.4 Irregularity

Steep-SS devices are susceptible to process variations. Due to the process variation in NCFET, a small change in threshold voltage results in a large variation in SS and off currents (leakage current) as compared to conventional MOSFETs. This is due to the presence of the FE material, which has a polycrystalline structure. This polycrystalline structure produces fluctuations in the placement of the ions. But some studies show that the NC phenomenon reduces the process variability in the FET structure such as FinFET [49]. In NCFET, surface potential is amplified when gate voltage is applied. But if the surface potential varies due to the process variation generated through FE materials, then there is a challenge in obtaining amplification in voltage.

9.6.5 Speed of NCFET

From the concept of NCFET, energy-efficient switching is possible. But in FE materials, ferroelectricity is generated through domain formation and displacement of ions, which restrict the frequency range up to the microwave frequency. The speed of NCFET is limited due to the polarization present in FE materials. To overcome it, dynamic analysis of NCFET should be done to predict the performance of NCFET. This can be done using L-K equations in a dynamic mode as given below:

$$\rho \frac{dp}{dt} = -\frac{dG}{dP}$$

where ρ is resistivity during switching and helps in finding the dynamic response of the FE materials. G is energy density as given in equation (9.3). ρ can be calculated using the transient analysis as given in Section 9.3. Multidomain effects can also be introduced by considering domain interaction energy. Studies shows that in NCFET, the operating frequency is obtained for less than 10 MHz, which is suitable only for low-speed devices. Some experimental studies also show hysteresis at a higher gate voltage sweep, which also satisfied the dynamic model of L-k [50]. ρ needs to be small for use in high-speed applications, and for this, the size of the sample taken for transient analysis should be in the micron range. There is a challenge of accurately characterizing a higher-frequency response of polarization and extract intrinsic dynamic parameters to estimate the operation speed of the NCFET.

9.6.6 Multidomain Effect

In FE materials, multidomains are present, but the analysis of NCFET is done by considering a single domain using the Landau free energy model. According to this model, NC is only present when there is strong interaction in different walls of domains so that they can be considered as the single domain. Through this, all characteristics of NCFET, like steep SS, DIBR, high on-to-off-current ratio, etc., are obtained. Nevertheless, it is natural that multidomains and antiparallel configuration are present in FE materials, and according to the physics of FE materials, multidomains should be considered for practical purposes.

9.6.7 Transient Analysis

NC is also obtained by transient analysis other than examining the S curve present in the P–V curve. Therefore, there is a need for a physical model other than the L–K model for the transient NC effect. Transient effect is also experimentally studied as defined in Section 9.3. There is a window where steep SS is obtained in transient analysis. There are two mechanisms that limit this window. First, when the sweeping time is very fast and polarization switching is unable to follow it, then there is no steep SS obtained in the sweeping time of gate voltage. Second, when switching time is slow, then free charge carriers are obstructed through polarization charges and no steep SS is obtained. But when both sweeping time and polarization switching time are comparable, then steep SS is obtained due to partial screening of the polarization charges. There are two models that are used for transient analysis. One is a time-dependent nonlinear capacitor and resistor model [51] and the other is a polarization switching delay model [52]. Both these models provide inappropriate quasistatic parameters for the Id–Vg test for fast-speed NCFET; therefore, there is a need for new test methods or models to estimate steep SS.

9.7 CONCLUSIONS

Negative capacitance has been experimentally seen in FE thin films, paving the way for revolutionary device technologies. NCFET is a symmetric device in terms of source and drain operation, which is advantageous in the development and manufacture of novel device technologies. Although NCFET is a good contender for a steep slope transistor, process integration has proven to be a challenge. Conventional FE materials including lead zirconate titanate (PZT) and barium titanate (BTO) as gate insulator materials must be a few hundreds of nanometers thick in order to balance the high polarization charge density and FET channel charge density, which is incompatible with advanced scaled CMOS front-end technologies. Furthermore, heavy metal contamination of the manufacturing line is a risk. Because NCFET may be made by any CMOS technology, from the most sophisticated to established fabrication methods, it can be one of the low-cost solutions for IoT power requirements without focusing on high-cost advance modern-day CMOS technologies. In this chapter, we discussed the need for NCFET over other types of FETs. After the basics of NCFET and its integration, its working principal are discussed in detail.

Capacitance matching, which is the main condition, of NCFET is discussed in detail. Experimental evidences of NCFET are also considered to be proof of concept and are on a transient basis. Advantages and applications area of NCFET are discussed over other FETs. The NC effect can be an effective performance booster at a low cost with the use of FE-HfO_2. Feasibility of NCFET is proved by the demonstration of 14 nm NC FinFET. Challenges regarding NCFET such as capacitance matching, reliability, SCEs, irregularity, operation speed, multidomain effect, and transient analysis are discussed here. For NCFET to become a manufacturable transistor solution to extend CMOS scaling, further research and development will be needed to address and tackle the challenges.

ACKNOWLEDGMENTS

The authors thank the Head, Department of Electronics and Communication Engineering, Malaviya National Institute of Technology, for providing the necessary support for carrying out this work.

REFERENCES

[1] Abdullah-Al-Shafi M, Bahar AN. Cloud computing: An aspect of information system. *International Journal of Applied Information Systems*. 2016;10(4):46–50.

[2] Arden WM. The international technology roadmap for semiconductors—perspectives and challenges for the next 15 years. *Current Opinion in Solid State and Materials Science*. 2002;6(5):371–7.

[3] Bohr M. A 30 year retrospective on Dennard's MOSFET scaling paper. *IEEE SolidState Circuits Society Newsletter*. 2007;12(1):11–3.

[4] Nilsson P. Arithmetic reduction of the static power consumption in nanoscale cmos. In: *2006 13th IEEE International Conference on Electronics, Circuits and Systems*. IEEE; 2006. p. 656–9.

[5] Sakurai T. Perspectives of low-power VLSI's. *IEICE Transactions on Electronics*. 2004;87(4):429–36.

[6] Amrouch H, Pahwa G, Gaidhane AD, Henkel J, Chauhan YS. Negative capacitance transistor to address the fundamental limitations in technology scaling: Processor performance. *IEEE Access*. 2018;6:52754–65.

[7] Yadav M, Bulusu A, Dasgupta S. Two dimensional analytical modeling for asymmetric 3T and 4T double gate tunnel FET in sub-threshold region: Potential and electric field. *Microelectronics Journal*. 2013;44(12):1251–9.

[8] Gopalakrishnan K, Griffin PB, Plummer JD. I-MOS: A novel semiconductor device with a subthreshold slope lower than kT/q. In: *Digest. International Electron Devices Meeting*, IEEE; 2002. p. 289–92.

[9] Salahuddin S, Datta S. Use of negative capacitance to provide voltage amplification for low power nanoscale devices. *Nano Letters*. 2008;8(2):405–10.

[10] Chaudhary S, Sahu C, Simulation study based performance projection of Negative capacitance FET. In: *2020 IEEE 17th India Council International Conference (INDICON)*. IEEE; 2020. p. 1–5.

[11] Lines ME, Glass AM. *Principles and Applications of Ferroelectrics and Related Materials*. Oxford University Press, Oxford, England; 2001.

[12] Tu L, Wang X, Wang J, Meng X, Chu J. Ferroelectric negative capacitance field effect transistor. *Advanced Electronic Materials*. 2018;4(11):1800231.

[13] Ota H, Ikegami T, Hattori J, Fukuda K, Migita S, Toriumi A. Fully coupled 3-D device simulation of negative capacitance FinFETs for sub 10 nm integration. In: *2016 IEEE International Electron Devices Meeting (IEDM)*. IEEE; 2016. p. 12–4.
[14] Ota H, Ikegami T, Fukuda K, Hattori J, Asai H, Endo K, et al. Multidomain dynamics of ferroelectric polarization and its coherency-breaking in negative capacitance fieldeffect transistors. In: *2018 IEEE International Electron Devices Meeting (IEDM)*. IEEE; 2018. p. 9–1.
[15] Yu T, Lu W, Zhao Z, Si P, Zhang K. Effect of different capacitance matching on negative capacitance FDSOI transistors. *Microelectronics Journal*. 2020;98:104730.
[16] Chaudhary S, Dewan B, Sahu C, Yadav M. Effect of negative capacitance in partially ground plane based SELBOX FET on capacitance matching and SCEs. *Silicon*. 2021.
[17] Khan AI, Yeung CW, Hu C, Salahuddin S. Ferroelectric negative capacitance MOSFET: Capacitance tuning & antiferroelectric operation. In: *2011 International Electron Devices Meeting*. IEEE; 2011. p. 11–3.
[18] Kim DJ, Jo JY, Kim YS, Chang YJ, Lee JS, Yoon JG, et al. Polarization relaxation induced by a depolarization field in ultrathin ferroelectric $BaTiO_3$ capacitors. *Physical Review Letters*. 2005;95(23):237602.
[19] Appleby DJR, Ponon NK, Kwa KSK, Zou B, Petrov PK, Wang T, et al. Experimental observation of negative capacitance in ferroelectrics at room temperature. *Nano Letters*. 2014;14(7):3864–8.
[20] Gao W, Khan A, Marti X, Nelson C, Serrao C, Ravichandran J, et al. Room-temperature negative capacitance in a ferroelectric–dielectric superlattice heterostructure. *Nano Letters*. 2014;14(10):5814–9.
[21] Kim YJ, Park MH, Lee YH, Kim HJ, Jeon W, Moon T, et al. Frustration of negative capacitance in $Al_2O_3/BaTiO_3$ bilayer structure. *Scientific Reports*. 2016;6(1):1–11.
[22] Sun FC, Kesim M, Espinal Y, Alpay S. Are ferroelectric multilayers capacitors in series? *Journal of Materials Science*. 2016;51(1):499–505.
[23] Krivokapic Z, Rana U, Galatage R, Razavieh A, Aziz A, Liu J, et al. 14nm ferroelectric FinFET technology with steep subthreshold slope for ultra low power applications. In: *2017 IEEE International Electron Devices Meeting (IEDM)*. IEEE; 2017. p. 15–1.
[24] Khan AI. *Negative Capacitance for Ultra-Low Power Computing*. University of California, Berkeley; 2015.
[25] Rahi SB, Asthana P, Gupta S. Heterogate junctionless tunnel field-effect transistor: future of low-power devices. *Journal of Computational Electronics*. 2017;16(1):30–8.
[26] Rahi SB, Ghosh B. High-k double gate junctionless tunnel FET with a tunable bandgap. *RSC Advances*. 2015;5(67):54544–50.
[27] Rahi SB, Ghosh B, Asthana P. A simulation-based proposed high-k heterostructure AlGaAs/Si junctionless n-type tunnel FET. *Journal of Semiconductors*. 2014;35(11):114005.
[28] Shafi N, Parmaar JS, Porwal A, Bhat AM, Sahu C, Periasamy C. Gate all around junctionless dielectric modulated BioFET based hybrid biosensor. *Silicon*. 2021;13(7):2041–52.
[29] Sun X, Moroz V, Damrongplasit N, Shin C, Liu TJK. Variation study of the planar ground-plane bulk MOSFET, SOI FinFET, and trigate bulk MOSFET. *IEEE Transactions on Electron Devices*. 2011;58(10):3294–9.
[30] Gundapaneni S, Ganguly S, Kottantharayil A. Bulk planar junctionless transistor (BPJLT): An attractive device alternative for scaling. *IEEE Electron Device Letters*. 2011;32(3):261–3.
[31] Avci UE, Morris DH, Young IA. Tunnel Field-effect transistors: Prospects and challenges. *IEEE Journal of the Electron Devices Society*. 2015;3(3):88–95.
[32] Awadhiya B, Kondekar PN, Meshram AD. Effect of ferroelectric thickness variation in undoped HfO_2-based negative-capacitance field-effect transistor. *Journal of Electronic Materials*. 2019;48(10):6762–70.

[33] Mehrotra S, Qureshi S. Analog/RF performance of thin (10nm) HfO_2 ferroelectric FDSOI NCFET at 20nm gate length. In: *2018 IEEE SOI-3D-Subthreshold Microelectronics Technology Unified Conference (S3S)*. IEEE; 2018. p. 1–3.
[34] Seo J, Lee J, Shin M. Analysis of drain-induced barrier rising in short-channel negative-capacitance FETs and its applications. *IEEE Transactions on Electron Devices*. 2017;64(4):1793–8.
[35] Scott JF, De Araujo CAP. Ferroelectric memories. *Science*. 1989;246(4936):1400–5.
[36] Liang Y, Li X, Gupta SK, Datta S, Narayanan V. Analysis of DIBL effect and negative resistance performance for NCFET based on a compact SPICE model. *IEEE Transactions on Electron Devices*. 2018;65(12):5525–9.
[37] George S, Aziz A, Li X, Sampson J, Datta S, Gupta SK, et al. NCFET based logic for energy harvesting systems. In: *SRC TECHCON 2015*; 2015.
[38] Mehrotra S, Qureshi S. Hysteretic behavior of PGP FDSOI NCFETs with hafnium oxide based ferroelectrics. In: *2019 Electron Devices Technology and Manufacturing Conference (EDTM)*. IEEE; 2019. p. 425–7.
[39] Meyer RB, Liebert L, Strzelecki L, Keller P. Ferroelectric liquid crystals. *Journal de Physique Lettres*. 1975;36(3):69–71.
[40] Nakamura K, Wada Y. Piezoelectricity, pyroelectricity, and the electrostriction constant of poly (vinylidene fluoride). *Journal of Polymer Science Part A-2: Polymer Physics*. 1971;9(1):161–73.
[41] Moazzami R, Hu C, Shepherd WH. Electrical characteristics of ferroelectric PZT thin films for DRAM applications. *IEEE Transactions on Electron Devices*. 1992;39(9):2044–9.
[42] Muller J, Boscke TS, Schroder U, Mueller S, Brauhaus D, Bottger U, et al. Ferroelectricity in simple binary ZrO_2 and HfO_2. *Nano Letters*. 2012;12(8):4318–23.
[43] Kobayashi M, Ueyama N, Jang K, Hiramoto T. Experimental study on polarization-limited operation speed of negative capacitance FET with ferroelectric HfO_2. In: *2016 IEEE International Electron Devices Meeting (IEDM)*. IEEE; 2016. p. 12–3.
[44] Kobayashi M, Hiramoto T. AIP Adv. 6, 025113 (2016).
[45] Zubko P, Wojdeł JC, Hadjimichael M, Fernandez-Pena S, Sené A, Luk'yanchuk I, et al. Negative capacitance in multidomain ferroelectric superlattices. *Nature*. 2016;534(7608):524–8.
[46] Stengel M, Vanderbilt D, Spaldin NA. Enhancement of ferroelectricity at metal–oxide interfaces. *Nature Materials*. 2009;8(5):392–7.
[47] Scott JF. *Ferroelectric Memories*. Springer-Verlag. Berlin. 2000.
[48] Muller J, Böscke TS, Müller S, Yurchuk E, Polakowski P, Paul J, et al. Ferroelectric hafnium oxide: A CMOS-compatible and highly scalable approach to future ferroelectric memories. In: *2013 IEEE International Electron Devices Meeting*. IEEE; 2013. p. 10–8.
[49] Lee HP, Su P. Suppressed Fin-LER induced variability in negative capacitance FinFETs. *IEEE Electron Device Letters*. 2017;38(10):1492–5.
[50] Zhou J, Wu J, Han G, Kanyang R, Peng Y, Li J, et al. Frequency dependence of performance in Ge negative capacitance PFETs achieving sub-30 mV/decade swing and 110mV hysteresis at MHz. In: *2017 IEEE International Electron Devices Meeting (IEDM)*. IEEE; 2017. p. 15–5.
[51] Kim YJ, Yamada H, Moon T, Kwon YJ, An CH, Kim HJ, et al. Time-dependent negative capacitance effects in $Al_2O_3/BaTiO_3$ bilayers. *Nano Letters*. 2016;16(7):4375–81.
[52] Jiang B, Computationally efficient ferroelectric capacitor model for circuit simulation. In: *1997 Symposium on VLSI Technology*. IEEE; 1997. p. 141–2.

10 Memory Designing Using Low-Power FETs for Future Technology Nodes

Young Suh Song
Korea Military Academy

Shiromani Balmukund Rahi
MCAET Ambedkar Nagar

Chandan Kumar Pandey
VIT-AP University

Shubham Tayal
SR University

Yunho Choi
University of Texas at Austin

Bijo Joseph
SRM Institute of Science and Technology

Tripuresh Joshi
Govind Ballabh Pant Engineering College

Daryoosh Dideban
University of Kashan

Suman Lata Tripathi
Lovely Professional University

DOI: 10.1201/9781003240778-10

CONTENTS

10.1 INTRODUCTION

In recent years, the IT market has been at the very important turning point. As fifth-generation (5G) technology has been gradually expanded, the demand for big data market has been gradually increased, and new mobile electronic devices such as Apple watches and Samsung Galaxy watches have become popular. Accordingly, memory devices are required to have (1) high data storage capacity and (2) low-power operation. In order to increase the data storage capacity, various new materials have been studied as charge trapping materials (CTMs) for storing information [1–4]. For low power consumption, the method of lowering the equivalent oxide thickness (EOT) has continuously contributed to lowering power consumption [5–7]. Specifically, new methodologies for designing memory devices have been broadly researched with high permittivity materials such as hafnium oxide (HfO_2), aluminum oxide (Al_2O_3, also known as alumina), nanolaminated hafnium-aluminum-oxide (HfAlO), and zirconium oxide (ZrO_2) [8–10].

In the present chapter, we address what kinds of performances are required as next-generation memory technology and what research has been conducted to achieve each performance. In addition, more importantly, trade-off issues in designing next-generation memory devices are also carefully and abundantly covered. After that, this chapter is concluded by discussing the remaining tasks and challenges that have to be solved.

This section is organized as follows. First, the needs for low-voltage and -subthreshold swing (SS) are introduced, and then the roadmap for future memory technology is addressed. Thereafter, the overall workflow of the study is briefly described with a visual flowchart for an intuitive and better understanding. Then, the optimization methodology for memory device is covered with stack-engineering and utilization of low-power FET (especially, tunnel FET). Finally, the remaining tasks and challenges are addressed from various perspectives. In all sections of this chapter, parentheses provide reader-friendly explanations.

10.1.1 Need for Low-Power and Low-Subthreshold Swing

In transistors (e.g., MOSFETs), their performance is generally evaluated by 'how well current is controlled by input gate voltage', which can be expressed as the inverse subthreshold slope or SS given by

$$\mathrm{SS} = \left(\frac{\sigma \log(I_{DS})}{\sigma V_{GS}} \right)^{-1}, \tag{6.1}$$

where V_{GS} is a potential difference between gate and source, and I_{DS} is the drain current [11]. It is desirable to lower SS as much as possible, since a lower SS generally leads to lower power consumption. Lower SS means there is a steep switching between on-state and off-state, and hence a lower on voltage (V_{ON}) and low-power operation can be simultaneously achieved. SS at room temperature can be expressed as follows:

$$\mathrm{SS} = 60 * \left(1 + \frac{C_{dep}}{C_{ox}} \right) [\mathrm{mV/dec}], \tag{6.2}$$

where C_{dep} is the depletion-layer capacitance and C_{ox} is the gate dielectric capacitance [11]. Therefore, SS has a lower limit of 60 mV/decade at room temperature, and it is desirable to lower C_{ox} by reducing EOT, which is given by,

$$\mathrm{EOT} = \sum_{n=1}^{i} \left(\text{Physical thickness} \times \frac{3.9\ \varepsilon_0}{\text{permittivity of each material}} \right), \tag{6.3}$$

where ε_0 is vacuum permittivity and 'i' is the number of gate dielectric layers [12].

Unlike logic devices, memory devices have bigger 'i' value in equation (6.3) because they have many gate dielectric stacks with tunneling oxide layer, charge trapping layer (CTL), and blocking oxide layer. That is to say, the memory device inevitably has a bigger EOT value due to its intrinsic structure with a high 'i' value in equation (6.3).

Therefore, the realization of low-power operation strongly depends on how well EOT is reduced by increasing permittivity of gate dielectric layers. Historically, the structure of memory cell has been continuously changed by reducing EOT [13,14]. In designing next-generation semiconductor devices, this tendency will remain unchanged.

10.1.2 Roadmap for Future Memory Technology with Low-Power Operation

The projected roadmap for future memory technology is illustrated in Figure 10.1. Traditionally, the SONOS structure has been broadly adopted after it was first developed by P. C. Y. Chen of Fairchild Camera and Instrument in 1977 [15]. This structure became the dominant cell design of charge trap flash (CTF) memory application, and it had been utilized for electrically erasable programmable read-only memory (EEPROM), flash memory, and thin-film-transistor (TFT) liquid-crystal display (LCD) displays [16]. Especially, semiconductor corporation of Toshiba, Macronix, United Microelectronics Corporation (also known as UMC), Floadia Corporation, and Cypress Semiconductor have offered various SONOS-based products [17].

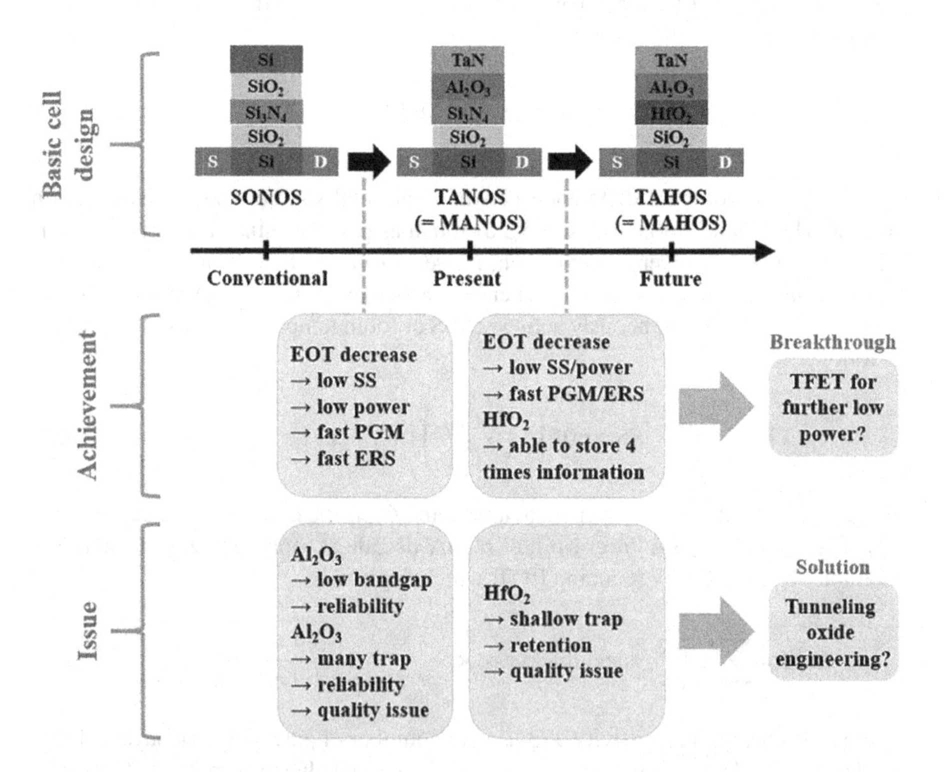

FIGURE 10.1 The projected roadmap for future memory technology. The terms SONOS, TANOS, and TAHOS are abbreviations for the following: SONOS: Si-SiO_2-Si_3N_4-SiO_2-Si, TANOS: TaN(or TiN)-Al_2O_3-Si_3N_4-SiO_2-Si, and TAHOS: TaN(or TiN)-Al_2O_3-HfO_2-SiO_2-Si. The symbolized terms of TaN (tantalum nitride), Si_3N_4 (silicon nitride), Si (silicon), S (source), D (drain), M (metal), PGM (programming), ERS (erasing), and TFET (tunnel FET) are used for better readability.

Later, the memory market required faster operation of memory cell: fast programming, fast erasing, and fast read operation. In addition, low-power consumption was also required as the mobile phone market expanded. According to this technological demand, the memory cell structure had been consistently required to reduce EOT. Consequently, the novel memory cell structure of TANOS was developed for faster programming/erasing/read operation [18–20]. This TANOS structure also had the virtue of reducing power consumption achieved by reduced EOT [18–20].

However, even though this TANOS structure successfully demonstrated the improvement of operation speed and reduced power consumption, quality issues inevitably and concomitantly arise due to Al_2O_3. Compared to conventional blocking oxide (SiO_2) in SONOS, the blocking oxide (Al_2O_3) in TANOS has lower bandgap and greater trap density [21–25]. From the lower bandgap, the increased electron tunneling resulted in degeneration of oxide quality [26]. Moreover, from the greater trap density, the increased electron trap is generated, and oxide quality of Al_2O_3 is also deteriorated [27]. Fortunately, these issues have been alleviated as fabrication equipment has developed, and this TANOS structure has become one of the dominant cell structures in modern memory technology.

Then, recently, as the demand for big data market has steadily increased, the methods of increasing data storage capacity are widely investigated [28–31]. Simply reducing the size of memory cell has been the most broadly discussed one because memory capacity could be increased without replacing the existing fabrication equipment. Namely, it has been possible to greatly increase the data storage capacity without a significant cost increase because the size of the memory device naturally decreases as extreme ultraviolet lithography (EUV) equipment is developed.

However, there is a physical limit to the scaling of memory devices. If the size of the memory cell becomes too small, normal operation becomes impossible due to the short-channel effects (SCEs). Therefore, a new method other than the existing scaling method is needed to increase the memory storage capacity. As a solution, using another material as CTL is suggested for the next-generation memory devices. Since HfO_2 has four times trap density compared to conventional CTL of Si_3N_4, HfO_2 CTL is possible to store four times the number of electrons compared to conventional Si_3N_4 CTL [32,33]. Namely, it might be possible to store four times information by utilizing HfO_2 as CTL. Surprisingly, since HfO_2 has significantly higher permittivity compared to Si_3N_4, the use of HfO_2 as CTL could also enable fast programming/erasing/read operation and low-power operation at the same time.

Therefore luckily, this TAHOS structure (which uses HfO_2 as CTL) arises as a viable solution for increasing data storage, and extensive research has been conducted for realization of the TAHOS structure. However, from various studies, it was found that one problem still remains, namely, retention problem [34–36]. Since HfO_2 has unstable trap distribution (namely, shallow trap), the retention of TAHOS becomes naturally poor. Namely, when we use a normal TAHOS structure, the memory cell cannot store the information (electron) for longer than 1 year. Considering the convention that we need to guarantee at least 10 years of stable operation in memory device, this basic TAHOS structure is hard to be commercially realized for the next-generation memory design.

Fortunately, previous researchers have successfully demonstrated the method of improving retention characteristics and guaranteeing high performance simultaneously.

Song *et al.* have demonstrated that the TAHOS structure can guarantee 10 years of stable operation by applying tunneling oxide engineering [37,38]. They also demonstrated that the TAHOS structure could lower power consumption by adopting low-power FETs, namely, tunnel FETs (TFETs) [37,38]. The detailed explanation of these state-of-art design methodologies will be addressed later in Sections 10.3.2 and 10.4.3.

10.1.3 Workflow and Scope

The workflow of this chapter is briefly shown in Figure 10.2 for a better understanding. In Section 10.2, the basic concept of memory devices including operation principle will be addressed. The physical principle of programming, erasing, and reading operation will be explained. Then, the main design object (design goal) in designing memory device will be described in terms of '3Ls: low power, large memory margin, and long retention'. In Sections 10.3 and 10.4, the optimization method will be addressed by gate stack optimization and doping optimization. Finally, in Sections 10.4 and 10.5, the design of future memory using low-power FET (TFET) and expected challenges will be addressed.

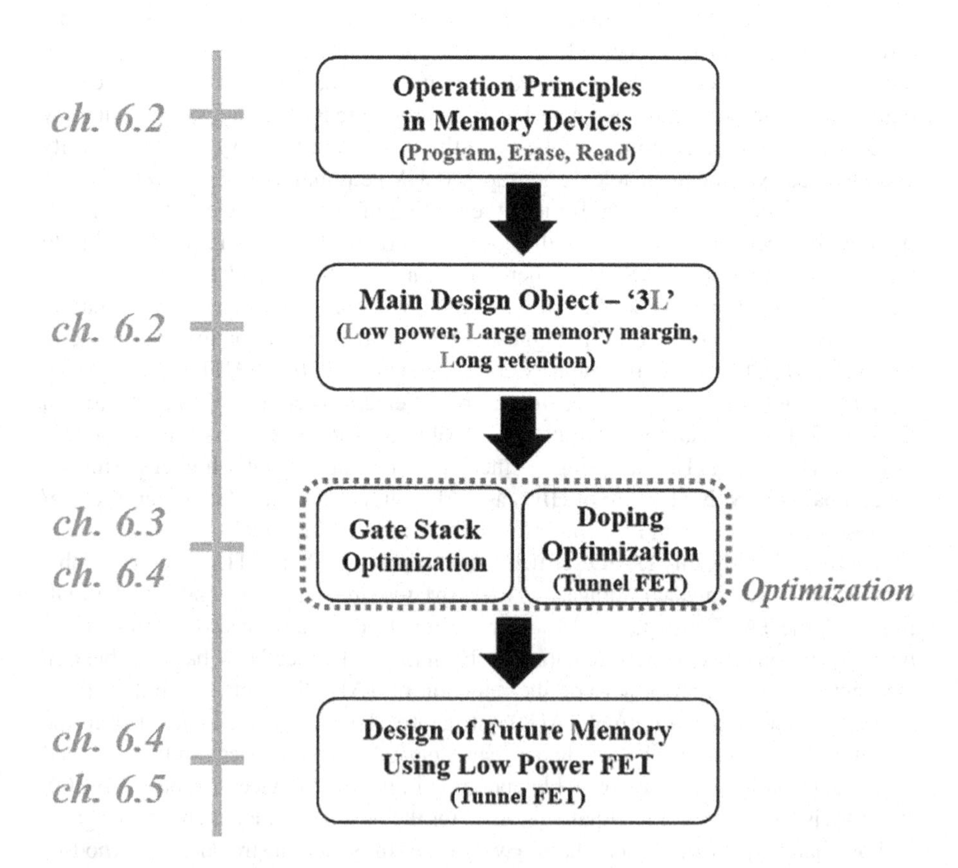

FIGURE 10.2 Illustration of workflow for a fast and better understanding.

10.2 NONVOLATILE MEMORY DEVICES (FLASH MEMORY)

10.2.1 Basic Structure of Nonvolatile Memory Devices

As mentioned previously in Section 10.1, the structure of memory cell has basically evolved from SONOS to TANOS, then to TAHOS. Interestingly, the names SONOS, TANOS, and TAHOS have something in common. They are common five-letter words. Even though the used materials in memory cell have changed over time, the memory cells are basically based on this five-layer structure. Let us take a look at what each of these five layers do.

The first letter of the names SONOS, TANOS, and TAHOS is the gate material where an external input voltage is applied. In some cases, doped silicon (usually doped by more than $10^{19}/cm^3$) is used as this gate material, and in other cases, metal is used. Traditionally, doped silicon was used as the gate material; however, a high-κ gate dielectric material with higher permittivity than that of SiO_2 was required to be used for lower EOT. However, in the case of most high-κ materials, doped silicon cannot be deposited on it for fabrication reasons. Specifically, if doped silicon is deposited on a high-κ insulator, the fabrication equipment that deposits doped silicon is contaminated by the high-κ material. Therefore, it is difficult to deposit the doped silicon on the top of the high-κ material, and only metal material can be deposited on it. As a result, metal gate has been adopted in TANOS and TAHOS technology.

The last letter 'S (silicon)' of the names SONOS, TANOS and TAHOS is the channel material where the current mainly flows. The principle of current-flow in the channel region is equally applied to the memory devices as in the logic devices.

The middle letters 'N (Si_3N_4)' and 'H (HfO_2)' are CTL, which stores electric charges. This middle layer is very important since it determines the information storage. When a very high voltage (e.g. 20 V) is applied to the gate, electrons in the channel region pass through SiO_2 and become finally trapped in this CTL. Usually, the electron is trapped by concentration of more than $10^{18}/cm^3$ after programming. When the information is stored (electron is trapped in CTL), the 0 state and 1 state are distinguished by whether there are electrons in the CTL or not. This process is called 'read operation', and this will be addressed later in Section 10.2.3.

The second letter 'O (SiO_2)' is a tunneling layer (also known as tunneling oxide) that fills between channel and CTL. This tunneling layer acts as a path for electrons to travel during programming. On the other hand, after programming, this layer acts as a blocker that prevents emission of trapped electrons in the CTL. Normally, the tunneling layer is composed of 3 nm SiO_2.

Finally, the fourth letter 'S (SiO_2)' and 'A (Al_2O_3)' is the blocking layer (also known as blocking oxide) that fills between CTL and gate material. This layer acts as a blocker that prevents trapped electrons in the CTL from passing to gate. Normally, the blocking layer is composed of insulator with a thickness of more than 6 nm.

Table 10.1 shows a brief summary. In essence, the memory cell consists of a total of five layers and can be summarized as a structure with CTL that stores information between gate and channel.

TABLE 10.1
Structure of SONOS, TANOS, TAHOS Memory Cell, and the Corresponding Materials in Each Layer

	SONOS (Conventional)	TANOS (Present)	TAHOS (Future)
Gate	S	T	T
Blocking layer (= Blocking oxide)	O	A	A
CTL	N	N	H
Tunneling layer (= Tunneling oxide)	O	O	O
Channel	S	S	S

The symbolized terms of S (Silicon), O (SiO_2), N (Si_3N_4), T (tantalum nitride or titanium nitride), A (Al_2O_3), and H (HfO_2) are used.

TABLE 10.2
Two Representative Methods for Programming: Fowler Nordheim (FN) Tunneling and Hot-Carrier Injection (HCI)

	FN Tunneling	HCI
V_{Gate}	16~20V	3~6V
V_{Drain}	0V	3~6V
Uniformity of trapped charge distribution (after programming)	Uniform	Concentrated on the drain side
Speed	Fast (1~100 μs)	Slow (0.1~10 ms)
Total power consumption	Low (10^{-11}~10^{-8} W)	High (10^{-5}~10^{-3} W)

10.2.2 Physical Principles: Program/Erase Operation

Based on the structure of memory cell described in the previous section, how the program/erase operation works will be explained in this section.

As shown in Table 10.2, there are two main ways of programming operation. Fowler Nordheim (FN) tunneling and hot-carrier injection (HCI) are the most common programming methods, and they have different characteristics.

In the case of FN tunneling, a very high voltage of 16 V or higher is applied to gate. Then, the energy band becomes sharp, especially in the upper part of the energy band diagram of tunneling oxide as shown in Figure 10.3. As a result, electron tunneling takes place and electrons move from channel to CTL. At this time, electrons are trapped uniformly throughout the CTL, since tunneling occurs in all areas of the tunneling oxide with high potential. Namely, the result of FN tunneling is the uniform distribution of trapped electrons in CTL.

On the other hand, in the case of HCI, 3–6 V is applied to the gate and drain. The differentiated characteristic of the HCI method is that a very high voltage is applied to the drain. This high drain voltage creates a fairly high electric field (E-field) in

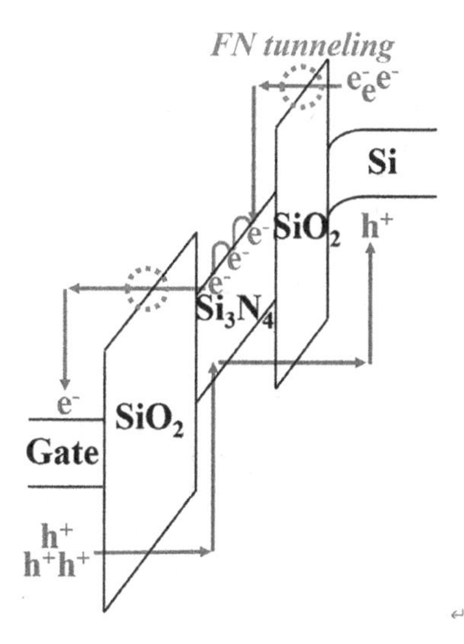

FIGURE 10.3 Energy band during program operation. The overall conduction band of gate stack becomes sharper, and electron moves through various tunneling mechanisms. As a result, many electrons are trapped in the charge trap layer.

the channel between source and drain. Thus, electrons moving through the channel are accelerated, and at some point, they reach their maximum speed (this is called 'velocity saturation'). After reaching the maximum speed, the total energy of electron continues to increase due to the E-field, resulting in electron–hole pairs. This generated electron is called hot electron, and this generated hole is called hot hole. Among the large number of hot electrons and hot holes created here, hot electrons are injected into the CTL region that has a relatively high potential compared to the channel region. These successive processes are called HCI.

In HCI, a very high voltage is applied to the drain, and a fairly high current (usually more than 100 μA) flows between the source and drain, which consumes very high power. This is a remarkable disadvantage of HCI. In addition, in the HCI method, electron is intensively trapped in the drain side of the CTL (namely, the uniformity of trapped charge distribution is poor), since many electron–hole pairs are generally formed in the channel region near the drain. However, despite these disadvantages, the HCI method is broadly utilized in certain memory designs because it has an advantage in the 'process of connecting thousands of memory cells into circuit (= array design)'. In this book, we will not address the array design in detail, since this book mainly focuses on semiconductor and memory devices. Readers who want to know more about array design, kindly refer to the book *Silicon Based Unified Memory Devices and Technology* (CRC Press) written by *A. Bhattacharyya* [39].

So far, we have dealt with the programming operation. In summary, there are two main methods of programming memory devices, FN tunneling and HCI. Even though the input voltage used in these two methods are different and the physical principle of programming is different, the result is the same as the injection of electrons into CTL. Namely, to put it simply, the program operation could be understood as 'the process of moving electrons from the channel to the CTL'.

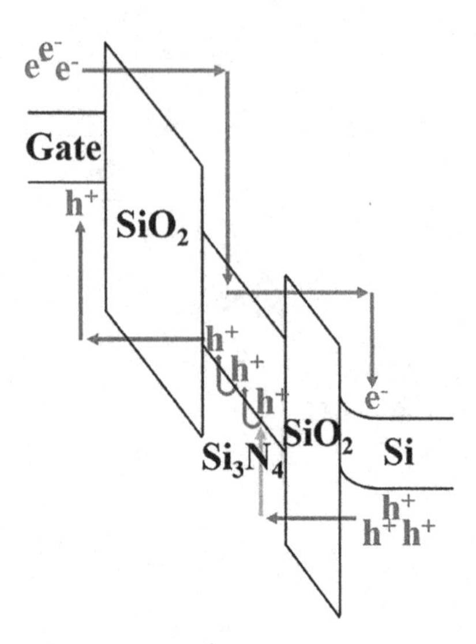

FIGURE 10.4 Energy band during erase operation. The overall valance band of gate stack becomes sharper, and hole moves through various tunneling mechanisms. Consequently, many holes are trapped in the charge trap layer.

Then, on the contrary, what about the erase operation? Considering the process of programming in reverse, the process of erasing could be inferred. In other words, erase operation could be understood as 'the process of moving electrons from the CTL to the channel'. Specifically, about −18 to −23 V is applied to the gate during erase operation. Then, the energy band of the tunneling layer becomes sharp, and electrons move from CTL to the channel, as illustrated in Figure 10.4. Namely, when a very low voltage is applied to the gate, the energy band changes due to the potential difference, and electrons move from the CTL to the channel. After the electrons exit the CTL into the channel, there are now significant holes in the CTL. This is because the hole moves in the opposite direction to the electron in the given electrical condition.

In summary, the program operation and erase operation could be identified by whether electrons or holes are consequently trapped in the CTL. If a significant number of electrons remain at the CTL of a certain cell, then the program operation might be previously held. On the other hand, if a significant number of holes remain at the CTL of a certain cell, the erase operation might be previously held in that cell.

10.2.3 Physical Principles: Read Operation

Then, how can we electrically distinguish the programmed cell from the erased cell? The method is simple. They can be distinguished by operating memory devices such as MOSFETs, as shown in Figure 10.5. For example, they can be distinguished by applying 1 V to the gate and drain. In the case of a programmed cell (= a cell with electron in the CTL), the trapped electron in the CTL will prevent the channel region from forming an inversion layer. In other words, in the programmed cell, the trapped electron in the CTL prevents the current from flowing in the channel region. Eventually, a low current consequently flows in the programmed cell.

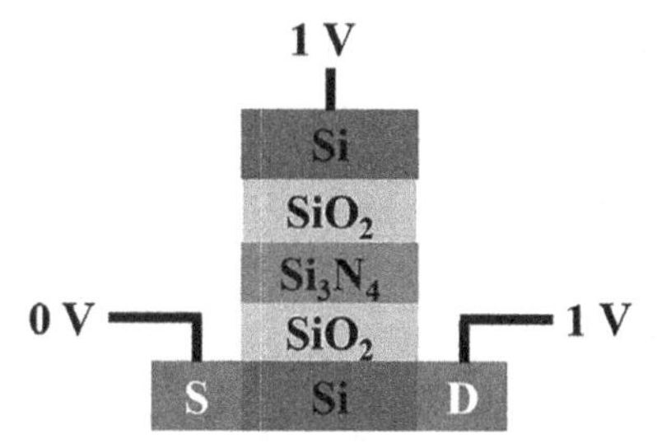

FIGURE 10.5 Read operation in memory device. The read operation is usually performed by applying a certain gate/source/drain voltage and comparing the amount of flowing current with a certain value.

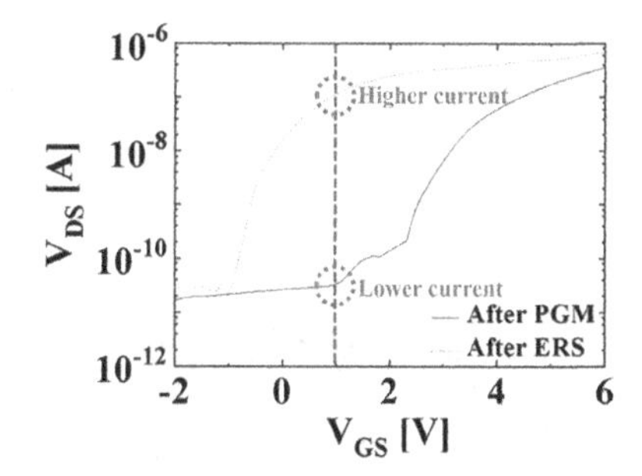

FIGURE 10.6 Illustration describing the process of read operation in memory device. After the gate/source/drain voltage is applied, the amount of flowing current is compared to the certain value. If the flowing current is greater, then the memory cell is recognized as 'erased'. On the other hand, if the flowing current is lower, then the memory cell is recognized as 'programmed'.

On the other hand, if 1 V is applied to the gate and drain of the erased memory cell (= a cell with hole in the CTL), the trapped hole in the CTL will help the channel to form an inversion layer. This is because the hole forms an E-field opposite to the electron. Eventually, a high current consequently flows in the erased cell.

In summary, the programmed cell and the erased cell can be distinguished by whether the current flows to a great extent or less when a certain voltage is applied to the gate and drain, as shown in Figure 10.6. If low current flows, that cell is identified as the programmed cell. On the contrary, if current flows to a great extent, that cell is identified as an erased cell. In this way, the 'read operation' of the memory devices is performed.

10.2.4 Main Design Goal '3Ls': Low-Power, Large Memory Margin, and Long Retention

We covered the basic structure and operation principles of memory devices previously in Sections 10.2.1–10.2.3. Then, what goals shall we accomplish when designing a memory device? First of all, it is obvious that power consumption should be lowered in designing memory devices to reduce battery consumption and ensure a longer use of the electronic device [40–46].

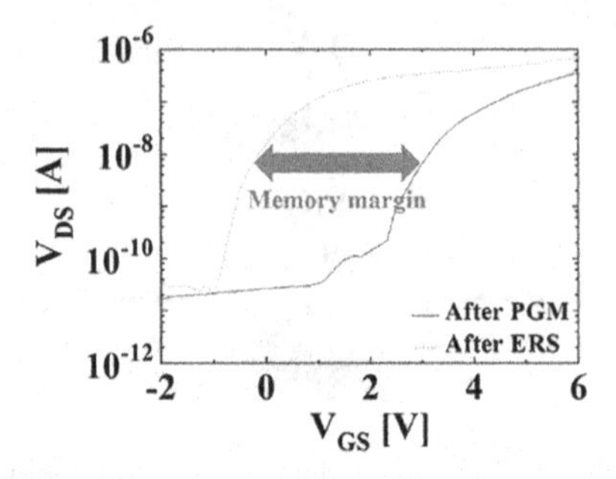

FIGURE 10.7 Illustration describing 'memory margin (also known as memory window)' in memory device. The difference between threshold voltage of the programmed cell and that of the erased cell is regarded as the memory margin.

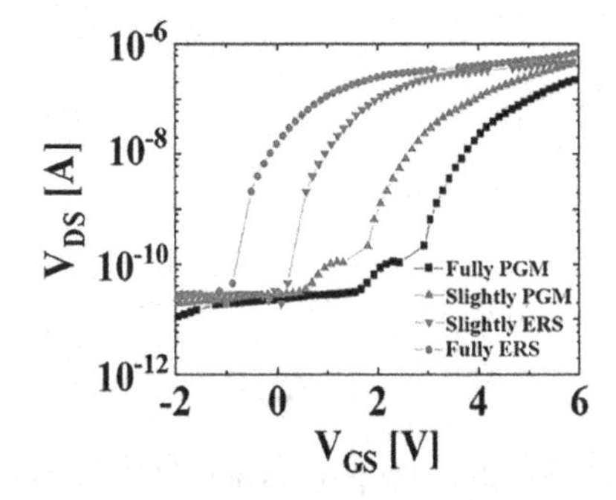

FIGURE 10.8 Illustration explaining multilevel cell (MLC). In MLC, four states could be separately identified. The fully programmed case could be treated as the 00-state, the slightly programmed case could be treated as the 01-state, the slightly erased case could be treated as the 10-state, and the fully erased case could be treated as the 11-state.

Another design goal is to increase the storage capacity. In order to design a memory cell with high memory capacity, the concept of memory margin (also known as memory window) is needed to be addressed first. Memory margin is the difference between the threshold voltage of the programmed cell and that of the erased cell (Figure 10.7).

The larger the difference between two threshold voltages , the more the information stored. Specifically, if the difference in the threshold voltage is large (= memory margin is large), the state of the memory device could be divided into four types. For example, as illustrated in Figure 10.8, a fully programmed case could be treated as the 00-state, the slightly programmed case could be treated as the 01-state, the slightly erased case could be treated as the 10-state, and the fully erased case could be treated as the 11-state [47]. This is called multilevel cell (MLC). Unlike the traditional single-level cell (SLC) that has only two states, MLC has four states; therefore, the storage capacity of memory devices could be doubled [48]. Namely, another important goal in designing a memory device is to increase the memory margin to enable the implementation of memory device with a larger number of bits.

Third, there is one more important goal in memory design. What if some memory device has a tremendous amount of memory storage capacity but it can maintain information for only 1 year? Users of that electronic device will have to format the electronic device every year. In other words, in designing a memory cell, its performance of storage must be maintained for a certain period or longer. This is technically called 'retention' [49]. In general, when designing a memory device, the retention characteristic should be guaranteed for at least 10 years [50–52].

It is clear that when designing a memory device, (1) low power consumption, (2) large memory margin, and (3) long retention characteristic must be accomplished. However, unfortunately, a trade-off issue between (2) large memory margin and (3) long retention characteristic makes memory designer challenging. When a memory device is technically designed to increase the memory margin, retention might be adversely affected, and this makes memory design difficult. For example, if HfO_2 is used as the CTL, the memory storage capacity could be theoretically increased by four times, but the retention characteristic is usually guaranteed for only 1 year. Then, the methodology to overcome this trade-off issue and improve both of them at the same time will be specifically discussed in the following section.

10.3 GATE STACK OPTIMIZATION FOR LOW-POWER OPERATION

10.3.1 Retention Issue in TAHOS Structure

In Section 10.1.2, the evolution of memory device has been addressed. It is explained that SONOS has been developed extensively in the past, then TANOS is utilized for its low-power and fast speed characteristics. For the next-generation memory device, extensive research has been conducted using HfO_2 as CTL (namely, TAHOS structure) because HfO_2 has four times charge trap density compared to the conventional CTL material Si_3N_4 [32,33]. Therefore, it is expected to be able to store four times as much information, and it might be possible to implement a triple-level cell (TLC) that stores more information beyond MLC [53].

Surely, HfO_2 is a very promising material. Its dielectric constant is about three times higher than the dielectric constant of Si_3N_4, and hence EOT and power consumption could be decreased by adopting HfO_2 as CTL [54–59]. Moreover, this decrease in EOT could also enable faster program/erase/read operation. In addition, the charge trap density of HfO_2 is four times compared to that of conventional CTL material (Si_3N_4), and much more information could be stored by applying HfO_2 as CTL [32,33]. However, there are some intrinsic characteristics of HfO_2 that make it difficult to commercialize HfO_2 CTL (TAHOS), which is the numerous shallow traps in HfO_2 [60]. In other words, in the case of HfO_2, the energy distribution of most charge trap is poor. Specifically, electron charge trap is located near the conduction band, and the hole charge trap is located near the valance band [60]. Therefore, the trapped electron or trapped hole in HfO_2 cannot stay in CTL for a long time and is easily emitted toward the channel. Namely, HfO_2 CTL has poor retention characteristics because there are numerous shallow traps. Therefore, additional engineering has been required to commercialize the TAHOS structure for low power/large memory storage capacity/long retention/fast operation speed.

10.3.2 Advanced TAHOS Structure: TAHOAOS

The solution for the retention issue in the TAHOS structure could be found in the history of memory devices. Memory devices have historically evolved by changing the material or the structure of the stack. Song *et al.* demonstrated the retention improvement in the TAHOS structure by applying a special design method to the tunneling layer [37,38]. The reason for applying engineering to the tunneling layer among five layers is that electron-escape mainly occurs by direct tunneling through the thin tunneling layer (usually 3 nm) compared to the thick blocking layer (usually 6–10 nm).

Figure 10.9 shows the structure of the tunneling oxide-engineered (TE) TAHOS structure [37,38]. In this TE-TAHOS (also called TAHOAOS), the three stacked layers of $SiO_2/Al_2O_3/SiO_2$ are used as tunneling oxide. From this specialized structure, the physical thickness of the tunneling oxide can be increased while maintaining the same EOT. For example, if a conventional tunneling layer of 3 nm-thick SiO_2 is replaced by SiO_2(1 nm)/Al_2O_3(2.3 nm)/SiO_2(1 nm), the total physical thickness of the tunneling layer could be increased from 3 to 4.3 nm [37,38]. Since direct tunneling (which usually contributes most to the emission of trapped electrons) is exponentially suppressed when the physical thickness of the tunneling layer becomes greater than 3 nm, the emission of electrons in the CTL could be significantly prevented by adopting the TAHOAOS structure [37,38].

As demonstrated in Figure 10.10, the retention characteristic could be remarkable improved when the TAHOAOS (TE-TAHOS) structure is applied [37,38]. In specific, the retention characteristics is significantly improved from 0.57 to 4.57 V after 10 years from programming and erasing [37,38]. Namely, through the TAHOAOS structure, it is possible to design a memory device with (1) low power, (2) large memory margin, and (3) long retention at the same time. Furthermore, this TAHOAOS technology has additional advantage that it does not require the introduction of

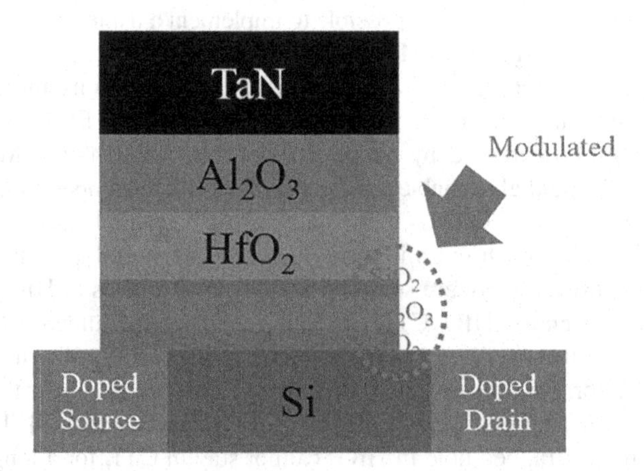

FIGURE 10.9 Schematic diagram of the TAHOAOS structure. In this TAHOAOS (also known as TE-TAHOS) structure, the advanced stacks of $SiO_2/Al_2O_3/SiO_2$ are adopted as the tunneling layer. By doing so, the physical thickness of the tunneling layer could be increased, whereas EOT of the tunneling layer is maintained. Consequently, the retention characteristic could be significantly improved.

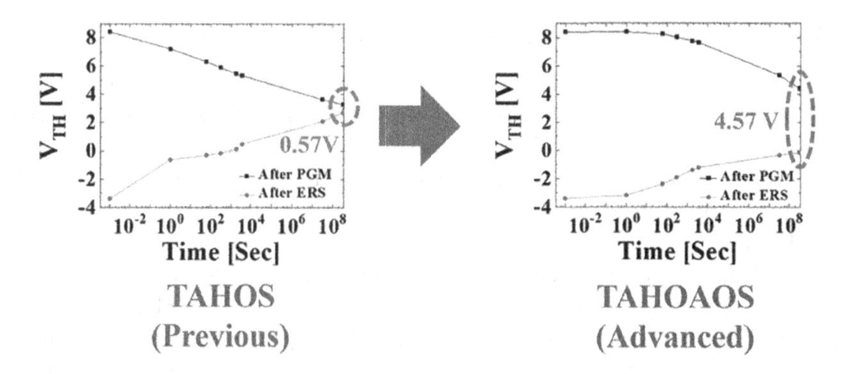

FIGURE 10.10 Improvement in retention characteristic by utilizing the TAHOAOS structure. The retention characteristic is calculated for 10 years after programming/erasing through Synopsys Sentaurus technology computer-aided design (TCAD) simulation.

additional fabrication equipment because the atomic layer deposition (ALD) equipment that deposits Al_2O_3 has been already used in TANOS technology.

10.4 DOPING OPTIMIZATION FOR LOW-POWER OPERATION

10.4.1 Tunnel FET for Low-Power Memory Device

In the previous Section 10.3, we covered the optimization method with stack engineering in memory design. In this section, another optimization method through doping technique will be addressed.

Basically, it is obvious that the power consumption could be reduced by applying the TAHOS or TAHOAOS structure. Interestingly, the power consumption could be further reduced utilizing the doping technique in TFETs [61–65]. Figure 10.11 shows that it is possible to significantly decrease the power consumption by utilizing the TFET structure in memory design. Specifically, in the case of the on-state, power consumption is expected to be reduced by about 100 times, and in the case of off-state, power consumption is expected to be reduced by about 300 times. The reason why the TFET structure flows less current compared to MOSFET was covered in the previous chapters and many previous papers [66].

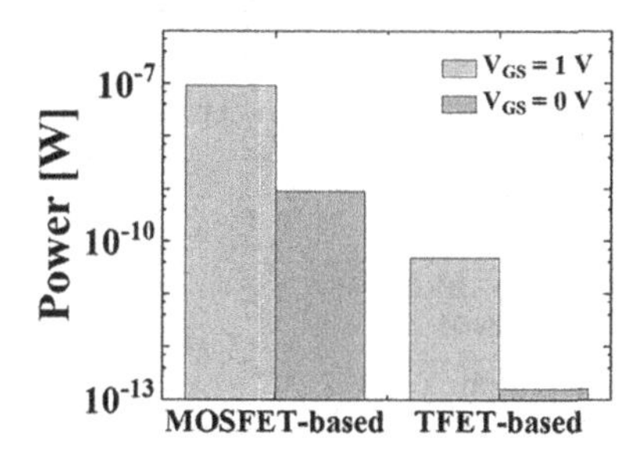

FIGURE 10.11 Comparison of power consumption between MOSFET-based memory and TFET-based memory. The TFET-based memory could significantly decrease the power consumption.

10.4.2 Additional Advantage of TFET: Fast Erase Speed

Essentially, TFET has different types of doping in source and drain, unlike MOSFET. TFET simultaneously has a p^+ doped region and an n^+ doped region, but this was not a big advantage for logic design (other than low-power characteristic and low SS of TFET). However, in memory devices, this special structure could act as a remarkable advantage for program/erase speed [66].

In the case of programming, the channel must supply electrons to CTL, and in the case of erasing, the channel must supply holes to CTL. However, there is a problem in one of the two when memory cell is designed from MOSFET. Specifically, if MOSFET has a $n^+/p/n^+$ structure, sufficient hole supply is not supplied during the erase operation. On the other hand, if MOSFET has a $p^+/n/p^+$ structure, sufficient electron is not supplied during the program operation.

In contrast to MOSFET, TFET has an n^+ region and a p^+ region at the same time. Therefore, sufficient electron supply is provided during program operation, and sufficient hole supply is also provided during erase operation (Figure 10.12) [66]. As demonstrated in Figure 10.13, the erase speed of TFET-based memory is 10,000

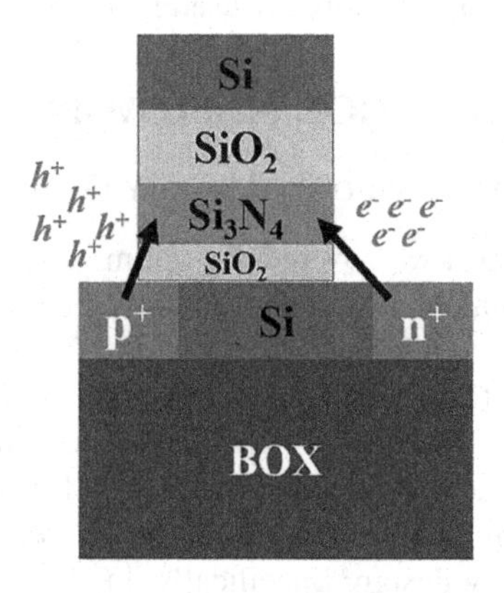

FIGURE 10.12 Illustration describing the additional advantage of the TFET-based memory device. The TFET-based memory cell could simultaneously have a hole supply depot and an electron supply depot.

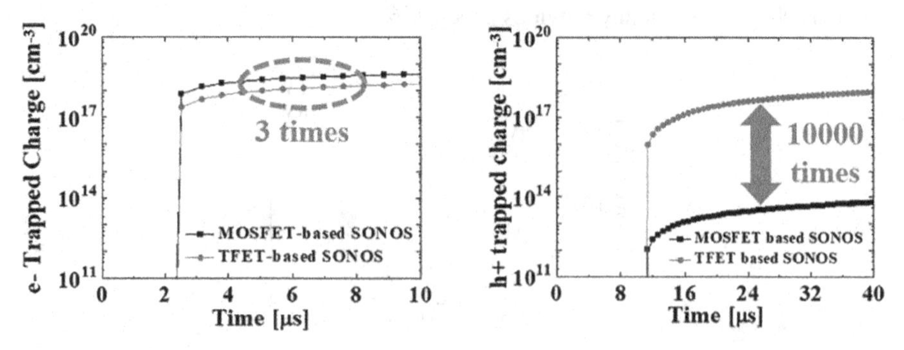

FIGURE 10.13 Comparison of program speed (left) and erase speed (right) between MOSFET-based memory and TFEET-based memory.

times faster than n-MOSFET-based memory, and the program speed of TFET-based memory is three times slower than n-MOSFET-based memory [66]. The reason why TFET-based memory has 10,000 times faster erase speed is that n-MOSFET-based memory has no hole supply depot whereas the TFET-based memory has one [66]. The reason why TFET-based memory has three times slower program speed is that n-MOSFET-based memory has an electron supply depot in both source and drain, whereas TFET-based memory has an electron supply depot in only one [66]. As a result, it could be said that the TFET-based memory device has an overall superior program/erase speed performance compared to the MOSFET-based memory device.

10.4.3 TFET TAHOAOS

Combining these two methods of stack optimization and doping optimization, TFET TAHOAOS structure could be designed [37,38]. This structure has the low-power advantage of TFET and '3Ls: low power, large memory margin, and long retention' advantage of TAHOAOS structure at the same time [37,38]. It is expected to be utilized as a very decisive structure in designing next-generation low-power memory devices, since the low-power goal could be realized from both the stack side and the doping side. However, since three tunneling layers are used in the TFET TAHOAOS structure, additional fabrication processes are required, and fabrication costs might be consequently increased. Moreover, the complicated fabrication process might cause some undesirable and unexpected issue. This will be specifically addressed in the next Section 10.5.

10.5 EXPECTED CHALLENGES

10.5.1 Abrupt Doping Profiles

So far, we have addressed the design method for low-power consumption and increased erase speed using TFET, which is one of the low-power FETs. In addition, to ensure the reliability of the memory device, the design methodology for retention improvement is addressed with stack engineering. The addressed design techniques could simultaneously guarantee the performance and reliability of memory device; however, there are some more issues to look at.

First of all, as when making a TFET logic device, an abrupt doping profile is required when fabricating a TFET-based memory device [61–65]. If the memory device is fabricated based on TFET, SS could be lowered and power consumption could be also reduced through band-to-band tunneling (BTBT), but the formation of an abrupt doping profile must be carried out. If a gradual doping profile rather than an abrupt doping profile is formed, it might show a worse SS than that of the MOSFET-based memory device, and the low-power characteristic might also be consequently weakened.

Specifically, unlike the MOSFET-based device, the performance of TFET-based devices has a much greater dependence on the doping profile because BTBT occurs between a highly doped region (source or drain) and a lowly doped region (channel region between source and drain) [61–65]. Therefore, in designing a TFET-based memory device, the main design purpose of 'low-power operation' could be well achieved when an abrupt doping profile is formed.

10.5.2 Increased Number of Photomasks

In both nonmemory semiconductor industry and memory semiconductor industry, the number of photomasks is a very decisive variable. Since the total manufacturing cost is heavily dependent on the number of photomasks, it is desirable to reduce the number of photomasks during fabrication.

Unfortunately, TFET-based devices usually use more photomasks than MOSFET-based devices. Specifically, in the case of the MOSFET-based device, since source and drain have the same doping type, the source and drain can be doped at the same time. On the other hand, in the case of TFET-based devices, the source and drain cannot be doped at the same time because source and drain have different doping types (technically speaking, MOSFET can be 'self-aligned' and TFET cannot be 'self-aligned'). Therefore, after doping at source (or drain), an additional series of the photolithography process (photoresist (PR) coating → soft bake → alignment with another photomask → light exposure → PR develop → hardbake) proceeds, and then doping at drain (or source) is performed. Therefore, the fabrication cost might increase.

For this reason, in designing and fabricating TFET-based memory devices, it is necessary to research how to optimize the total number of photomasks.

10.5.3 Variability

The third issue to be considered for the TFET-based memory device is related to the issue of 'increased number of photomasks' in the previous section. In the case of TFET-based memory devices, since more photo processes are performed, there is a high probability of variability issue. In other words, as the number of process step increases, variability problems among wafers might arise when the memory devices are actually fabricated. Eventually, this leads to a device yield issue; therefore, it is important to consider how to improve the variability in fabricating TFET-based memory devices.

10.5.4 Array Design

In the case of memory devices, the integration of memory and the performance of memory greatly depend on how the memory devices are connected in circuit. The method of connecting memory cells for circuit is called 'array design', and there are two main array designs in the memory industry: NAND array and NOR array [67,68].

Figure 10.14 schematically illustrates NAND array and NOR array. Each of these two design techniques has advantages and disadvantages, respectively. As shown in Table 10.3, NAND flash has advantages in terms of integration, memory capacity, erase speed, write speed, and manufacturing cost [39]. However, when memory cells are connected by the NAND flash design technique, there are disadvantages in terms of standby power, reliability, read speed, and code execution [39]. On the other hand, in the case of NOR flash, it has advantages of low standby power, high reliability, fast read speed, and

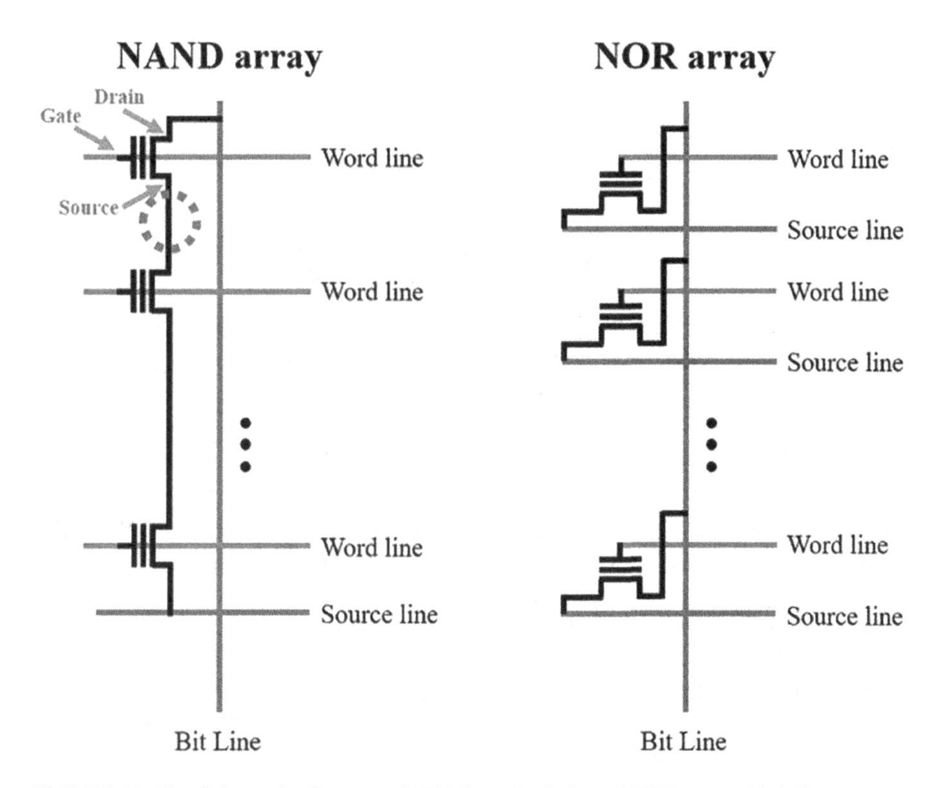

FIGURE 10.14 Schematic diagram of NAND array (left) and NOR array (right).

TABLE 10.3
Comparison between NAND Flash and NOR Flash

Design Consideration	NAND Flash	NOR Flash
Integration (cell size)	O ($4F^2$)	X ($10F^2$)
Memory capacity	O (high)	X (low)
Erase time	O (1 ms/sector)	X (1 s/sector)
Write speed	O (high)	X (low)
Cost per bit	O (low)	X (high)
Standby power	X (high)	O (low)
Reliability	X (high)	O (low)
Read speed	X (20 µs)	O (<80 ns)
Code execution	X (hard)	O (easy)

Symbol 'O' means good, and symbol 'X' means bad.

easy code execution [39]. At the same time, NOR flash has disadvantages in terms of low density, slow erase speed, slow write speed, and high manufacturing cost [39].

In the past, NOR flash was used extensively, but in recent years, NAND flash is widely used due to its high density (high storage capacity) and low manufacturing

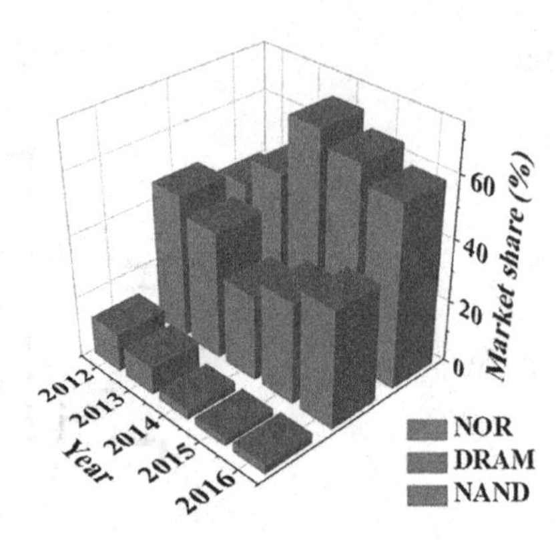

FIGURE 10.15 Percent market share of NOR, DRAM, and NAND in Micron Technology.

cost [67–69]. According to *R. Castellano* (president of The Information Network), total sales of NOR flash in Micron Technology (MU) have remained less than 30% of NAND flash sales (Figure 10.15).

Unfortunately, however, NAND flash is hard to be designed with TFET-based memory cells. As shown in the orange dotted line in Figure 10.14, in NAND flash, source and drain must have the same type of doping [38,39]. However, the TFET-based memory cell has different types of doping in source and drain, and it is difficult to apply TFET-based memory cells to NAND array. Therefore, TFET-based low-power memory devices have been studied extensively for the NOR flash market.

Like this, several trade-off issues exist between device design and circuit design. Therefore, it is important to design the whole by simultaneously considering device characteristics and circuit characteristics. This method of designing the whole by taking these two things into account is called 'device/circuit co-design' (also known as design technology co-optimization (DTCO)) [70]. In making memory electronic devices based on TFET [71–73], it is also essential to design them with device/circuit co-design so that overall good performance is achieved. Therefore, the collaboration between the device design engineer and the circuit design engineer might be paramount.

10.6 CONCLUSIONS

In the first part of this chapter, the projected roadmap for future memory technology is shown and explained. The conventional structure of SONOS had evolved to TANOS structure for lower EOT, which enables low-subthreshold swing, low-power operation, fast program operation, and fast erase operation. Then, this TAHOS structure has been recently suggested for viable candidates for replacing TANOS structure for better memory storage capacity and reduced EOT. However, the TAHOS structure has quality issue, such as it is difficult to guarantee stable operation of 10 years.

For the solution and further breakthrough, two optimization methodologies are suggested for next-generation memory devices. The proposed optimization methodologies utilize the low-power characteristics of TFET and increased physical thickness of tunneling layers (while EOT is unchanged) from stack engineering. By doing so, the three parameters required for memory design—low power, large memory margin, and long retention (3Ls)—could be achieved. In addition, the TFET-based memory design could improve erase speed as well, since TFET structure simultaneously has p^+ region and n^+ region whereas the MOSET structure does not.

In the final part of this chapter, four possible issues in these optimization techniques are addressed. For better performance, abrupt doping profiles are critically required, and possible variability issues from the increased number of photomasks should be researched. Most importantly, the device/circuit co-design shall be adopted for solving trade-off issues between device design and circuit design, and the collaboration between the device design engineer and the circuit design engineer would be paramount.

ACKNOWLEDGMENTS

This chapter has been written utilizing information obtained from various studies, and the sources of the information are indicated in a reference format. Some figures in this chapter were written by re-editing and utilizing the figures from the open access journal, after obtaining permission from all authors of the corresponding manuscripts (references 37 and 38). In addition, several plagiarism checks (Turnitin) have been conducted in writing this chapter. As a result, 3%–5% similarity has been confirmed in this chapter (except for the 'reference list part'). The author is grateful to previous studies and the plagiarism screening program (Turnitin) for providing a variety of information.

REFERENCES

1. H.-W. You and W.-J. Cho, "Charge Trapping Properties of the HfO_2 Layer with Various Thicknesses for Charge Trap Flash Memory Applications," *Applied Physics Letters*, vol. 96, pp. 093506, Mar. 2010, doi: 10.1063/1.3337103.
2. D. Yun, H. Kang and S. Yoon, "Process Optimization and Device Characterization of Nonvolatile Charge Trap Memory Transistors Using In–Ga–ZnO Thin Films as Both Charge Trap and Active Channel Layers," *IEEE Transactions on Electron Devices*, vol. 63, no. 8, pp. 3128–3134, Aug. 2016, doi: 10.1109/TED.2016.2580220.
3. H. Yin, S. Kim, C. J. Kim, I. Song, J. Park, S. Kim and Y. Park, "Fully Transparent Nonvolatile Memory Employing Amorphous Oxides as Charge Trap and Transistor's Channel Layer," *Applied Physics Letters*, vol. 93, pp. 172109, Oct. 2008, doi: 10.1063/1.3012386.
4. S. Spiga, F. Driussi, A. Lamperti, G. Congedo and O. Salicio, "Effects of Thermal Treatments on the Trapping Properties of HfO2 Films for Charge Trap Memories," *Applied Physics Express*, vol. 5, no. 2, pp. 021102, Jan. 2012, doi: 10.1143/APEX.5.021102.
5. S. Tayal and A. Nandi, "Study of 6T SRAM Cell Using High-K Gate Dielectric Based Junctionless Silicon Nanotube FET," *Superlattices and Microstructures*, vol. 112, pp. 143–150, Dec. 2017, doi: 10.1016/j.spmi.2017.08.061.
6. U. Ganguly, T. Guarini, D. Wellekens, L. Date, Y. Cho, A. Rothschild and J. Swenberg, "Impact of Top-Surface Tunnel-Oxide Nitridation on Flash Memory Performance and Reliability," *IEEE Electron Device Letters*, vol. 31, no. 2, pp. 123–125, Feb. 2010, doi: 10.1109/LED.2009.2036577.

7. P. K. Asthana, Y. Goswami, S. Basak, S. B. Rahi and B. Ghosh, "Improved Performance of a Junctionless Tunnel Field Effect Transistor with a Si and SiGe Heterostructure for Ultra Low Power Applications," *RSC Advances*, vol. 5, no. 60, pp. 48779–48785, Apr. 2015, doi: 10.1039/C5RA03301B.
8. Z. Tang, X. Zhu, H. Xu, Y. Xia, J. Yin, Z. Liu, A. Li and F. Yan, "Impact of the Interfaces in the Charge Trap Layer on the Storage Characteristics of ZrO_2/Al_2O_3 Nanolaminate-Based Charge Trap Flash Memory Cells," *Materials Letters*, vol. 92, pp. 21–24, Feb. 2013, doi: 10.1016/j.matlet.2012.10.024.
9. S. Tayal and A. Nandi, "Effect of High-K Gate Dielectric In-Conjunction with Channel Parameters on the Performance of FinFET Based 6T SRAM," *Journal of Nanoelectronics and Optoelectronics*, vol. 13, no. 5, pp. 768–774, May 2018, doi: 10.1166/jno.2018.2287.
10. S. Tayal and A. Nandi, "Performance Analysis of Junctionless DG-MOSFET-Based 6T-SRAM with Gate-Stack Configuration," *Micro & Nano Letters*, vol. 13, no. 6, pp. 838–841, June 2018, doi: 10.1049/mnl.2017.0702.
11. T.-J. K. Liu and K. Kuhn, "Tunnel Transistors," in *CMOS and Beyond: Logic Switches for Terascale Integrated Circuits*, 1st ed. Cambridge University Press, Cambridge, England; ch. 6, pp. 117–143, 2015.
12. H. Harris, K. Choi, N. Mehta, A. Chandolu, N. Biswas, G. Kipshidze, S. Nikishin, S. Gangopadhyay and H. Temkin, "HfO_2 Gate Dielectric with 0.5 nm Equivalent Oxide Thickness," *Applied Physics Letters*, vol. 81, pp. 1065, July 2002, doi: 10.1063/1.1495882.
13. J.-Y. Wu, Y.-T. Chen, M.-H. Lin and T.-B. Wu, "Ultrathin HfON Trapping Layer for Charge-Trap Memory Made by Atomic Layer Deposition," *IEEE Electron Device Letters*, vol. 31, no. 9, pp. 993–995, July 2010, doi: 10.1109/LED.2010.2052090.
14. S. Maikap, S. Z. Rahaman and T. C. Tien, "Nanoscale (EOT = 5.6 nm) nonvolatile memory characteristics using n-Si/SiO_2/HfAlO nanocrystal/Al_2O_3/Pt capacitors," *Nanotechnology*, vol. 19, no. 43, pp. 435202, Sep. 2008, doi: 10.1088/0957–4484/19/43/435202.
15. P. C. Y. Chen, "Threshold-alterable Si-gate MOS devices," *IEEE Transactions on Electron Devices*, vol. 24, no. 5, pp. 584–586, May 1977, doi: 10.1109/T-ED.1977.18783.
16. S.-C. Chen, T.-C. Chang, P.-T. Liu, Y.-C. Wu, P.-S. Lin, B.H. Tseng, J.-H. Shy, S. M. Sze, C.-Y. Chang and C.-H. Lien, "A Novel Nanowire Channel Poly-Si TFT Functioning as Transistor and Nonvolatile SONOS Memory," *IEEE Electron Device Letters*, vol. 28, no.9, pp. 809–811, Sep. 2007, doi: 10.1109/LED.2007.903885.
17. M. Chen, H. Y. Yu, N. Singh, Y. Sun, N. S. Shen, X. Yan, G.-Q. Lo and D.-L. Kwong, "Vertical-Si-Nanowire SONOS Memory for Ultrahigh-Density Application," *IEEE Electron Device Letters*, vol. 30, no. 8, pp. 879–881, Aug. 2009, doi: 10.1109/LED.2009.2024442.
18. A. Padovani, L. Larcher, D. Heh and G. Bersuker, "Modeling TANOS Memory Program Transients to Investigate Charge-Trapping Dynamics," *IEEE Electron Device Letters*, vol. 30, no. 8, pp. 882–884, Aug. 2009, doi: 10.1109/LED.2009.2024622.
19. M. F. Beug, T. Melde, M. Czernohorsky, R. Hoffmann, J. Paul, R. Knoefler and A. T. Tilke, "Analysis of TANOS Memory Cells with Sealing Oxide Containing Blocking Dielectric," *IEEE Transactions on Electron Devices*, vol. 57, no. 7, pp. 1590–1596, May 2010, doi: 10.1109/TED.2010.2049217.
20. C. M. Compagnoni, A. Mauri, S. M. Amoroso, A. Maconi and A. S. Spinelli, "Physical Modeling for Programming of TANOS Memories in the Fowler–Nordheim Regime," *IEEE Transactions on Electron Devices*, vol. 56, no. 9, pp. 2008–2015, Sep. 2009, doi: 10.1109/TED.2009.2026315.
21. J. Kolodzey, E.A. Chowdhury, T. N. Adam, G. Qui, I. Rau, J. O. Olowolafe, J. S. Suehle and Y. Chen, "Electrical Conduction and Dielectric Breakdown in Aluminum Oxide Insulators on Silicon," *IEEE Transactions on Electron Devices*, vol. 47, no. 1, pp. 121–128, Jan. 2000, doi: 10.1109/16.817577.

22. Y. N. Tan, W. K. Chim, W. K. Choi, M. S. Joo and B. J. Cho, "Hafnium Aluminum Oxide as Charge Storage and Blocking-Oxide Layers in SONOS-type Nonvolatile Memory For High-Speed Operation," in *IEEE Transactions on Electron Devices*, vol. 53, no. 4, pp. 654–662, Apr. 2006, doi: 10.1109/TED.2006.870273.
23. S. B. Rahi and P. Rastogi, "A Review Report for Next Generation of CMOS Technology as Spintronics: Fundamentals, Applications and Future," *International Journal of Advances in Engineering & Technology*, vol. 6, no. 2, pp. 696–702, May 2013.
24. Y.-N. Tan, W. -K. Chim, B. J. Cho and W.-K. Choi, "Over-Erase Phenomenon in SONOS-Type Flash Memory and Its Minimization Using a Hafnium Oxide Charge Storage Layer," in *IEEE Transactions on Electron Devices*, vol. 51, no. 7, pp. 1143–1147, July 2004, doi: 10.1109/TED.2004.829861.
25. S. B. Rahi, B. Ghosh and P. Asthana, "A Simulation-Based Proposed High-k Heterostructure AlGaAs/Si Junctionless n-type Tunnel FET," *Journal of Semiconductors*, vol. 35, no. 11, pp. 114005, Nov. 2014, doi: 10.1088/1674–4926/35/11/114005.
26. Y. S. Song, S. Kim, G. Kim, H. Kim, J.-H. Lee, J. H. Kim and B.-G. Park, "Improvement of Self-Heating Effect in Ge Vertically Stacked GAA Nanowire pMOSFET by Utilizing Al2O3 for High-Performance Logic Device and Electrical/Thermal Co-Design," *Japanese Journal of Applied Physics*, vol. 60, no. SC, pp. SCCE04, Mar. 2021, doi: 10.35848/1347–4065/abec5c.
27. B. E. Davis and N. C. Strandwitz, "A Systematic Investigation of Aluminum Oxide Passivating Tunnel Layers for Titanium Oxide Electron-Selective Contacts," *2020 47th IEEE Photovoltaic Specialists Conference (PVSC)*, pp. 1557–1561, 2020, doi: 10.1109/PVSC45281.2020.9300502.
28. R. Chen, Z. Qin, Y. Wang, D. Liu, Z. Shao and Y. Guan, "On-Demand Block-Level Address Mapping in Large-Scale NAND Flash Storage Systems," *IEEE Transactions on Computers*, vol. 64, no. 6, pp. 1729–1741, 1 June 2015, doi: 10.1109/TC.2014.2329680.
29. C. Du, Y. Yao, J. Zhou and X. Xu, "VBBMS: A Novel Buffer Management Strategy for NAND Flash Storage Devices," *IEEE Transactions on Consumer Electronics*, vol. 65, no. 2, pp. 134–141, May 2019, doi: 10.1109/TCE.2019.2910890.
30. V. KhademHosseini, D. Dideban, M. T. Ahmadi and R. Ismail, "An Analytical Approach to Model Capacitance and Resistance of Capped Carbon Nanotube Single Electron Transistor," *AEU-International Journal of Electronics and Communications*, vol. 90, pp. 97–102, June 2018, doi: 10.1016/j.aeue.2018.04.015.
31. H. Tasdighi and D. Dideban, "Analysis and Modeling of the Effect of Statistical Fluctuations on (6T) nano-CMOS SRAM Cell Stability," *Journal of Electron Devices*, vol. 21, pp. 1834–1841.
32. J. Fu, N. Singh, C. Zhu, G. Lo and D. Kwong, "Integration of High-k Dielectrics and Metal Gate on Gate-All-Around Si-Nanowire-Based Architecture for High-Speed Nonvolatile Charge-Trapping Memory," *IEEE Electron Device Letters*, vol. 30, no. 6, pp. 662–664, June 2009, doi: 10.1109/LED.2009.2019254.
33. F. Driussi, S. Spiga, A. Lamperti, G. Congedo and A. Gambi, "Simulation Study of the Trapping Properties of HfO2-Based Charge-Trap Memory Cells," *IEEE Transactions on Electron Devices*, vol. 61, no. 6, pp. 2056–2063, June 2014, doi: 10.1109/TED.2014.2316374.
34. E. Gnani, A. Gnudi, S. Reggiani, G. Baccarani, J. Fu, N. Singh, G.Q. Lo and D.L. Kwong, "Performance Analysis of Nonvolatile Gate-All-Around Charge-Trapping TAHOS Memory Cells," *2009 International Semiconductor Device Research Symposium*, pp. 1–2, 2009, doi: 10.1109/ISDRS.2009.5378209.
35. T. Ali, K. Mertens, R. Olivo, M. Rudolph, S. Oehler, K. Kühnel, D. Lehninger, F. Müller, M. Lederer, R. Hoffmann and P. Schramm, "A Novel Hybrid High-Speed and Low Power Antiferroelectric HSO Boosted Charge Trap Memory for High-Density Storage," *2020 IEEE International Electron Devices Meeting (IEDM)*, pp. 18.3.1–18.3.4, 2020, doi: 10.1109/IEDM13553.2020.9371980.

36. G. Congedo, A. Lamperti, L. Lamagna and S. Spiga, "Stack Engineering of TANOS Charge-Trap Flash Memory Cell Using High-κ ZrO_2 Grown by ALD as Charge Trapping Layer," *Microelectronic Engineering*, vol. 88, no. 7, pp. 1174–1177, July 2011, doi: 10.1016/j.mee.2011.03.066.
37. Y. S. Song, T. Jang, K. K. Min, M.-H. Baek, J. Yu, Y. Kim, J.-H. Lee and B.-G. Park, "Tunneling Oxide Engineering for Improving Retention in Nonvolatile Charge-Trapping Memory with $TaN/Al_2O_3/HfO_2/SiO_2/Al_2O_3/SiO_2/Si$ Structure," *Japanese Journal of Applied Physics*, vol. 59, pp. 061006, June 2020, doi: 10.35848/1347–4065/ab8275.
38. Y. S. Song and B.-G. Park, "Retention Enhancement in Low Power NOR Flash Array with High-κ–Based Charge-Trapping Memory by Utilizing High Permittivity and High Bandgap of Aluminum Oxide," *Micromachines*, vol. 12, no. 3, pp. 328, Mar. 2021, doi: 10.3390/mi12030328.
39. A. Bhattacharyya, "Historical Progression of NVM Devices," *Silicon Based Unified Memory Devices and Technology*, 1st ed. Boca Raton, FL: CRC Press, ch. 2, pp. 13–34, 2017.
40. Y. S. Song, J. H. Kim, G. Kim, H.-M. Kim, S. Kim and B.-G. Park, "Improvement in Self-Heating Characteristic by Incorporating Hetero-Gate-Dielectric in Gate-All-Around MOSFETs," *IEEE Journal of the Electron Devices Society*, vol. 9, pp. 36–41, 2021, doi: 10.1109/JEDS.2020.3038391.
41. Y. Choi, K. Lee, K.Y. Kim, S. Kim, J. Lee, R. Lee, H.M. Kim, Y.S. Song, S. Kim, J.H. Lee, and B.G. Park, "Simulation of the Effect of Parasitic Channel Height on Characteristics of Stacked Gate-All-Around Nanosheet FET," *Solid-State Electronics*, vol. 164, pp. 107686, Feb. 2020, doi: 10.1016/j.sse.2019.107686.
42. Y. S. Song, S. Hwang, K. K. Min, T. Jang, Y. Choi, J. Yu, J.-H. Lee and B.-G. Park, "Electrical and Thermal Performances of Omega-Shaped-Gate Nanowire Field Effect Transistors for Low Power Operation," *Journal of Nanoscience and Nanotechnology*, vol. 20, no. 7, pp. 4092–4096, July 2020, doi: 10.1166/jnn.2020.17787.
43. Y. S. Song, H. Kim. J. Yu and J. Lee, "Improvement in Self-Heating Characteristic by Utilizing Sapphire Substrate in Omega-Gate-Shaped Nanowire Field Effect Transistor for Wearable, Military, and Aerospace Application," *Journal of Nanoscience and Nanotechnology*, vol. 21, no. 5, pp. 3092–3098, May 2021, doi: 10.1166/jnn.2021.19149.
44. T. Kim, Y. S. Song and B.-G. Park, "Overflow Handling Integrate-and-Fire Silicon-on-Insulator Neuron Circuit Incorporating a Schmitt Trigger Implemented by Back-Gate Effect," *Journal of Nanoscience and Nanotechnology*, vol. 19, no. 10, pp. 6183–6186, Oct. 2019, doi: 10.1166/jnn.2019.17004.
45. J. H. Kim, H. W. Kim, Y. S. Song, S. Kim and G. Kim, "Analysis of Current Variation with Work Function Variation in L-Shaped Tunnel-Field Effect Transistor," *Micromachines*, vol. 11, no. 8, pp. 780, Aug. 2020, doi: 10.3390/mi11080780
46. H. B. Joseph, S. K. Singh, R. M. Hariharan, Y. Tarauni and D. J. Thiruvadigal, "Simulation Study of Gated Nanowire InAs/Si Hetero p Channel TFET and Effects of Interface Trap," *Materials Science in Semiconductor Processing*, vol. 103, pp. 104605, Nov. 2019, doi: 10.1016/j.mssp.2019.104605.
47. F. Bedeschi, R. Fackenthal, C. Resta, E.M. Donze, M. Jagasivamani, E.C. Buda, F. Pellizzer, D.W. Chow, A. Cabrini, G.M.A. Calvi and R. Faravelli, "A Bipolar-Selected Phase Change Memory Featuring Multi-Level Cell Storage," *IEEE Journal of Solid-State Circuits*, vol. 44, no. 1, pp. 217–227, Jan. 2009, doi: 10.1109/JSSC.2008.2006439.
48. T. Cho, Y.T. Lee, E.C. Kim, J.W. Lee, S. Choi, S. Lee, D.H. Kim, W.G. Han, Y.H. Lim, J.D. Lee and J.D. Choi, "A Dual-Mode NAND Flash Memory: 1-Gb Multilevel and High-Performance 512-Mb Single-Level Modes," *IEEE Journal of Solid-State Circuits*, vol. 36, no. 11, pp. 1700–1706, Nov. 2001, doi: 10.1109/4.962291.
49. M. H. White, Y. Yang, A. Purwar and M. L. French, "A Low Voltage SONOS Nonvolatile Semiconductor Memory Technology," *IEEE Transactions on Components, Packaging, and Manufacturing Technology: Part A*, vol. 20, no. 2, pp. 190–195, June 1997, doi: 10.1109/95.588573.

50. S. -I. Minami and Y. Kamigaki, "A Novel Monos Nonvolatile Memory Device Ensuring 10-Year Data retention After 10/Sup 7/Erase/Write Cycles," *IEEE Transactions on Electron Devices*, vol. 40, no. 11, pp. 2011–2017, Nov. 1993, doi: 10.1109/16.239742.
51. Z. Wei, T. Takagi, Y. Kanzawa, Y. Katoh, T. Ninomiya, K. Kawai, S. Muraoka, S. Mitani, K. Katayama, S. Fujii and R. Miyanaga, "Demonstration of High-Density ReRAM Ensuring 10-year Retention at 85°C Based on a Newly Developed Reliability Model," *2011 International Electron Devices Meeting*, 2011, pp. 31.4.1–31.4.4, doi: 10.1109/IEDM.2011.6131650.
52. C.H. Lai, C.C. Huang, K.C. Chiang, H.L. Kao, W.J. Chen, A. Chin and C.C. Chi, "Fast High-k AIN MONOS Memory with Large Memory Window and Good Retention," *63rd Device Research Conference Digest, 2005. DRC'05.*, pp. 99–100, doi: 10.1109/DRC.2005.1553074.
53. T. Ali, K. Mertens, R. Olivo, M. Rudolph, S. Oehler, K. Kühnel, D. Lehninger, F. Müller, R. Hoffmann, P. Schramm and K. Biedermann, "Impact of the Nonlinear Dielectric Hysteresis Properties of a Charge Trap Layer in a Novel Hybrid High-Speed and Low-Power Ferroelectric or Antiferroelectric HSO/HZO Boosted Charge Trap Memory," *IEEE Transactions on Electron Devices*, vol. 68, no. 4, pp. 2098–2106, April 2021, doi: 10.1109/TED.2021.3049758.
54. T. Joshi, Y. Singh and B. Singh, "Extended-Source Double-Gate Tunnel FET With Improved DC and Analog/RF Performance," *IEEE Transactions on Electron Devices*, vol. 67, no. 4, pp. 1873–1879, April 2020, doi: 10.1109/TED.2020.2973353.
55. S. L. Tripathi, R. Mishra and R. A. Mishra, "Characteristic Comparison of Connected DG FINFET, TG FINFET and Independent Gate FINFET on 32 nm Technology," *2012 2nd International Conference on Power, Control and Embedded Systems*, pp. 1–7, 2012, doi: 10.1109/ICPCES.2012.6508037.
56. S. L. Tripathi, R. Mishra and R. A. Mishra, "Multi-Gate MOSFET Structures with High-k Dielectric Materials," *Journal of Electron Devices*, vol. 16, pp. 1388–1394, Mar. 2012.
57. S. L. Tripathi, R. Mishra, V. Narendra and R. A. Mishra, "High Performance Bulk FinFET with Bottom Spacer," *2013 IEEE International Conference on Electronics, Computing and Communication Technologies*, pp. 1–5, 2013, doi: 10.1109/CONECCT.2013.6469282.
58. T. Joshi, B. Singh and Y. Singh, "Controlling the Ambipolar Current in Ultrathin SOI Tunnel FETs Using the Back-Bias Effect," *Journal of Computational Electronics*, vol. 19, pp. 658–667, Mar. 2020, doi: 10.1007/s10825-020-01484–8.
59. T. Joshi, Y. Singh and B. Singh, "Dual-Channel Trench-Gate Tunnel FET for Improved ON-Current and Subthreshold Swing," *Electronics Letters*, vol. 55, no. 21, pp. 1152–1155, Oct. 2019, doi: 10.1049/el.2019.2219.
60. I. Z. Mitrovic, Y. Lu, O. Buiu and S. Hall, "Current Transport Mechanisms in (HfO_2) x(SiO_2) 1−x/SiO_2gate Stacks," *Microelectronic Engineering*, vol. 84, no. 9–10, pp. 2306–2309, Sep. 2007, doi: 10.1016/j.mee.2007.04.087.
61. S. B. Rahi, P. Asthana and S. Gupta, "Heterogate Junctionless Tunnel Field-Effect Transistor: Future of Low-Power Devices," *Journal of Computational Electronics*, vol. 16, no. 1, pp. 308–38, Mar. 2017, doi: 10.1007/s10825-016-0936–9.
62. P. K. Asthana, B. Ghosh, S. B. Rahi and Y. Goswami, "Optimal Design for a High Performance H-JLTFET Using HfO_2 as a Gate Dielectric for Ultra Low Power Applications," *RSC Advances*, vol. 4, no. 43, pp. 22803–22807, Mar. 2014, doi: 10.1039/C4RA00538D.
63. A. Singh, S. Chaudhury, C. K. Pandey, S. M. Sharma and C. K. Sarkar, "Design and Analysis of High k Silicon Nanotube Tunnel FET Device," *IET Circuits, Devices & Systems*, vol. 13, no. 8, pp. 1305–1310, Nov. 2019, doi: 10.1049/iet-cds.2019.0230.
64. C. K. Pandey, D. Dash and S. Chaudhury, "Approach to Suppress Ambipolar Conduction in Tunnel FET Using Dielectric Pocket," *Micro & Nano Letters*, vol. 14, no. 1, pp. 86–90, Jan. 2019, doi: 10.1049/mnl.2018.5276.

65. H. B. Joseph, S. K. Singh, R.M. Hariharan, P. A. Priya, N. M. Kumar and D. John Thiruvadigal, "Hetero Structure PNPN Tunnel FET: Analysis of Scaling Effects on Counter Doping," *Applied Surface Science*, vol. 449, pp. 823–828, Aug. 2018, doi: 10.1016/j.apsusc.2018.01.274.
66. Y. S. Song, T. Jang, H.-M. Kim, J.-H. Lee and B.-G. Park, "Erase Speed Enhancement with Low Power Operation by Incorporating Boron Doping," *Journal of Semiconductor Technology and Science*, vol. 21, no. 2, pp. 92–100, Apr. 2021, doi: 10.5573/JSTS.2021.21.2.092.
67. J.-D. Lee, S.-H. Hur and J.-D. Choi, "Effects of Floating-Gate Interference on NAND Flash Memory Cell Operation," in *IEEE Electron Device Letters*, vol. 23, no. 5, pp. 264–266, May 2002, doi: 10.1109/55.998871.
68. C. Lee, S. H. Baek and K. H. Park, "A Hybrid Flash File System Based on NOR and NAND Flash Memories for Embedded Devices," *IEEE Transactions on Computers*, vol. 57, no. 7, pp. 1002–1008, July 2008, doi: 10.1109/TC.2008.14.
69. K. Kim and J. Choi, "Future Outlook of NAND Flash Technology for 40nm Node and Beyond," *2006 21st IEEE Non-Volatile Semiconductor Memory Workshop*, pp. 9–11, 2006, doi: 10.1109/.2006.1629474.
70. P. K. Pal, B. K. Kaushik and S. Dasgupta, "Asymmetric Dual-Spacer Trigate FinFET Device-Circuit Codesign and Its Variability Analysis," *IEEE Transactions on Electron Devices*, vol. 62, no. 4, pp. 1105–1112, April 2015, doi: 10.1109/TED.2015.2400053.
71. Kumar D., "Performance Evaluation of Double Gate Tunnel FET Based Chain of Inverters and 6-T SRAM Cell," *Engineering Research Express*, vol. 1, no. 2, pp. 025055, Dec. 2019, doi: 10.1088/2631–8695/ab5f16.
72. Kumar D., and Jain P. "Double Gate Tunnel Field Effect Transistor with Extended Source Structure and Impact Ionization Enhanced Current," *Intelligent Communication, Control and Devices, Advances in Intelligent Systems and Computing*, vol. 624, pp. 973–980, Apr. 2018, doi: 10.1007/978–981-10–5903-2_102.
73. S. Ahmad, N. Alam and M. Hasan, "Robust TFET SRAM cell for ultra-low power IoT application," *2017 International Conference on Electron Devices and Solid-State Circuits (EDSSC)*, pp. 1–2, 2017, doi: 10.1109/EDSSC.2017.8333263.

11 TFET-Based Flash Analog-to-Digital Converter

Asra Ansari and Naushad Alam
ZHCET, AMU

CONTENTS

11.1 INTRODUCTION

In accordance with Moore's law, the number of transistors doubles every 2 years because the size of MOSFET is reduced with successive technology nodes to design faster and functionality--rich integrated circuits (ICs). However, increased leakage power poses a significant challenge while designing an IC. Along with the reduction of channel length, leakage current increases, and short channel effects (SCEs) came into picture. Therefore, to reduce the SCEs, new designs and technologies are being explored that include FinFET, junctionless transistor, carbon nanotube FET, TFET, etc. [1]. Among the emerging future technologies, TFET is a promising device in terms of low-voltage, low-power applications. In this work, we observed the

DOI: 10.1201/9781003240778-11

characteristics of TFET and conclude that it is a promising device and a better alternative for low-voltage low-power IC design with a steep subthreshold slope (SS) that can go below 60 mV/decade, ultra-low-power and low voltage, negligible SCEs, extremely low leakage current, and high I_{on}/I_{off} current ratio [2]. The operation of TFET is based on band-to-band tunneling (BTBT), that is, carriers tunnel through the barrier, which is a bandgap present between the valence band and the conduction band, for the flow of electron from source to drain. Besides the many attractive features of TFET, there are some TFET device characteristics such as PIN forward bias current, low ON current, and unidirectional current conduction (i.e., the source–drain terminals are asymmetric) that pose challenges in the circuit design. These characteristics of TFET differ from those of conventional MOSFET characteristics; therefore, TFET implementation of existing MOSFET circuits requires some modifications [2].

From the above discussion, it is concluded that TFET is a potential substitute for conventional MOSFET for designing low-power, low-voltage, low leakage current, and energy-efficient circuits. In this work, we present the implementation of a 4-bit flash ADC using TFET with a 20 nm channel length. Ultra-low-power ADC attributed to its TFET implementation would be very useful in communication systems, sensors, biomedical devices, and other hand-held battery operated devices. Flash ADC is the fastest ADC available among various others ADCs, but its power consumption is high [3]. Therefore, this work aims at developing ultra-low-voltage designs with ultralow power consumption for high-speed and medium-resolution flash ADC to improve energy efficiency. We also compare CMOS-based flash ADC with the TFET-based flash ADC.

11.2 OVERVIEW OF TUNNEL FIELD EFFECT TRANSISTOR

CMOS is the workhorse of the present day semiconductor industry. But due to the scaling down of channel length, leakage current and SS have significantly increased. This is attributed to the various SCEs that come into picture at reduced channel lengths, which ultimately reduce the overall performance of CMOS devices. Therefore, we need a device that is capable of having a lower SS below 60 mV/decade, thereby supporting voltage scaling. TFET offers a high I_{on}/I_{off} current ratio at low voltages, high transconductance in the subthreshold region, and high output resistance as compared to CMOS, which translates into high gain [4].

Structurally, TFET is a gated p-i-n diode that involves BTBT mechanism for current conduction. The electrostatic potential of the intrinsic region is controlled by a gate terminal.

Figure 11.1 shows a schematic of TFET in which the source is of p-type, channel is of n-type, and drain is of n-type. When no V_{gs} is applied, there will be no conduction. The valence band of the source is lower than the conduction band of the channel; therefore, no electrons will tunnel through it. When a positive V_{gs} is applied, then the conduction band of the channel will go below the valence band of the source. Therefore, electrons will tunnel through it, and conduction will take place. When a negative V_{gs} is applied, then the valence band of the channel goes above the conduction band of the drain. Therefore, conduction takes place through tunneling again. However, in this case, a channel drain junction is formed and the channel is changed

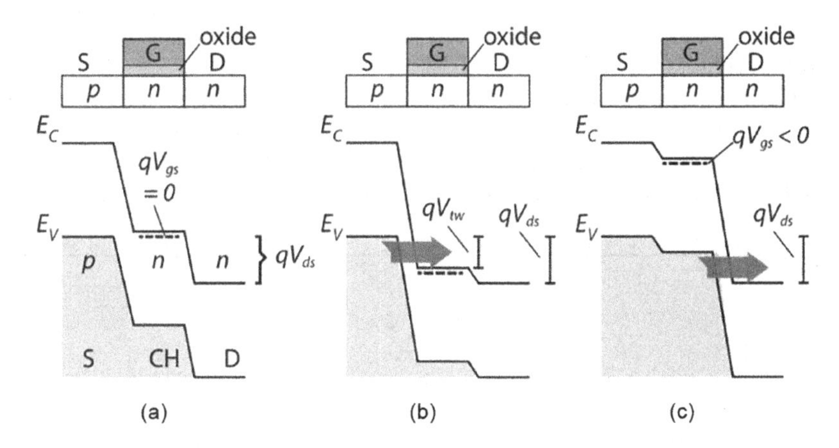

FIGURE 11.1 Energy band diagram of tunnel field effect transistor: (a) OFF state, (b) ON state, and (c) ambipolar state [1].

from one carrier type to another carrier type leading to the nomenclature ambipolar. This region seriously impacts the performance of a TFET device, which leads to poor stability of the circuit. The overall working region of a TFET is divided into four regions labeled as the tunneling region (Kane-Sze), diode region, ambipolar region, and the negative differential resistance (NDR) region. It is important to make sure that TFET works in the Kane-Sze region, otherwise, other regions of operation will seriously impact its performance and lead to instability of the device and compromised circuit performance [1].

11.2.1 TFET and Its Characteristics

In n-type TFET, the source has p+ doping, the channel has n-type doping, and the drain has n+ doping. On the other hand, in p-type TFET, the source has n+ doping, the channel has n-type doping, and the drain has p+ doping. In this work, double-gate InAs TFET is used for the implementation of 4-bit flash ADC. Some of the device characteristics are presented in Table 11.1.

Simulation of the NTFET device has been performed using the Verilog-A model with the HSPICE tool. Characteristics such as I_d vs V_{ds} and I_d vs V_{gs} have been studied,

TABLE 11.1
Process Parameters of the InAs TFET Model

Parameter	Value
Gate length, L_g (nm)	20
Equivalent oxide thickness, ETO (nm)	0.2
Channel thickness, Tch (nm)	5
Threshold voltage, V_{th} (V)	0.145
Semiconductor band gap, E_g (eV)	0.354

and it is observed that TFET demonstrates unidirectional current flow, that is, there is no drain current for $V_{ds} < 0\,V$ for NTFET.

In Figure 11.2, the characteristics of drain current vs drain–source voltage are plotted, and it is observed that the current increases on increasing the gate voltage. In Figure 11.3, drain current vs gate–source voltage is plotted, and we observe that NTFET starts conducting when V_{gs} becomes greater than V_{th}. In Figure 11.4, we take

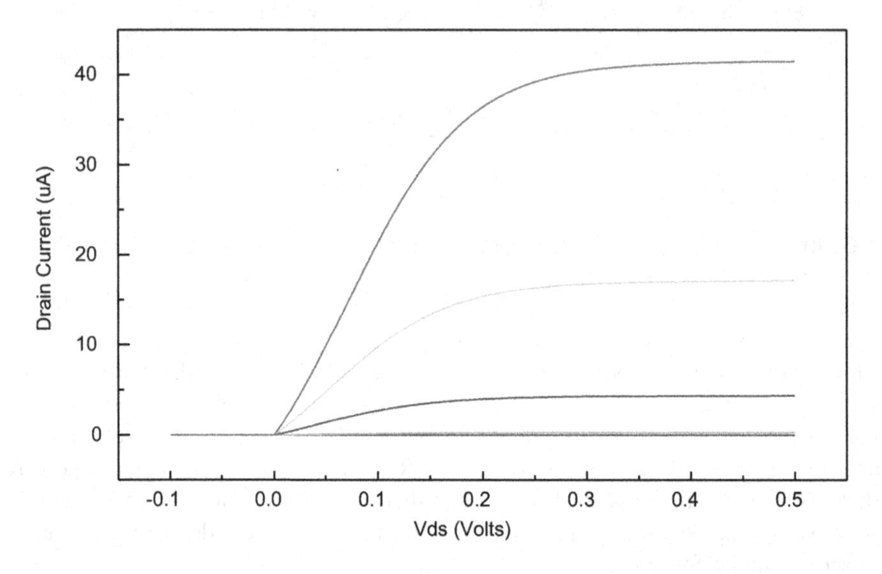

FIGURE 11.2 I_D vs V_{DS} varying from −0.1 to 0.5 V with a gate voltage sweep of −0.1 to 0.4 V in a step of 0.1 V.

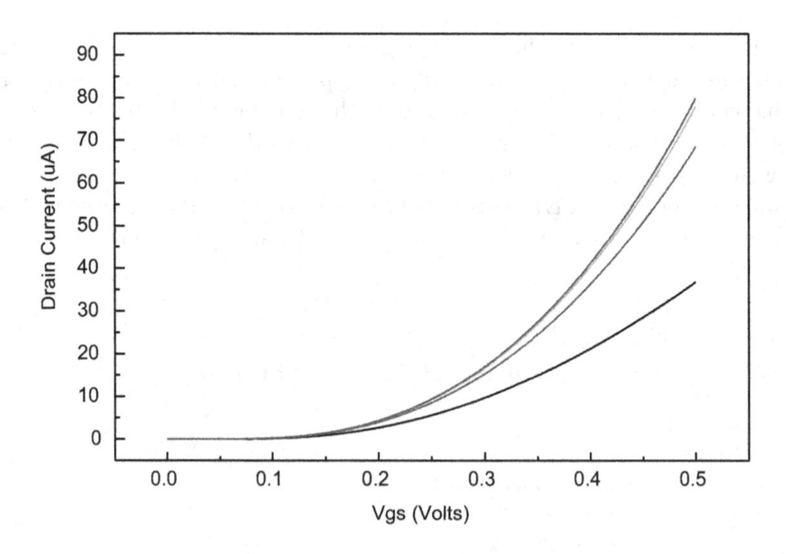

FIGURE 11.3 I_D vs V_{GS} varying from 0 to 0.5 V with a drain voltage sweep of 0–0.5 V in a step of 0.1 V.

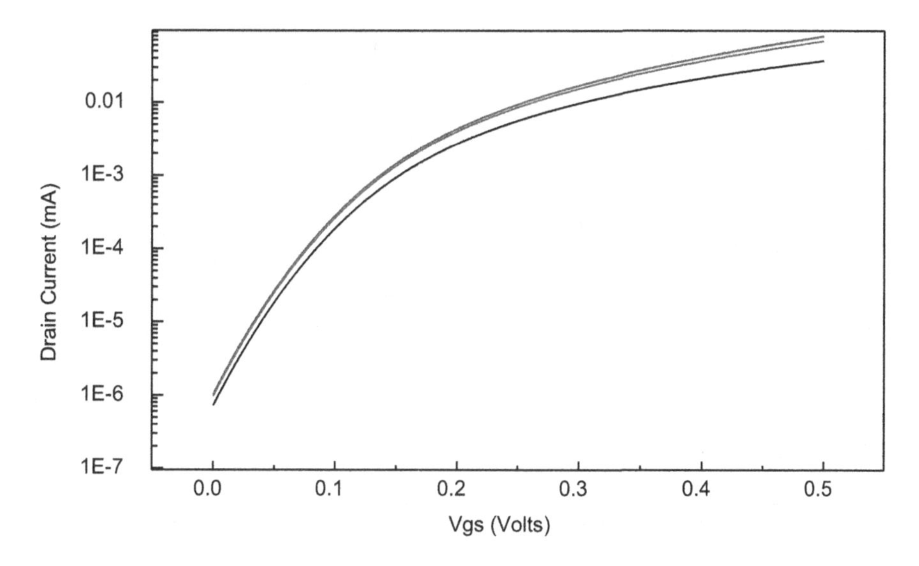

FIGURE 11.4 Plot between log I_d and V_{gs} for the calculation of SS.

the logarithmic of the drain current and calculate its SS from the graph, which is found to be less than 60 mV/decade, which is approximately equal to 45.7 mV/decade. This SS indicates that TFET has a low leakage current and supports better voltage scaling as compared to MOSFET and is good for low-voltage and low-power applications.

11.3 TFET-BASED DESIGN CONSIDERATION

11.3.1 TFET-Based Inverter

This is the basic building block using TFET, and its performance can be studied by performing simulation using HSPICE. This circuit would be later used for the implementation of comparator and encoder. TFET inverter consists of one NTFET and one PTFET, input is given to the gate terminal of TFETs, and output is taken from the drain terminal of the TFETs (Figure 11.5).

11.3.1.1 Simulation of TFET Inverter

DC analysis has been performed using HSPICE simulation. For this, input is varied from 0 to 0.9 V, and the supply voltage is 0.9 V. In Figure 11.6, it is observed that the voltage transfer characteristic (VTC) curve of TFET inverter has a straight line and a gentle slope.

11.3.2 TFET-Based Sample and Hold Circuit

Sample and hold circuit is an important building block in data converter systems, which helps to minimize errors that may occur due to delays in internal operation of the converter. This circuit will be later used in the design of voltage comparator. In the conventional S/H circuit, when MOSFET is replaced by TFET, it results in a

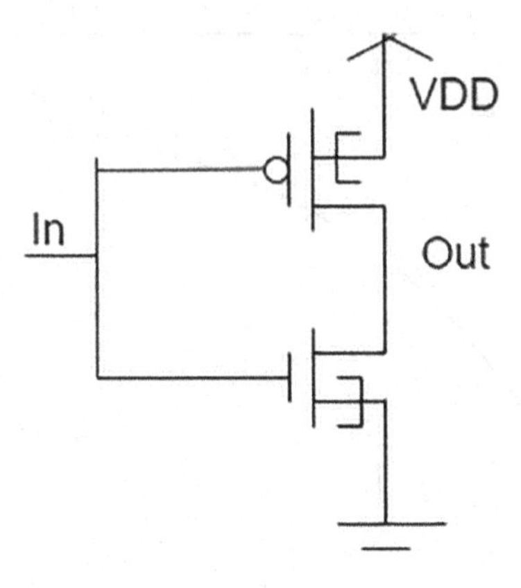

FIGURE 11.5 Schematic diagram of a TFET inverter.

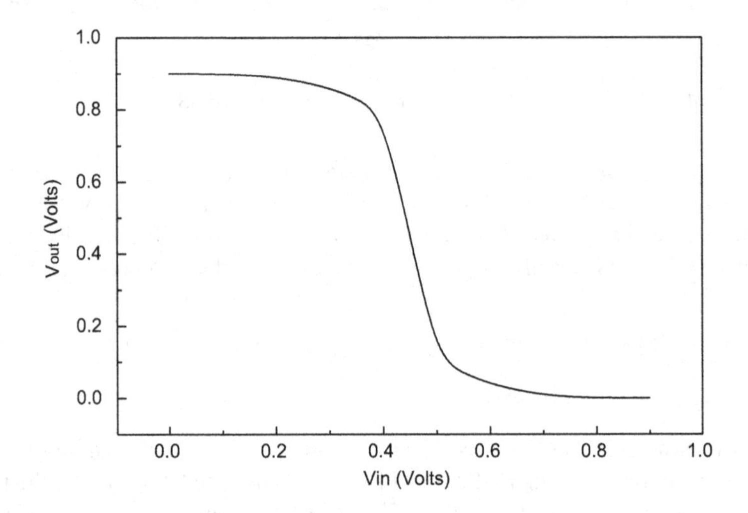

FIGURE 11.6 Voltage transfer characteristics of TFET inverter.

circuit as shown in Figure 11.7 (*hereafter referred to as circuit 1*). Simulated waveforms of circuit 1 have been observed, and we found that there was some trouble and the signal was not properly sampled during fall transition as shown in Figure 11.8. This is attributed to the PIN forward bias effect, and the gate terminal loses control over the device and current flows between drain and source. In order to eliminate this limitation, the conventional S/H circuit 1 has been improved to obtain proper sampled signals, and the improved circuit is referred to as circuit 2 as shown in Figure 11.9. Circuit 2 has been verified for various sampling frequencies, and for different amplitudes of input signals as shown in Figure 11.10.

Various performance metrics of sample and hold circuit 2 have been measured and are tabulated in Table 11.2.

The droop rate is zero, and also the hold mode settling time is zero. It is also observed that by decreasing the value of the capacitor, the droop rate increases and

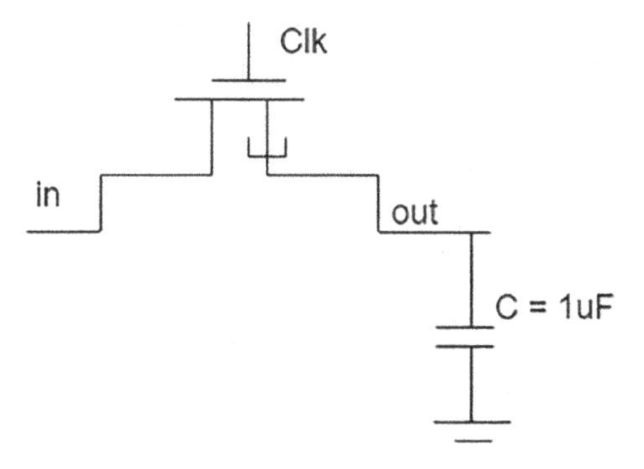

FIGURE 11.7 Conventional S/H circuit implemented using TFET and referred to as circuit 1.

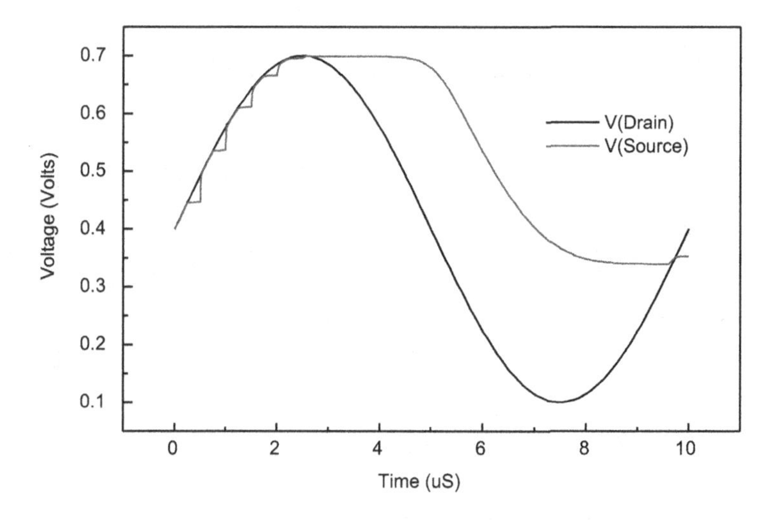

FIGURE 11.8 Behavior of circuit 1 for the input voltage of 0.3 V peak to peak and sampling frequency of 100 KHz with load capacitance of 1 pF.

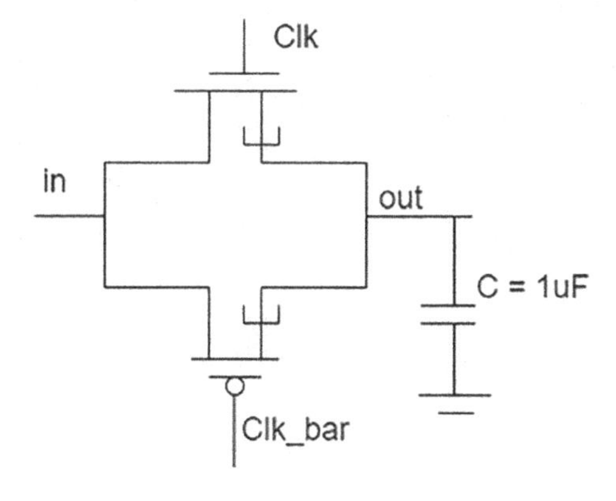

FIGURE 11.9 Modified S/H circuit referred to as circuit 2.

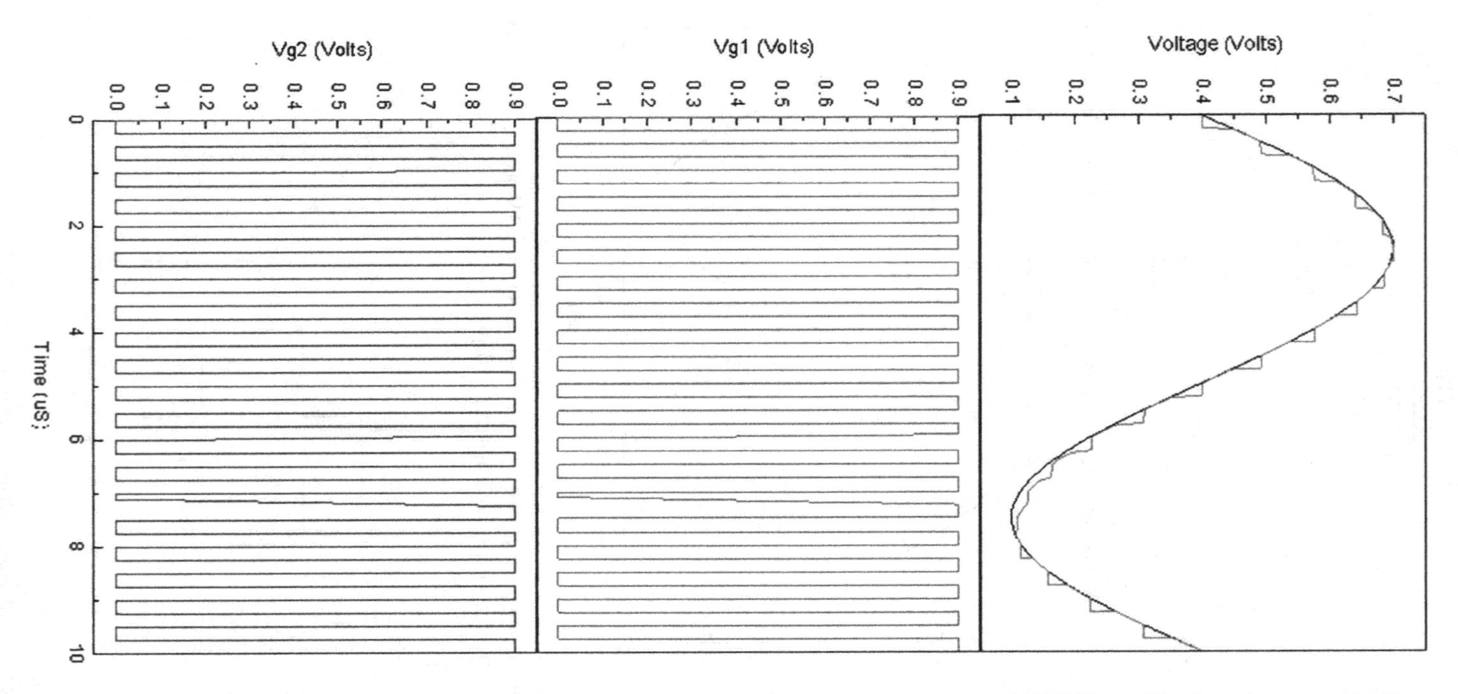

FIGURE 11.10 Behavior of circuit 2 for the input voltage of 0.3 V peak to peak, sampling frequency of 100 KHz, and load capacitance of 1 pF.

TABLE 11.2
Performance Metrics of Sample and Hold Circuit

Actual sampling time (μs)	Time to go from hold to sample (μs)	Acquisition time (μs)
0.519	0.501	0.018
1.54	1.5	0.04
2.05	2	0.05

acquisition time decreases. Therefore, TFET is a promising device option for the design of data converters. Performance parameters of the sample and hold circuit are improved with some modification in terms of droop rate, acquisition time, and hold mode settling time. TFET is also compatible with any digital/analog, internet of things (IoT), and biomedical applications. Comparator and encoder have been designed using TFET for converting the analog signal to digital signal properly, with the help of building blocks discussed in previous sections.

11.4 TFET-BASED COMPARATOR AND ENCODER DESIGN FOR FLASH ADC

11.4.1 Introduction to Comparator

Comparator is an important block of any ADC, and its design is also very important for the overall performance of any ADC. Speed of operation, power consumption, and the sensitivity of the input signal are important performance parameters while designing a comparator. The primary purpose of comparator in Figure 11.11 is to compare the input signal with the reference voltage and generate high- or low-output signal depending upon the comparison as shown in Figure 11.12. Basically, it is a decision-making circuit.

If the input voltage is greater than the reference voltage, then the output will be logic high, and if the input voltage is less than the reference voltage, the output will be logic low.

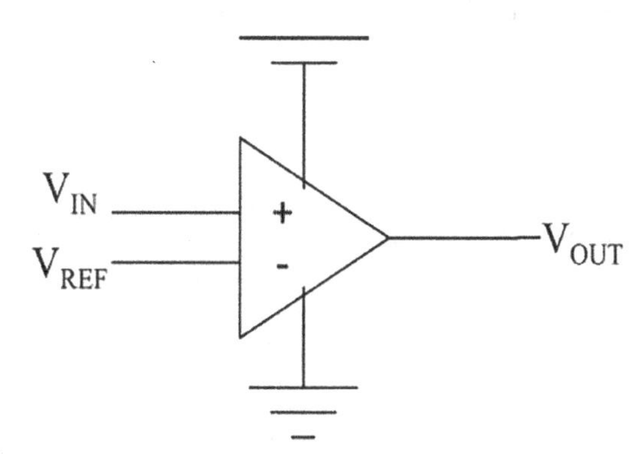

FIGURE 11.11 Basic diagram of comparator.

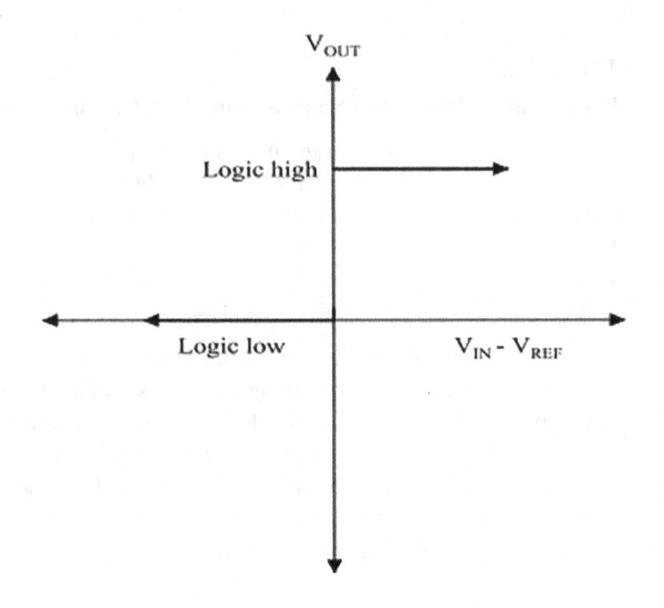

FIGURE 11.12 Comparator's ideal transfer characteristics.

This comparator is an inverter-based comparator designed using a TFET device that samples the signal only on a clock edge and proves to have more speed and less power dissipation as compared to its MOSFET counterpart.

11.4.2 TFET-Based Comparator Design

Figure 11.13 shows an inverter-based comparator that is based on the functioning of the clock signal, that is, positive and negative edges of the clock. When the clock signal is high, an input signal is sampled through the sample and hold circuit, S1. When the clock signal is low/CLK bar is high, a reference voltage is passed through the sample and hold circuit, S2. The inverter acts as a comparator, the latch (I3 and I4 constitute latch) as a memory element. To boost the gain of the comparator, an additional buffer has been used.

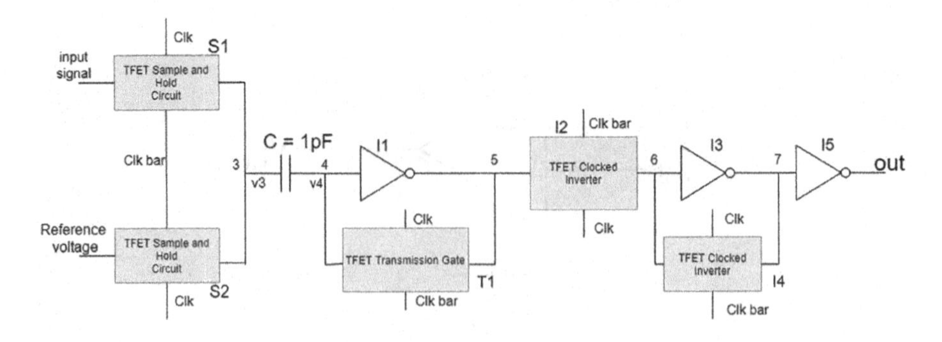

FIGURE 11.13 Block diagram of TFET inverter-based comparator.

TABLE 11.3
Value of Reference Voltages that are Used for the Implementation of TFET-Based Flash ADC

Reference Voltage	Value (V)
Comparator 1	0.1
Comparator 2	0.14
Comparator 3	0.18
Comparator 4	0.22
Comparator 5	0.26
Comparator 6	0.3
Comparator 7	0.34
Comparator 8	0.38
Comparator 9	0.42
Comparator 10	0.46
Comparator 11	0.5
Comparator 12	0.54
Comparator 13	0.58
Comparator 14	0.62
Comparator 15	0.66

As this comparator will be used in the implementation of the 4-bit flash ADC, 15 such comparators are required corresponding to which there are 15 reference voltages as shown in Table 11.3.

11.4.2.1 Working

During the sampling mode (when the clock is high), S1, T1, and I4 are closed and S2 and I2 are open. In this mode, an input signal is sampled with the help of a capacitor, C. During hold mode, when the clock bar is high, a reference voltage is passed through the sample and hold circuit. In this mode, S2 and I2 are closed and others are open. The inverter I1 acts as a comparator, and I3 and I4 act as a latch and a memory element, respectively.

During the sample mode, the input is sampled as the S1 is open using capacitor C. In this mode, T1 is closed, which makes nodes 4 and 5 short-circuited. The voltages of nodes 4 and 5 are equal, which helps the inverter to regain its switching threshold voltage, which is VDD/2. Actually, the inverter I1 is designed in such a way that the switching threshold voltage has to be VDD/2. As I2 is open, I3 and I4 act as a latch, keeping the previously stored value. Therefore, node 4 is has a voltage VDD/2, and node 3 is equal to Vin.

$$\text{Voltage of capacitor, } Vc = V4 - V3$$

$$Vc = VDD/2 - Vin$$

During the hold mode, when CLK bar is high, S2 and I2 are closed and the others are open. Constant reference voltages are passed through the sample and hold circuit, S2, and are accumulated over the capacitor; subsequently, the voltage at node 4 would be for comparison.

$$Vc = V4 - V3$$

$$Vc = V4 - Vref$$

$$V4 = Vc + Vref$$

$$V4 = VDD/2 - Vin + Vref$$

$$V4 = VDD/2 - (Vin - Vref)$$

Therefore, from the above equation, it is clear that V4 is the voltage that is present at the input of the inverter, I1. If Vin > Vref, the input at inverter I1 would be less than VDD/2, which makes the output of the inverter logic high (VDD), which is 0.9 V. Similarly, for Vin < Vref, the voltage of node 4 would be greater than VDD/2. This implies that the input of the inverter has a voltage greater than VDD/2, which leads to having an output voltage corresponding to logic 0(0V). There is a possibility that the circuit may not generate proper high and low voltages at the output of inverter I1–I3 that helps in getting full voltage levels of 0.9 and 0 V corresponding to logic high and logic low. I2 separates the latch and the comparator during sample mode, while I3 acts as a buffer for hold mode.

11.4.2.2 Components of TFET Comparator

11.4.2.2.1 TFET Sample and Hold Circuit

This circuit is implemented using NTFET and PTFET with their sources and drains connected together and clock and clock_bar applied at the gate terminals, respectively, as shown in Figure 11.14. Input is applied at the drain terminals of TFETs, while the output is taken from the source terminals of TFETs as shown in Figure 11.14.

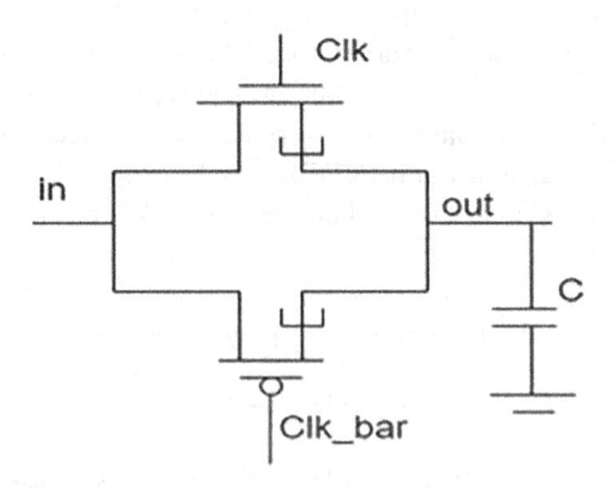

FIGURE 11.14 TFET sample and hold circuit.

11.4.2.2.2 TFET Transmission Gate

A TFET-based transmission gate is shown in Figure 11.15. Source terminals of NTFET and PTFET are connected together. Input is applied at the source terminal, and output is taken from the drain terminal. When the clock is high, the input signal is passed to the output terminal.

11.4.2.2.3 TFET Clocked Inverter

It is sensitive to a clock which means that it is working with the appropriate clock levels. In Figure 11.16, when the clock is low, it will work normally as an inverter, otherwise the source of PTFET is not connected to VDD and the source of NTFET is not connected to the ground, that is, the output node is floating.

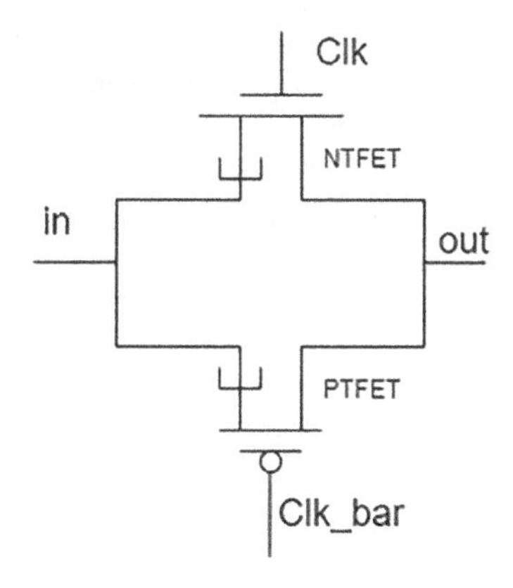

FIGURE 11.15 TFET transmission gate circuit.

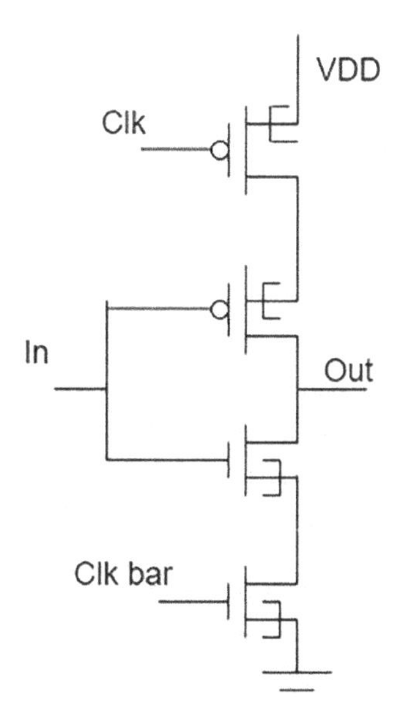

FIGURE 11.16 TFET clocked inverter circuit.

11.4.3 Design of Encoder for TFET-Based Flash ADC

11.4.3.1 Introduction to Encoder Design

ADC is used to convert analog signal into digital signal for communication, signal processing, System on Chips (SOCs), mixed-signal design, and other applications. Flash ADC provides high-speed and moderate resolution. Flash ADC of N bit requires (2^N-1) comparators, 2^N resistances, and an encoder. Outputs of comparators form a pattern of output in terms of consecutive 0s and 1s, which is known as the thermometer code. The thermometer code has to be converted into binary values, which can be possible using the thermometer-to-binary-code converter.

11.4.3.2 Architectures of Encoders

In a flash ADC, an input signal is compared with a reference voltage at the input of the comparator, and depending on the input signal voltage, the output of the comparator is either '0' or '1'. Subsequently, the generated thermometer code is applied to the input of the thermometer-to-binary-code converter. For flash ADC, delay and power are the important performance parameters for its functioning. Therefore, delay and power should be minimum for the IC design.

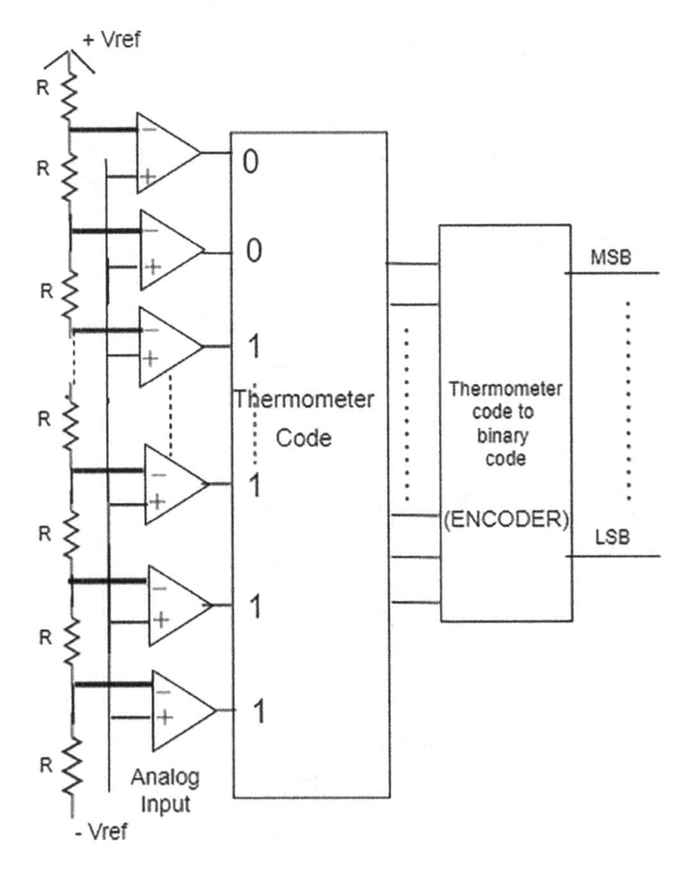

FIGURE 11.17 N bit flash ADC [3].

Flash ADC should be designed in such a way that it gives high speed and low power. There are some architectures of encoders that are used in flash ADC that are discussed below. These architectures are taken from existing literature and are compared with the one that is used in this work (Figure 11.17).

11.4.3.2.1 ROM-Based Encoder

This encoder has two stages: the first one is to convert the thermometer code to a one-hot code. The one-hot code provides the address location to the ROM in the second stage. At the memory location of ROM, its equivalent binary values are present. Therefore, it helps to convert the thermometer code to binary values. Disadvantages of ROM-based encoder are high power consumption and large delay [5].

11.4.3.2.2 Wallace Tree Encoder

This encoder counts the number of 1s at the output of comparators (thermometer code). But it has the disadvantage of high power and delay [6] (Figure 11.18).

11.4.3.2.3 Fat Tree Encoder

Fat tree encoder offers smaller delay, consumes lesser area as compared to ROM-based encoder and Wallace tree encoder. It comprises two stages. Initially, the input thermometer code is converted into a one-hot code. Subsequently, the one-hot code is converted into binary code using multiple trees of OR gates [7] (Figure 11.19).

11.4.3.2.4 MUX-Based Encoder

Multiplexer-based encoder is designed using 2:1 MUX and XOR gate [8]. This encoder is implemented in 180nm CMOS technology, with a 1.8V supply voltage [8]. In this

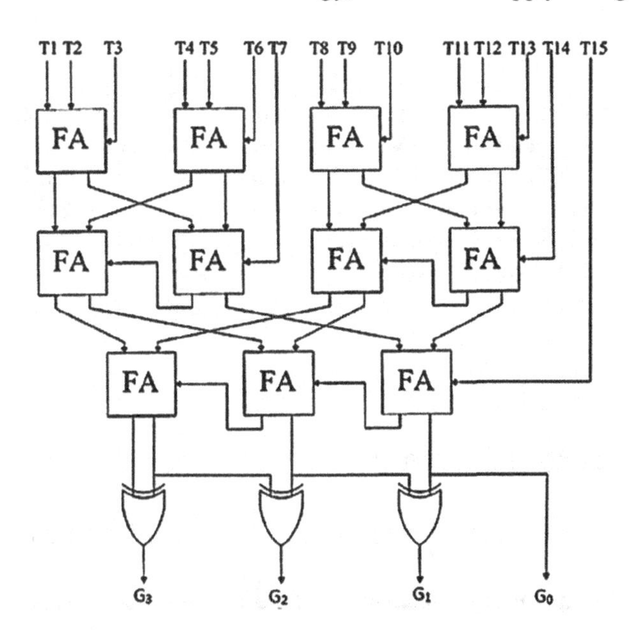

FIGURE 11.18 Wallace tree-based encoder.

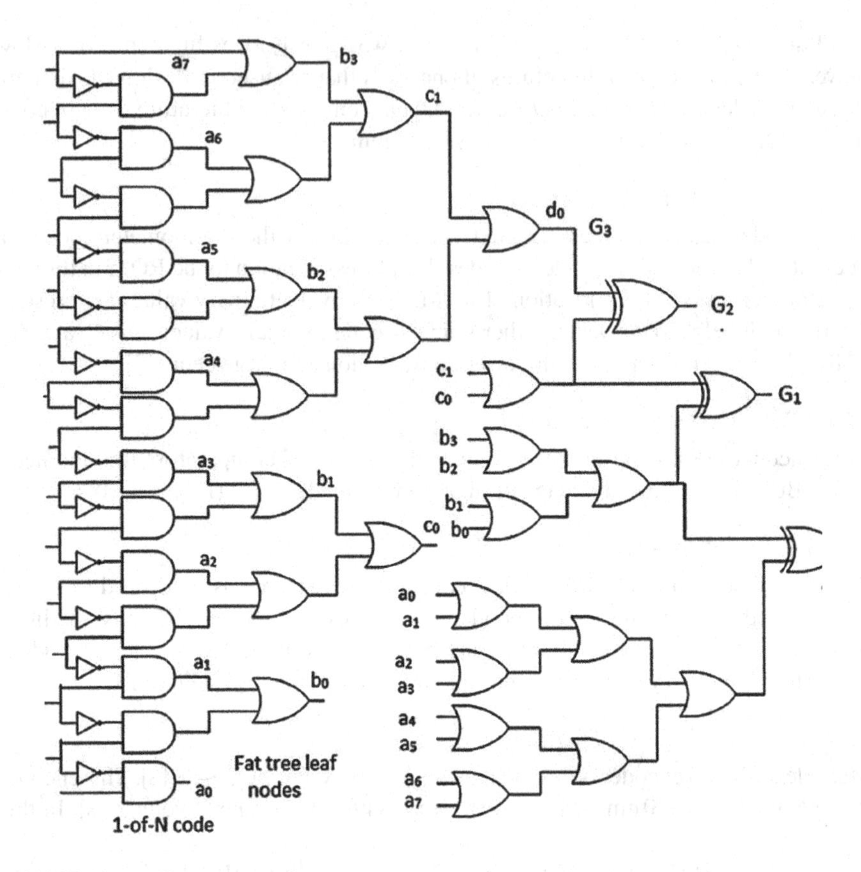

FIGURE 11.19 Fat tree encoder.

work, this encoder circuit has been implemented using 20nm TFET technology. This architecture of the encoder consumes less power and offers smaller delay. First, thermometer codes are passed through 2:1 MUX, and this converts it into Gray codes. Gray codes are converted into binary codes using XOR gates. Expressions between the thermometer code and Gray code are evaluated using Table 11.4 (Figure 11.20).

Using Table 11.4, expressions are derived as follows:

$$G4 = T8$$

$$G3 = T4.\overline{T12}$$

$$G2 = T2.\overline{T6} + T6.T10.T14$$

$$G1 = T1.\overline{T13} + T3.\left(T5.\ \overline{T7} + T7.\left(T9.\overline{T11} + T11.\left(T13.\overline{T15}\right)\right)\right)$$

The Gray codes are implemented using 2:1 MUX as shown in Figure 11.21, and for the binary values, the XOR gate-based circuit converts Gray code to binary code. 2:1 MUX and XOR gates have been designed using standard NAND gates implemented in the TFET technology (Figure 11.22).

TABLE 11.4
Truth Table for a 4-Bit Encoder

Thermometer Code															Gray code			
T15	T14	T13	T12	T11	T10	T9	T8	T7	T6	TS	T4	T3	T2	Tl	G4	G3	G2	Gl
0	0	0	0	0	0	0	0	0	0	0	0	0	0	0	0	0	0	0
0	0	0	0	0	0	0	0	0	0	0	0	0	0	1	0	0	0	1
0	0	0	0	0	0	0	0	0	0	0	0	0	1	1	0	0	1	1
0	0	0	0	0	0	0	0	0	0	0	0	1	1	1	0	0	1	0
0	0	0	0	0	0	0	0	0	0	0	1	1	1	1	0	1	1	0
0	0	0	0	0	0	0	0	0	0	1	1	1	1	1	0	1	1	1
0	0	0	0	0	0	0	0	0	1	1	1	1	1	1	0	1	0	1
0	0	0	0	0	0	0	0	1	1	1	1	1	1	1	0	1	0	0
0	0	0	0	0	0	0	1	1	1	1	1	1	1	1	1	1	0	0
0	0	0	0	0	0	1	1	1	1	1	1	1	1	1	1	1	0	1
0	0	0	0	0	1	1	1	1	1	1	1	1	1	1	1	1	1	1
0	0	0	0	1	1	1	1	1	1	1	1	1	1	1	1	1	1	0
0	0	0	1	1	1	1	1	1	1	1	1	1	1	1	1	0	1	0
0	0	1	1	1	1	1	1	1	1	1	1	1	1	1	1	0	1	1
0	1	1	1	1	1	1	1	1	1	1	1	1	1	1	1	0	0	1
1	1	1	1	1	1	1	1	1	1	1	1	1	1	1	1	0	0	0

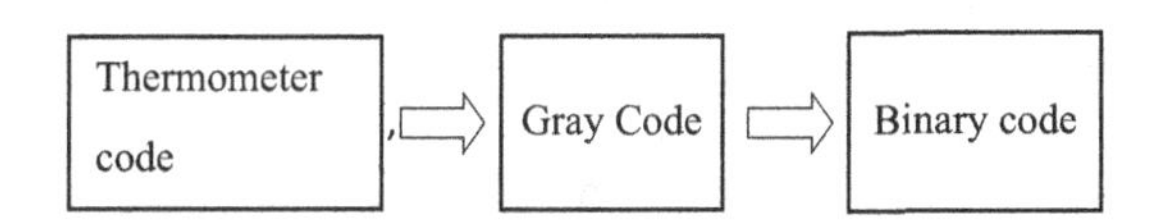

FIGURE 11.20 Block diagram of the thermometer-to-binary-code converter.

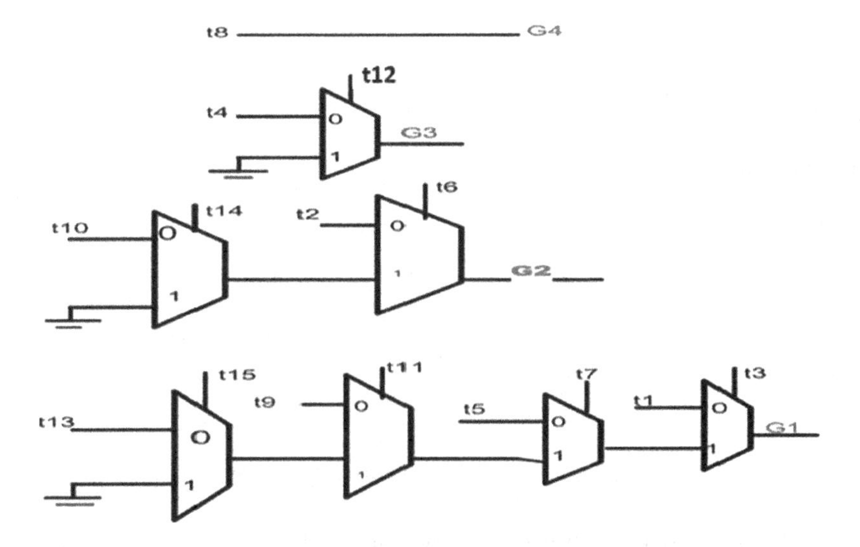

FIGURE 11.21 Implementation of Gray codes from thermometer codes using 2:1 MUX.

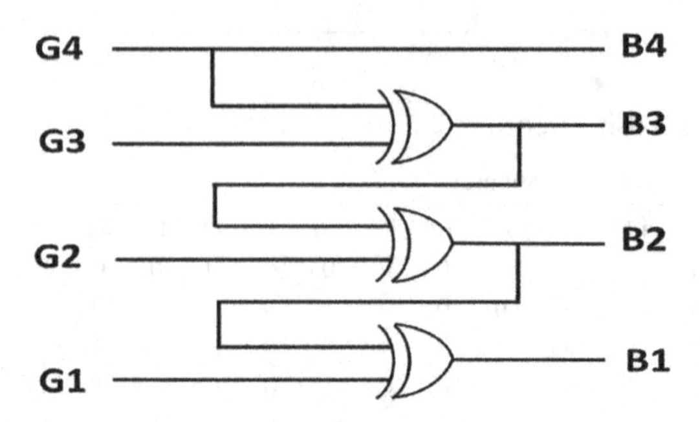

FIGURE 11.22 Gray code to binary code conversion using XOR gate.

In Figure 11.21, G4 is equivalent to T8. G3 is generated through 2:1 MUX with T4 as the input and T14 as the select line. Similarly, G2 is generated through two 2:1 MUX, T10 and T2 as the input signal, and T14 and T6 as the select lines. Subsequently, for G1, T13, T9, T5, and T1 are used as the input signal to the MUXes, and T15, T11, T7, and T3 as the select lines. Thereafter, the Gray code is converted into binary code using XOR gate-based circuits. B4 is equivalent to G4, B3 is the XOR of B4 and G3, B2 is the XOR of B3 and G2, and B1 is the XOR of B2 and G1. This design has a self-reconfigurable property 2:1 MUX and XOR gates are implemented using the universal NAND gates as shown in Figures 11.23 and 11.24.

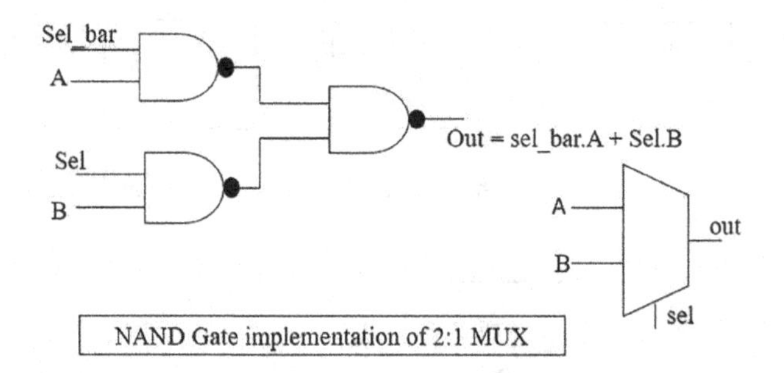

FIGURE 11.23 NAND gate implementation of 2:1 MUX.

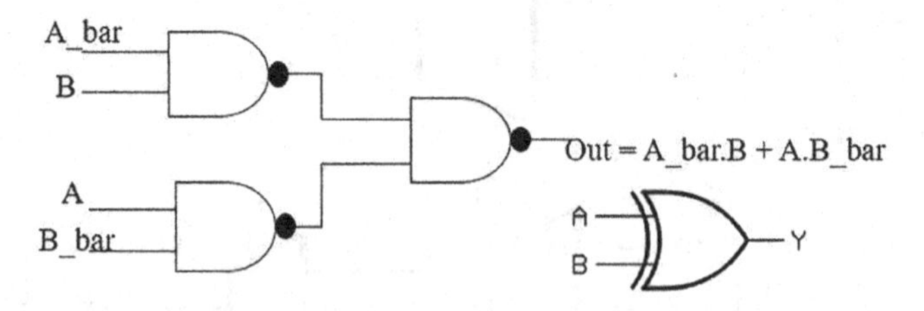

FIGURE 11.24 NAND gate-based implementation of XOR gate.

TABLE 11.5
Parameters for Thermometer-to-Binary-Code Converter (16:4 Encoder)

Device Technology	Power (μW)	Delay
180 nm CMOS [8]	25.64	0.518 ns
20 nm TFET [This Work]	0.328	10.67 ps

NAND gate is implemented using four TFET devices with an aspect ratio of 600/20 nm of both NTFET and PTFET devices. Sel_bar, A_bar, and B_bar are also implemented using the NAND gates.

Power and delay are estimated using HSPICE simulations for this encoder and are reported and compared in Table 11.5. It can be observed from Table 11.5 that the TFET-based thermometer-to-binary-code converter is faster and consumes lesser power.

11.4.4 Implementation of TFET-Based 4-Bit Flash ADC

This work targets the ADC of 4-bit resolution using a TFET device. In the previous sections, we discussed various components of flash ADC and its implementation using TFET, along with its performance parameters. In this section, all the building blocks are combined to implement the 4-bit flash ADC [9].

In Figure 11.25, the basic flow and conversion of analog signal to digital signal is explained pictorially. Reference voltages varying from 100 to 660 mV are generated using R ladder with a resolution of 40 mV.

11.5 SIMULATION RESULTS AND DISCUSSION

11.5.1 Simulation Results of TFET-Based Comparator

TFET-based comparator is implemented and simulated using HSPICE. This proposed comparator has been verified for various clock frequencies up to 2 GHz. Also, any input signal can be selected while considering the Nyquist criterion. As the supply voltage is 0.9 V, any reference voltage can be taken from 0 to 0.85 V. This comparator is verified for sampling frequencies up to Gigahertz. In Figure 11.26, an input

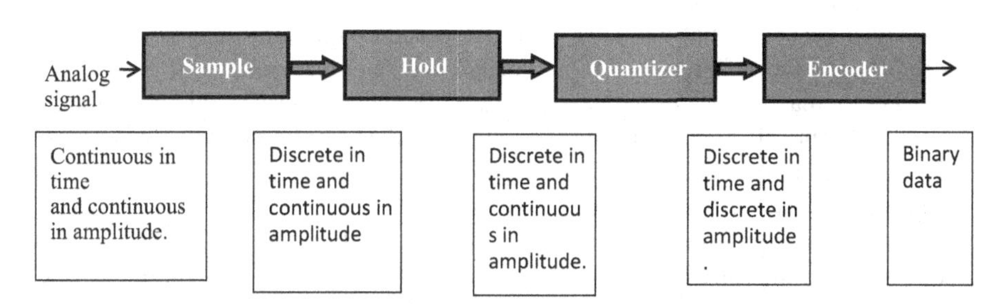

FIGURE 11.25 Basic components of ADC.

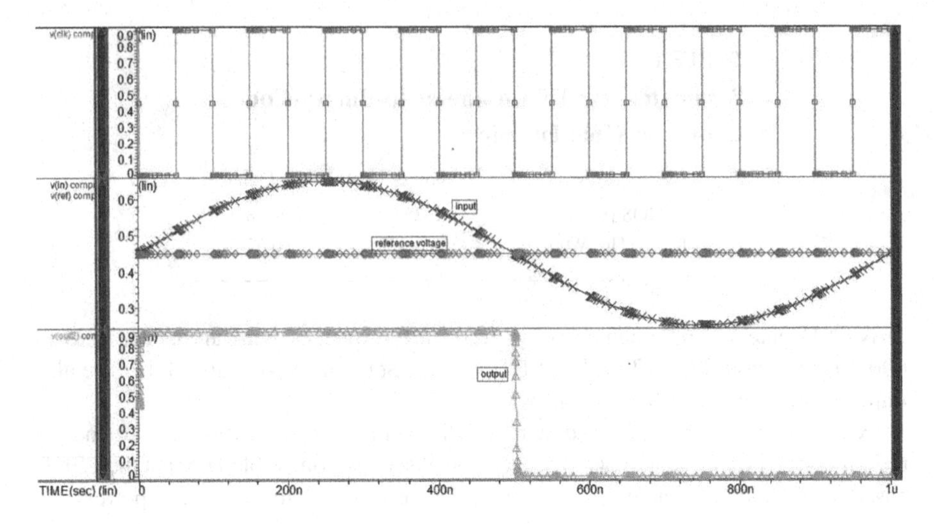

FIGURE 11.26 Response of TFET-based comparator for the input signal of 0.4 V peak to peak having a frequency of 1 MHz. Clock frequency is 10 MHz, and reference voltage is 0.45 V.

TABLE 11.6
Performance Metrics of TFET Comparator

Parameter	Values
Technology (nm)	20
Maximum operating frequency (GHz)	2
Reference voltage (V)	0.1–0.85
Average power dissipation (uW)	<50
Input range (V)	0–0.9

signal of 0.4 V p-p and a reference voltage of 0.45 V have been used. When the input signal is greater than the reference voltage, the output of the comparator will be 0.9 V. However, when the input signal is lower than the reference voltage, the output of the comparator will be 0 V. This circuit is suitable as a comparator, and its performance metrics are listed in Table 11.6.

It is also observed that the delay of TFET-based comparator depends on offset, where offset is the difference of input voltage and reference voltage. It is observed that delay increases with an increase in offset. We have observed this variation by having different offsets for different reference voltages as shown in Figures 11.27–11.29 for a sampling frequency of 10 MHz.

Variation of delay by varying the offset has been observed. Therefore, the delay depends on how much input voltage and reference voltages are apart. Hence, this work concludes that this TFET-based comparator is suitable for designing an ADC that can be used for communication systems, IoT, and biomedical applications. It is also observed that the speed of the comparator is fast, having only a few picoseconds.

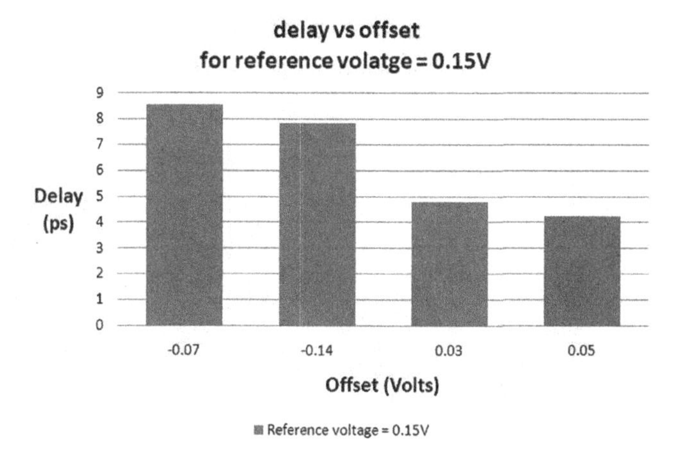

FIGURE 11.27 Variation of delay with offset for a reference voltage of 0.15 V.

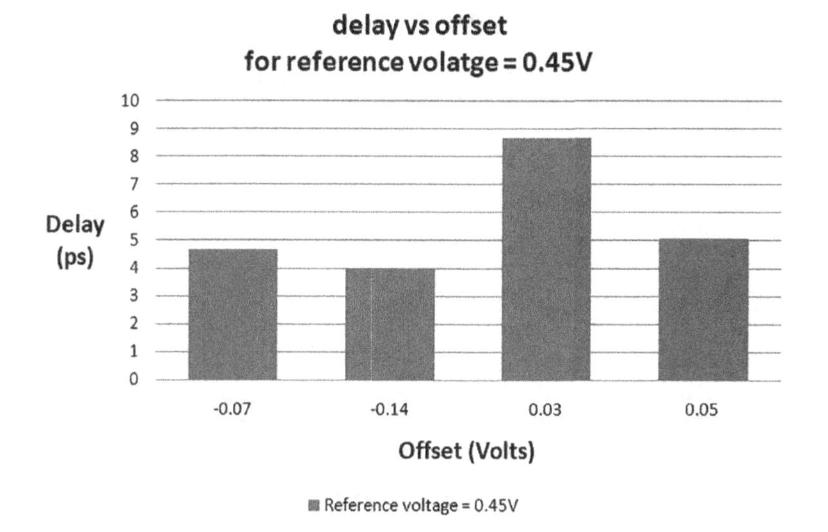

FIGURE 11.28 Variation of delay with offset for a reference voltage of 0.45 V.

The TFET-based comparator performance is satisfactory in terms of power (<50 μW), speed, operating frequency (<2 GHz), and reference voltage range. The TFET-based comparator is suitable for lower technology nodes that can provide minimum resolution and a wide dynamic range for ADC with less power dissipation.

11.5.2 Simulation Results of Flash ADC

Implementation of 4-bit flash ADC design has been carried out using the HSPICE tool with a 20 nm channel length TFET device with 0.9 V of supply voltage. First, the analog signal is given to the input pins of comparators, and another input pin has

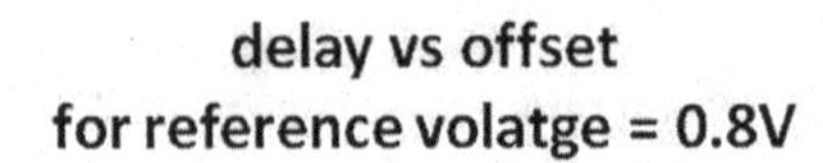

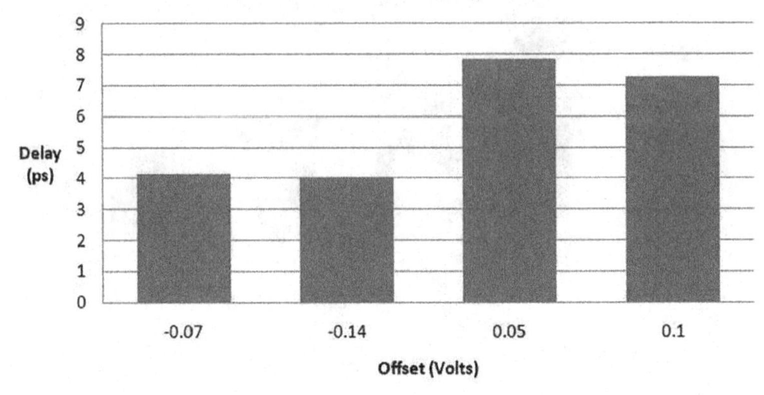

FIGURE 11.29 Variation of delay with offset for a reference voltage of 0.8 V.

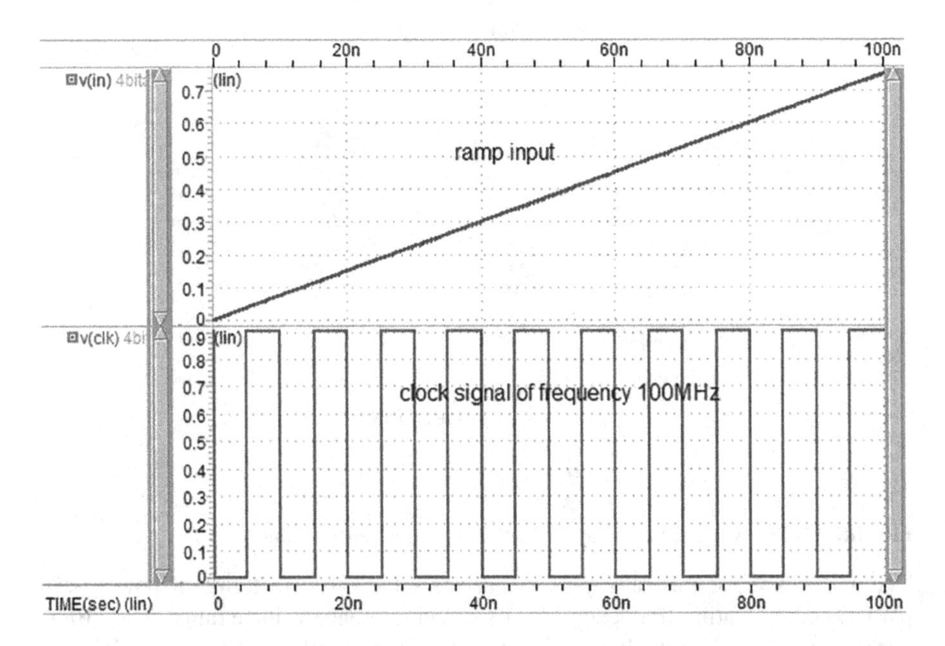

FIGURE 11.30 Ramp input signal and clock signal are given at the inputs of comparator.

reference voltages, which are generated using R ladder. Figure 11.30 shows the ramp input signal varying from 0 to 0.75 V with a clock frequency of 100 MHz following the Nyquist criterion. For the 4-bit ADC, 15 reference voltages are required for 15 comparators used in ADC. Reference voltages vary from 0.1 to 0.75 V at a difference of 40 mV within two consecutive reference voltages, for example, reference 1 is 0.1 mV, reference 2 is 0.14 mV, reference 3 is 0.18 mV, and so on as shown in Figure 11.31.

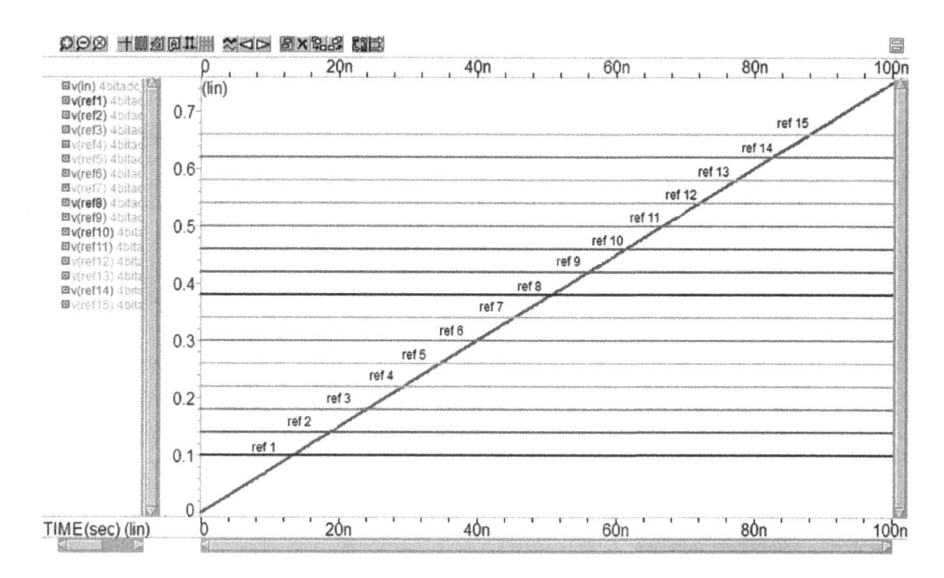

FIGURE 11.31 Reference voltages and input signal for an ADC.

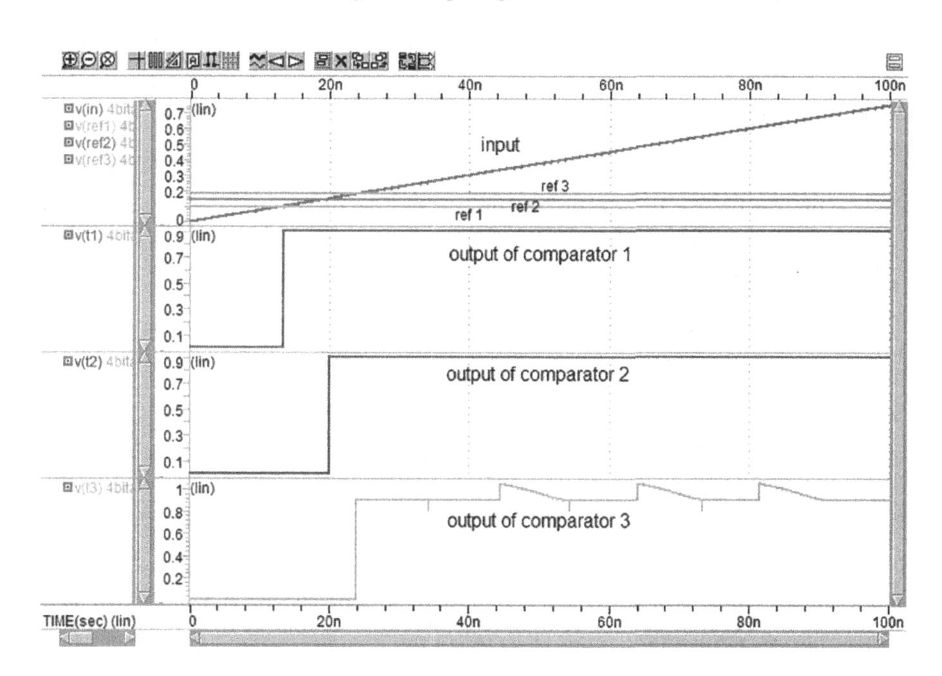

FIGURE 11.32 Input signal with reference voltages and its corresponding output of comparators 1, 2, and 3.

In Figures 11.32 and 11.33, output of comparator has been shown along with the corresponding reference voltages applied at the input of comparators. The output of comparator remains low when the input signal is lower than reference voltages that are given to comparators. Likewise, reference 1 is of 0.1 V and the corresponding

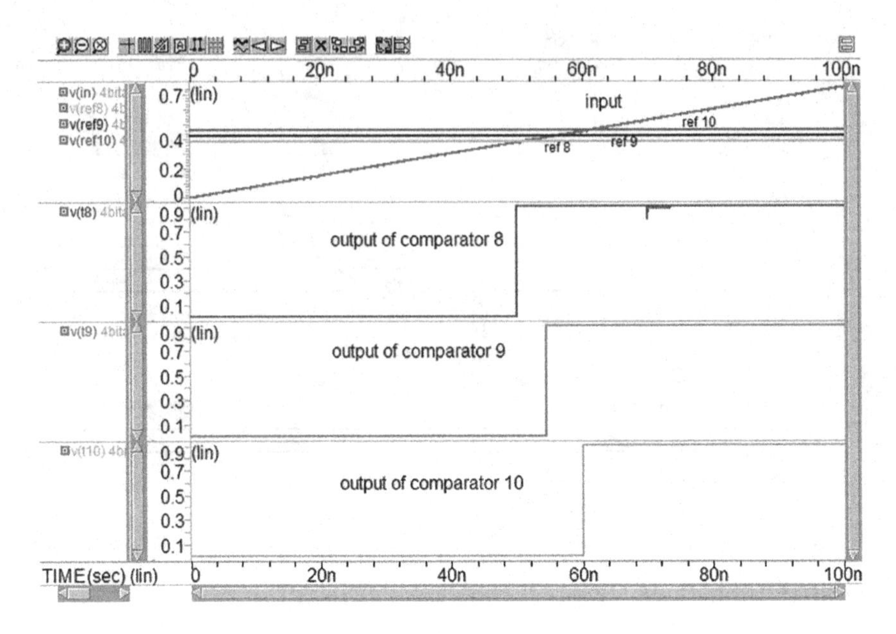

FIGURE 11.33 Input signal with reference voltages and its corresponding output of comparators 8, 9, and 10.

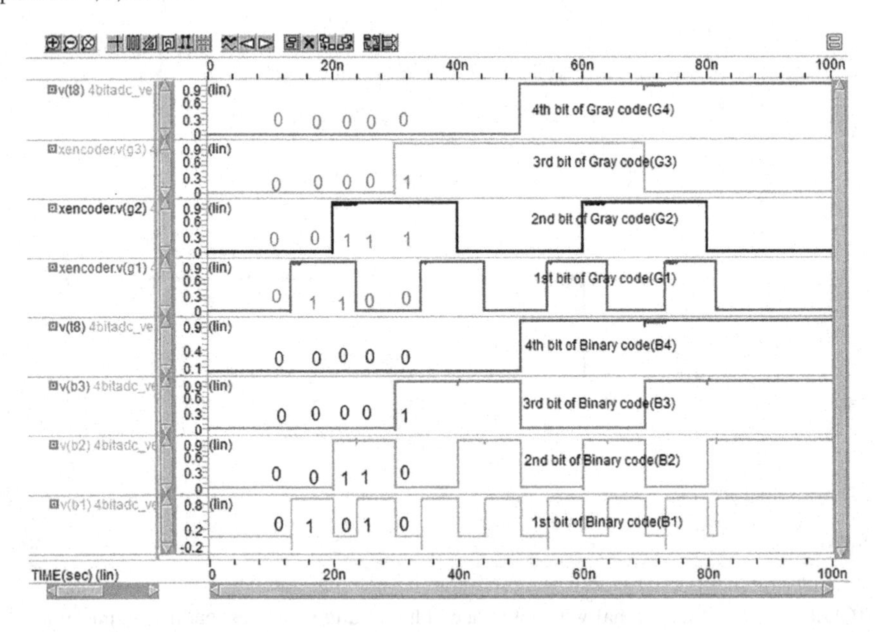

FIGURE 11.34 Output of encoder: Gray to binary conversion.

output of comparator is shown in Figure 11.32. Similarly, other comparators 2, 3, 4, 5, 6, 7, 8, 9, 10, 11, 12, 13, 14, and 15 function properly and give desirable results.

In Figure 11.34, the thermometer code is first converted into Gray code with the help of multiplexers, and then into binary code using XOR gates. Gray code and

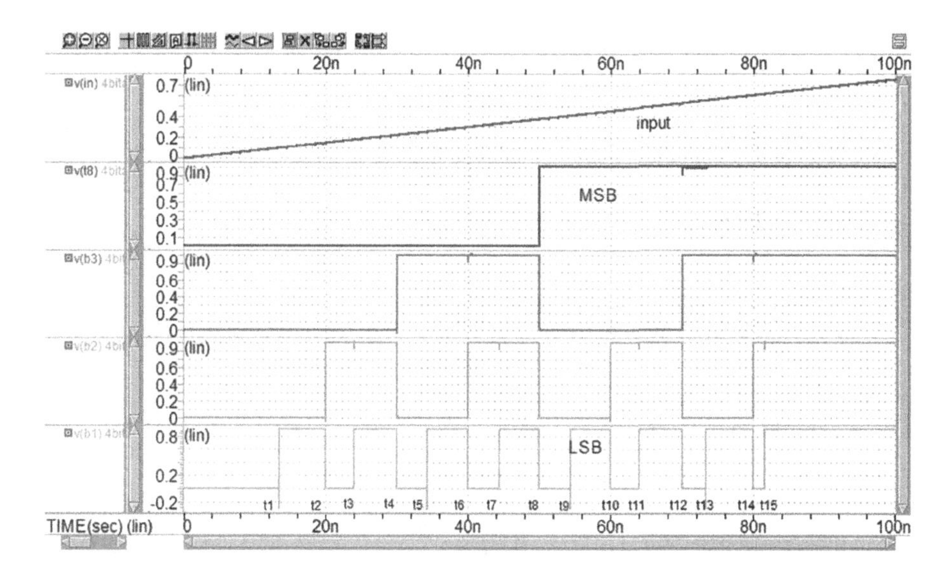

FIGURE 11.35 Output of 4-bit flash ADC for a ramp input.

its corresponding binary code have been shown in Figure 11.34. Initially, when the Gray code is 0000, then the binary value will be 0000. When ramp input is increased slightly above 0.1 V, the Gray code will be 0001, and its corresponding binary values would be 0001. When an input is above reference 2, 0.14 V, then the Gray code would be 0011, and its corresponding binary values would be 0010 (decimal value is 2). Similarly, when the Gray code is 0110, its binary value would be 0100 as shown clearly in Figure 11.34. G4, G3, G2, and Gl are the fourth, third, second, and first bits, respectively, of Gray codes. B4, B3, B2, and Bl are the fourth, third, second, and first bits, respectively, of binary codes.

In Figure 11.35, binary values are shown for an input signal from 0 to 0.75 V. Binary values are 0000 till 't1' because input is less than the reference voltage 1, and then it would be 0001 till 't2', when reference voltage 1 becomes lower than the input signal. Afterward, binary values become 0010→0011→0100→0101→0110→0111→1000→1001→1010→1011→1100→1101→1110→1111. All the bits would be high when input signal reaches about 0.65 V or above.

Implementation and simulation of TFET-based 4-bit flash ADC has been done successfully as shown in Figure 11.35. To check the proper functioning of 4-bit flash ADC, various input signals are taken for different amplitudes and of different frequencies, keeping the Nyquist criterion for selecting the sampling frequency. A sinusoidal input signal has been selected to pass through ADC and is converted into digital signals for the purpose of communication/transmission of data.

Sinusoidal input signal of frequency 10 MHz, 0.3V p-p, for a clock frequency of 100 MHz is shown in Figure 11.36, and its corresponding output binary values are shown in Figure 11.37. Initially, digital output is 1000, because input signal has an offset of 0.4 V and it is greater than reference 4; therefore, its corresponding output would be 1000. Then, as the signal increases, digital output would be 1001→1010 and so on.

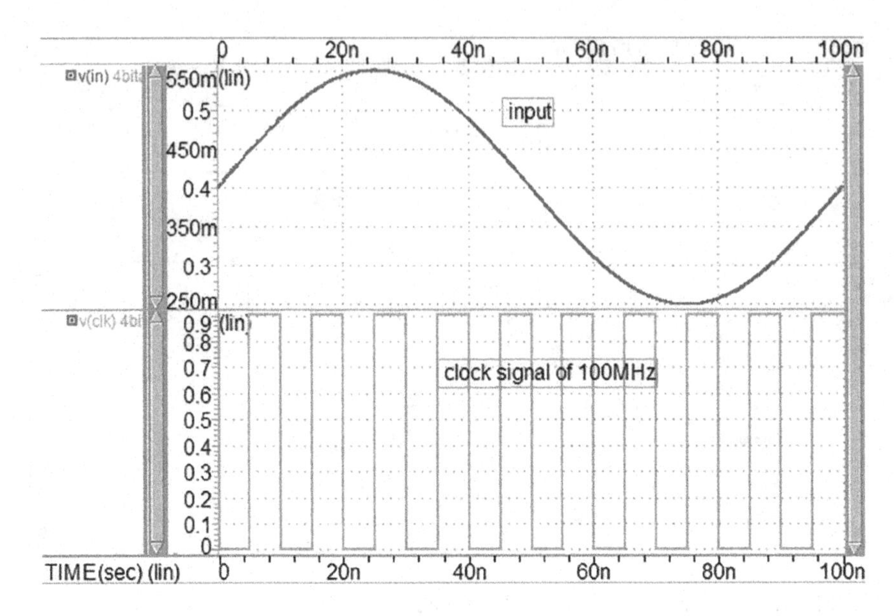

FIGURE 11.36 Sinusoidal input signal of frequency 10 MHz, 0.3V p-p, and a sampling frequency of 100 MS/s.

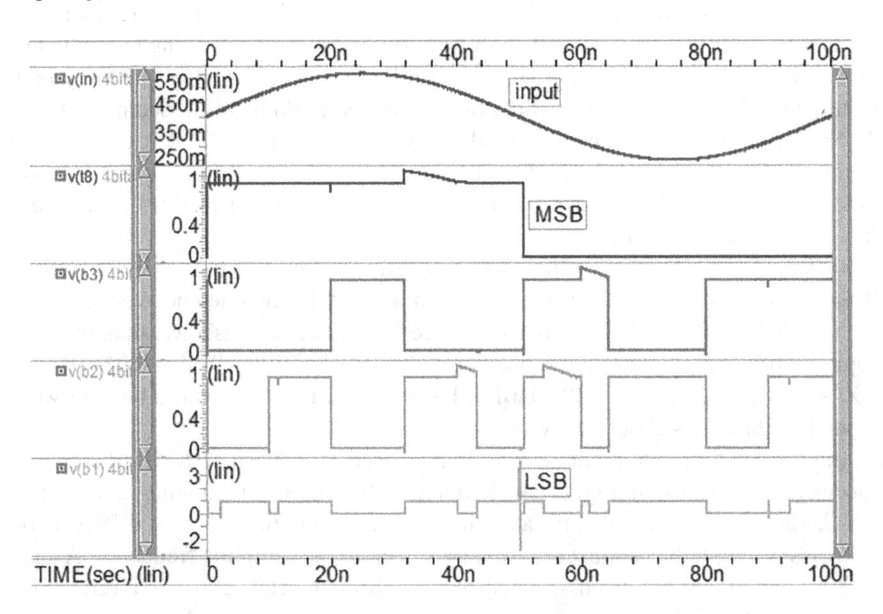

FIGURE 11.37 Sinusoidal input signal of frequency 10 MHz and binary output.

This system is able to respond to up to 2 GS/s sampling frequency and an input frequency of 500 MHz. In this work, we have calculated ADC's performance parameters such as DNL, INL, SNDR (signal-to-noise-distortion ratio), ENOB (effective number of bits), power, delay, and figure of merit (FoM) to check the performance of 4-bit flash ADC. Calculated performance metrics are shown in Table 11.7. We also observed these parameters by varying temperature and study its impact upon the

TABLE 11.7
Measured Parameters of TFET-Based 4-Bit Flash ADC

Parameters	Values
Resolution	4
Supply voltage	0.9
Sampling frequency (maximum) (GS/s)	2
ENOB	3.18
Average power (maximum)	259 μW
Delay	2 ns @ 500 MHz clock frequency
Power delay product	0.518 fWs
DNL, INL	<0.5 LSB

TABLE 11.8
Variation of Performance Parameters of ADC w.r.t. Temperature for a Sampling Frequency of 100 MS/s

Temperature (°C)	Power (uW)	Delay (ns)	SNDR	ENOB
−40	289.73	10	21.67	3.3
25	250.68	10	20.918	3.18
50	250	10	20.94	3.18
75	243.53	10	21.12	3.21
125	234	10	21.3435	3.25

ADC's parameters. This work recorded the readings at temperatures −40°C, 25°C, 50°C, 75°C, and 125°C.

$$\text{ENOB} = (\text{SNDR} - 1.76)/6.02$$

$$\text{FoM} = \frac{\text{Power dissipated}}{2^{ENOB} \times \text{sampling frequency}}$$

In this work, the circuit works on a 0.9 V supply voltage for a 4-bit resolution. The acceptable sampling frequencies vary from 50 MS/s to 2 GS/s, with the input frequency up to 500 MS/s. Maximum power consumption of 0.259 mW is observed. Delay varies according to the sampling frequency; for example, for the sampling frequency of 100 MHz, the delay is found to be 10 ns. Delay for 500 MHz sampling frequency is found to be 2 ns and so on. DNL and INL are found to be less than 0.5 LSB, and LSB is of 40 mV.

TFET-based 4-bit flash ADC shows lesser power consumption, and delay is also less, which implies that speed is fast with a considerable FoM of 0.0143 pJ/conversion step. At the output of 4-bit flash ADC, considerable ENOBs of about 3.18 are obtained. ENOB is calculated after passing the binary output through ideal DAC. Ideal DAC is performed using Verilog-A code, and we observe its waveform and then take a FFT of the output of DAC along with SNDR and ENOB in the HSPICE. These performance parameters are also observed by varying the temperature as shown in Table 11.8.

Table 11.8 enables us to see the variation in power, delay, SNDR, and ENOB with respect to temperature variation. As observed from Table 11.8, the power decreases on increasing the temperature, delay is almost constant with respect to temperature, SNDR also slightly increases with respect to temperature, and ultimately ENOBs vary within 3.18–3.3. For better visualization, these variations are shown using bar graphs in Figures 11.38–11.40.

As power decreases on increasing the temperature, power is decreased from 25°C to 125°C by about 6.65%. This can be due to TFET current characteristics'

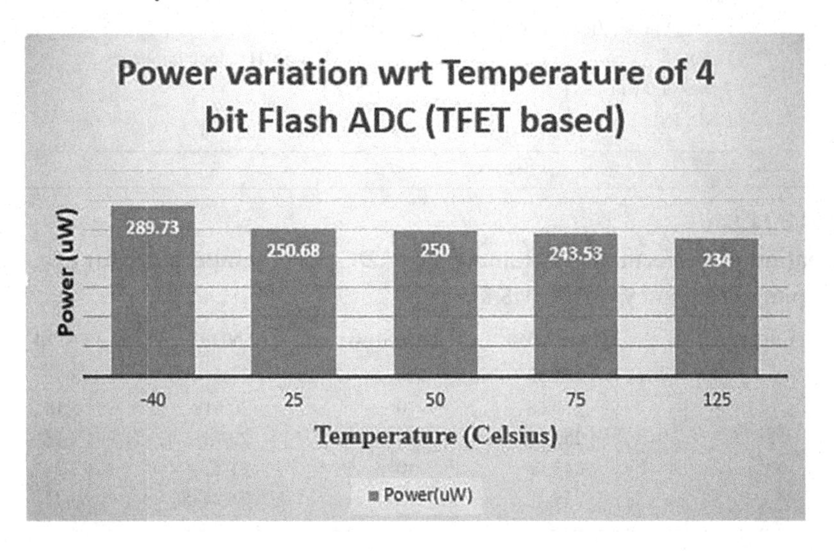

FIGURE 11.38 Graphical representation of power w.r.t temperature for a sampling frequency of 100 MS/s.

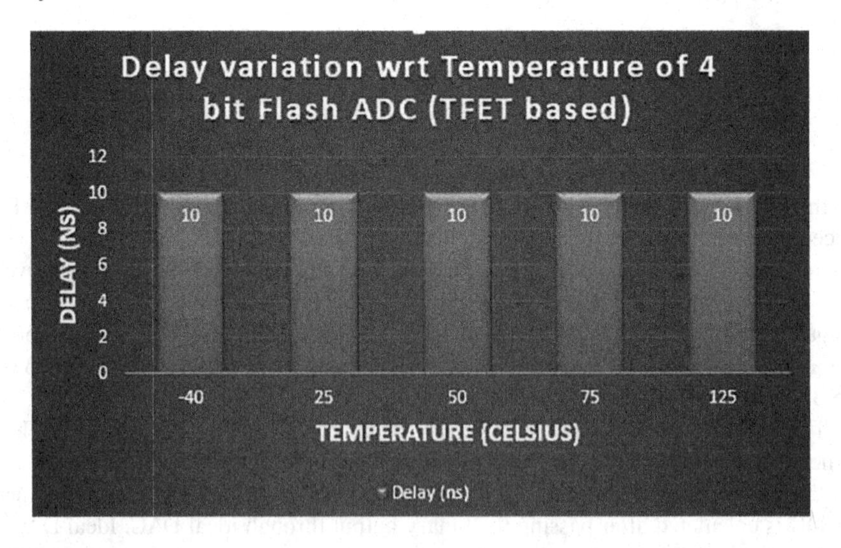

FIGURE 11.39 Graphical representation of delay w.r.t temperature for a sampling frequency of 100 MS/s.

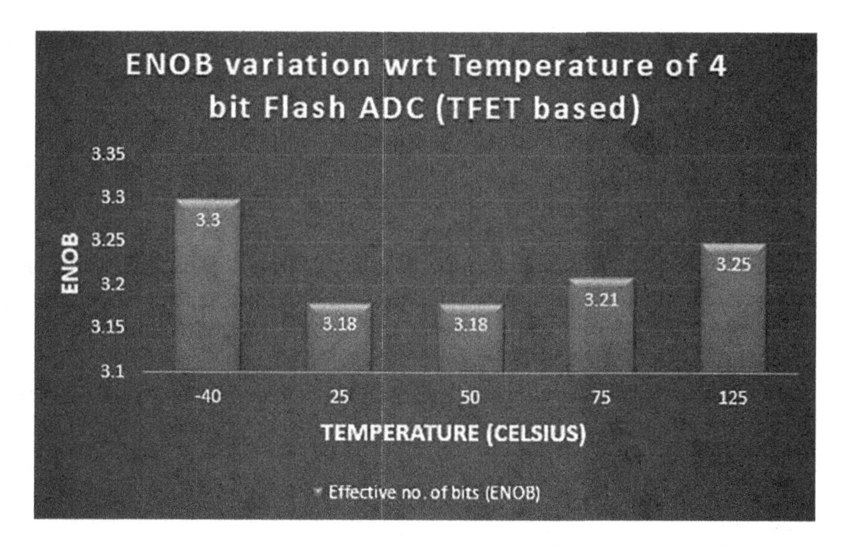

FIGURE 11.40 Graphical representation of ENOB w.r.t temperature for a sampling frequency of 100 MS/s.

TABLE 11.9
Comparison of This Work with CMOS Flash ADC

Parameters	Reference [11]	Reference [12]	Reference [13]	Reference [14]	This work
Technology	45 nm CMOS	180 nm CMOS	65 nm CMOS	45 nm CMOS	20 nm TFET
Resolution	4	4	4	5	4
Supply voltage (V)	1	1.8	1	0.4	0.9
Sampling frequency (MS/s)	1	100	1	0.25	2,000
Power (mW)	2.95958	1.08	0.7	0.2761×10^{-6}	0.259
DNL	0.4	1.04	0.2 LSB	Nil	0.42 LSB
INL	0.38	1.36	0.2 LSB	Nil	0.45 LSB
SNDR (dB)	24.8636	23.59	22.3	Nil	20.918
ENOB (bits)	3.83	3.62	3.42	4.01	3.18
FoM (pJ/step)	1.284	0.87	0.065	0.0000685	0.0143

dependence on temperature, as the off-state current is sensitive to temperature [10]. If we consider the trend of delay w.r.t temperature, it is independent of temperature variation, which can be because delay is dependent on ON current of TFET. As from the characteristics of TFET, it is concluded that ON current is independent of temperature variation [10].Therefore, delay is does not vary by varying the temperature; it is constant for a particular sampling frequency. ENOBs slightly vary with temperature.

TFET-based 4-bit flash ADC is compared with the reported flash ADC designed using 45, 65, and 180 nm CMOS technology. Table 11.9 compares the performance of this work with the existing work.

In this work, TFET-based 4-bit flash ADC has been designed and compared with recently reported CMOS-based flash ADCs.

- TFET-based flash ADC exhibits less power consumption as compared to [11–13] and offers about 91.22% decrease in power as compared to reference [11].
- It can support a sampling frequency of up to 2GS/s by maintaining the Nyquist criterion for input frequency.
- In reference [14], power consumption is lesser than that in our work but reference [14] can support only a 0.25MS/s sampling frequency, while the presented TFET-based work can be used at the sampling frequency of up to 2GS/s.
- ENOB in the TFET-based 4-bit flash ADC is also comparable and it gives desirable output bits that can be properly transmitted and received by a transmitter/receiver.
- FoM of this work is better than those of other references available in the literature.

11.6 CONCLUSIONS

This chapter presents the TFET implementation of a high-speed 4-bit flash ADC. Flash ADC is successfully designed with the power dissipation of 0.259 mW and has the highest sampling frequency of 2GS/s with a power supply of 0.9 V.

With the simulation of NTFET and PTFET and observing their characteristics, we can conclude that it is good for low-power, low-voltage applications with little leakage current. The SS of NTFET is found to be 45.7 mV/dec. Comparator (TFET-based) and thermometer-to-binary-code converters (TFET-based) are implemented using TFET at a 20 nm technology node using the HSPICE tool of SYNOPSYS. Comparators, thermometer-to-binary-code converter, and resistor ladder are combined, which result in flash ADC. In the thermometer-to-binary-code converter, first, the thermometer code is converted into Gray code and then into binary code. Capacitors are present at every node of the resistor ladder to reduce the clock feedthrough. Various performance parameters of the TFET-based flash ADC have been discussed. Finally, the proposed work is compared with the previous work on CMOS-based flash ADC, and various conclusions are drawn based on the sampling frequency, ENOB, power, supply voltage, resolution, and others. ADC is the significant block in digital ICs, and its performance needs to be improved in terms of conversion speed, power, area, and ability to do more complex operations.

REFERENCES

[1] Hao, L., Y. Trond, and S. Alan. Universal TFET model implementation in Verilog-A. (2015).
[2] Turcane, S. M., and A. K. Kureshi. Review of tunnel field effect transistor. *International Journal of Applied Engineering Research* 11, no. 7 (2016): 4922–4929.
[3] Lakshmi, Taninki Sai, and Avireni Srinivasulu. A low power encoder for a 5-GS/s 5-bit flash ADC. In *2014 Sixth International Conference on Advanced Computing (ICoAC)*, pp. 41–46. IEEE. 2014.

[4] Yasir, Mohd, and Naushad Alam. Design of TFET based Op-Amp and CCII for low voltage and low power applications. *International Journal of Electronics* 108, no. 10 (2021): 1733–1753.

[5] Agrawal, Niket, and Roy Paily. An improved ROM architecture for bubble error suppression in high speed flash ADCs. In *Proc. of AISPC*, pp. 1–5. 2008.

[6] Wallace, Christopher S. A suggestion for a fast multiplier. *IEEE Transactions on Electronic Computers*, EC-13, no. 1 (1964): 14–17.

[7] Lee, Daegyu, Jincheol Yoo, Kyusun Choi, and Jahan Ghaznavi. Fat tree encoder design for ultra-high speed flash A/D converters. In *The 2002 45th Midwest Symposium on Circuits and Systems, 2002. MWSCAS-2002.*, vol. 2, pp. II-87–II-90. IEEE, 2002.

[8] Mayur, S. Marinaik. Design of novel multiplexer based thermometer to binary code encoder for 4 bit flash ADC. In *2017 2nd IEEE International Conference on Recent Trends in Electronics, Information & Communication Technology (RTEICT)*, pp. 1006–1009. IEEE, 2017.

[9] Kalyani, Nayana, and M. Monica. Design and analysis of high speed and low power 6-bit flash ADC. In *2018 2nd International Conference on Inventive Systems and Control (ICISC)*, pp. 742–747. IEEE, 2018.

[10] Lin, Zhiting, Panpan Chen, Le Ye, Xu Yan, Lanzhi Dong, Shuguang Zhang, Zhou Yang, Chunyu Peng, Xiulong Wu, and Junning Chen. Challenges and solutions of the TFET circuit design. *IEEE Transactions on Circuits and Systems I: Regular Papers* 67, no. 12 (2020): 4918–4931.

[11] Malathi, D., R. Greeshma, R. Sanjay, and B. Venkataramani, A 4 bit medium speed flash ADC using inverter based comparator in 0.18 μm CMOS. In *19th International Symposium on VLSI Design and Test*, pp. 1–5. IEEE, 2015.

[12] Soman, Vanitha, and Sudhakar S. Mande. A 4-Bit 1GS/s folding flash ADC using 45-nm CMOS technology. *Microprocessors and Microsystems* (2020): 103512.

[13] Nasri, Bayan, Sunit P. Sebastian, Kae-Dyi You, RamKumar RanjithKumar, and Davood Shahrjerdi. A 700 μW 1GS/s 4-bit folding-flash ADC in 65nm CMOS for wideband wireless communications. In *2017 IEEE International Symposium on Circuits and Systems (ISCAS)*, pp. 1–4. IEEE, 2017.

[14] George, Glyny, and AV Jos Prakash. Design of ultralow-voltage high speed flash ADC in 45nm CMOS technology. In *2018 3rd IEEE International Conference on Recent Trends in Electronics, Information & Communication Technology (RTEICT)*, pp. 920–924. IEEE, 2018.

12 Demand of Low-Power-Driven FET as Biosensors in Biomedical Applications

Mekonnen Getnet
Delhi Technological University
Debre Tabor University

Rishu Chaujar
Delhi Technological University

CONTENTS

12.1 INTRODUCTION: BACKGROUND AND DRIVING FORCES

Modern business operations and consumer needs for data acquisition, processing, and entertainment drive the growing industrial need for creating, acquiring, storing, processing, and transmitting information. The problem started with the introduction of the electronic calculator, personal computer, and microprocessor-driven games,. The increased need for computing, combined with a growing demand, has resulted in integrated circuits (ICs) that are faster, smaller, more dependable, and less powerful. The technology accompanying these advancements can be incorporated into the combatant's electronic systems to improve functionality and reduce battery consumption. The industry has embarked on a long-term strategy to reduce electronic

DOI: 10.1201/9781003240778-12

devices' size and operating voltages while increasing device and circuit integration. In technological advancement, we are studying the National Technology Roadmap for Semiconductors (NTRS). ICs further lowered power usage by allowing multiple devices to be packed on a single chip, eliminating the interconnection constraint and minimizing power consumption (1). "Invented complementary metal-oxide-semiconductor (CMOS) integrated circuit was the crucial low-power CMOS device used as building blocks in most ICs, significantly lowering IC power consumption" (2). Future social life quality depends on the perspective improvements of information technology. In the field of sensing technology, highly desirable products like house monitoring (air pollution, humidity, fire, temperature, human health, and security), traffic monitoring (like car speed, traffic jams, and accidents), agriculture monitoring (air pollution, temperature, humidity, wind, lightning, rain, and storms), space monitoring (stars, sun, meteorites, moon, and other astronomical phenomena), security monitoring, and so on are created in many companies.

Three primary applications that drive the demand for low-power FET electronic devices are as follows.

i. Portable electrical goods including wristwatches, cardiac pacemakers, personal digital assistants, hearing aids, pocket calculators, etc.
ii. Improving performance by increasing the package density of components in ICs while countermeasuring power consumption restrictions; and
iii. power preservation in desktop computers with a modest life cycle of cost-to-performance ratio, which necessitates low power for less power supply and cooling expenses (2). Low utilization of power results in lower chip temperatures, higher stability, and less expensive plastic packages (3).

The power consumption of dynamic random access memory (DRAM) circuits can be significantly lower. Power dissipation challenges in electronic systems are treated at many levels. Out of the different hierarchies in electronic systems, only device issues relating to low-power electronics are discussed here, such as circuits, materials, devices, and systems.

To study low-power electronics, one must first study the components of ICs or semiconductor devices. MOS-field-effect transistor (MOSFET) is the backbone of CMOS used in low-power circuits (4). Silicon CMOS technology utilizes both P-type and N-type transistors (5). In recent years, the leading semiconductor technology is CMOS technology for the application-specific IC microprocessors and memory (4).

12.2 ALTERNATIVE STRUCTURES FOR SCALING CMOS DEVICES

According to the new International Technology Roadmap System (ITRS) for "Semiconductors, the rapid scalability of metal-oxide-semiconductor field-effect transistor (MOSFET) is speeding up the introduction of new technologies to extend complementary MOS (CMOS) down to, and possibly beyond, the 22-nm node" (1,6–8). This speeding up necessitates the industry's intensification of study on two extremely hard drives. The first "crucial issue of scaling CMOS into an increasingly difficult manufacturing domain much below the 90-nm node is a high performance, reasonably low power, and low reliable power implementations, and the second is an exciting opportunity to develop fundamentally novel approaches" (9). Scaling challenges involve monitoring short-channel effects (SCEs) and tunnel currents in digital applications, increasing drain saturation current while significantly plummeting the power supply voltage, and sustaining device parameter consistency, thereby overpowering leakage current, drain-induced barrier lowering, threshold voltage instability, and impact ionization effects across the chip and from chip to chip. In order to minimize challenges under CMOS scaling, "new materials utilized in the gate stack (high-dielectrics and metal electrode materials) in the conducting channel leads to enhanced carrier transport capabilities including improved device performance and minimizing scaling related problems" (10,11). New transistor structures focus on improving the MOSFET's electrostatics, providing a platform for introducing new materials, and meeting new material integration requirements (12). In transport-enhanced MOSFETs, boosting the average velocity of carriers in the channel can result in an augmented transistor drive current for advanced circuit performance. According to scaling technology, double and triple hybrid metal gate devices would considerably lower CMOS circuit leakage currents (13,14). 3D-FETs, such as FinFETs and TG MOSFETs, are being intensively investigated for boosting circuit performance and suppressing SCEs, as well as low power consumption (7,15–17). The contribution of the top surface of the fin of the trigate MOSFET to the channel current improves drivability because it serves as an additional channel in the case of low-power circuit operation (18). Therefore, the "silicon-on-insulator FinFET triple hybrid metal gate simulation result is much more improved than the planar double/single gate silicon-on-insulator metal oxide semiconductor field-effect transistors" (SOI-MOSFETs) (15).

12.3 LOW-POWER FET AS BIOSENSORS IN BIOMEDICAL APPLICATIONS

In recent years, in biomedical applications, low-power FET biosensors (bio-FET) have seen rapid advancements with improvements in FET electrical properties and changes to bioreceptor designs (19–21). Due to their unlimited advantages, CMOS-FET biosensors are predicted to continue advancing as one of the most promising technologies for biomedical applications (19). FET-based biosensors have attracted scientific attention in advanced technology due to their extremely high sensitivity, speed, cost-effective nature, selective, simple, easy signal readout, and label-free assay capabilities (22,23). Due to their low cost, fast speed, and accessibility, bio-FETs appear to be one of the most promising choices among a wide variety of electrical detecting devices (23). We address the fabrication of high-field gated bio-FET to detect target analytes within physiological conditions without requiring widespread sample pretreatments (24). For instance, protein detection in human serum blood (25), nucleotide detection in buffer, whole cell-based biosensor, and recognition of biological tissues (26) are only a few of the biomedical applications of NWFET (NWFET)-based biosensors highlighted in this review from the original research study. Also, pH sensing (27), ion sensing (19), gas sensing (28), and biomolecule sensing (29) are additional applications that have been demonstrated in current studies. A few examples of low-power bio-FET, such as compound semiconductor FETs (21), silicon nanowire FETs (SiNWFETs) (30,31), and graphene FETs (32), have been used to develop different low-power NWFET biosensor devices. NWFET-based biosensors are implemented in biomedical applications since they effectively interpret biological interactions at the device surface into easily legible electrical signals (33,34) that can be treated rapidly. For instance, silicon nanowire-based bio-FET has been shown to have excellent sensitivity for label-free and real-time detection. It can be highly commercialized due to its mass-production capability in the semiconductor industry (33). Subsequently, SiNWFET-based biosensors have been effectively used to sense a wide range of biological and chemical molecules such as viruses, nucleic acids, proteins, and DNA beleaguered substances; there have been numerous strategies for improving their detecting capability in the subthreshold regime, using frequency domain measurement, and augmenting surface amendment with a variety of surface modification techniques (35). Because of the rapid growth of solid-state technologies, an interesting sensing technique, namely, the NWFET-based biosensor, has been proposed and has become an emerging field in biosensing mechanisms due to most biomolecules having electrostatic charges and bioactivities that require potential electrical changes (36); therefore, NWFET-based biosensors are a suitable candidate for applications demanding high sensitivity and speedy response times (19). NWFET-based biosensors, for example, have been utilized to detect amino acids, nucleotides, and biological cells. Biological biomarkers such as protein, DNA, Uricas, APTES, and ChO_X are commonly used to diagnose cardiac, diabetes, inflammatory, cancer (37,39), kidney injury, and infectious diseases (33,39,40). NWFET-based biosensors have gained popularity in the early diagnosis, drug screening, and disease screening fields because of their improved sensitivity, selectivity, and ability to be easily functionalized in rural and urban areas. For instance, "Microfluidic cartridges for sample preparation can also be used in combination with FET-based sensing devices to provide completely automated testing" (19).

12.4 FET-BASED BIOSENSOR WORKING PRINCIPLES

An active source, drain, and gate device comprises FETs and critical electrical components (7,38,41). The gate connects electrodes between the drain and source electrodes and modulates the channel conductance at the arrival and beginning point of charge carriers. Charge carriers (electrons in n-type FETs, holes in p-type FETs) flow from source to drain through the gate, and the amount of charge carriers can be increased or lowered by the voltage between the gate and the source. External power sources affect the voltage applied to the gate in conventional FETs, while the concentration and species of biomolecules chemically attached to the gate influence the voltage in bio-FETs (42,43). Due to the induced electric field of the semiconductor channel, the metal potential (ψ_M) is changed accordingly, and the band bending is created. The band bending results in channel carrier concentration changes, such as inversion, depletion, or accumulation (44). In addition to gate voltage ($V_{GS} > V_{th}$, used to turn ON the transistor devices), solution potentials of the gate, for instance, a charge of biomolecules or pH value, can be other issues that impact the status of channel carriers and make the current–voltage characteristics change (shift) positively or negatively because the conductivity of the nanowire channel is affected by the charge carriers from analytes due to attractive or repulsive electrostatic forces, "charges from analytes in the solution influence the degree of depletion within a channel and causes high electric field leads to depletion or growth of charge carriers in the channel" (19), and a change in electrostatic charge or charge transfer from biomolecules to the transducing nanoparticle produces a change in the gate voltage environment and causes a conductance/current shift. The small change of biomolecule charge would be amplified by threshold voltage (V_{th}). This characteristic difference of threshold voltage (ΔV_{th}) indicates the sensitivity of semiconductor-FET devices (45). Finally, the target is recognized because a shift in conductance is a function of the target biomolecule concentration.

12.4.1 Some Examples of NWFET-Based Biosensor Working Principles

The typical design of SiNW biosensors for biomolecule recognition is shown in Figure 12.1. A receptor molecule that identifies the target is mounted on the SiNW surfaces to recognize a particular target (30,46), as illustrated in Figure 12.1a. Specific binding causes a reduction in conductance when the surface receptors are subjected to the target molecules, which have negative charges, as shown in Figure 12.1b.

Figure 12.2 shows a schematic architecture for label-free multiplexed biomarker detection. APTES and glutaraldehyde were used to modify the SiNW surface, leading

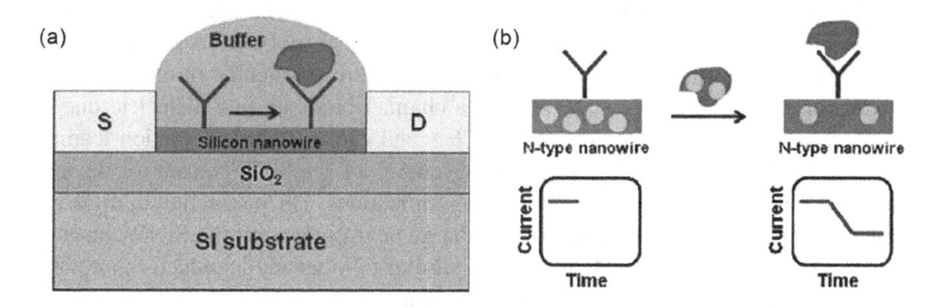

FIGURE 12.1 SiNW biosensor working principle. (a) Receptor molecules on the SiNW channel surface. (b) Interface of negatively charged target molecules with negatively charged receptor molecules leading to a drop in conductance for an n-type doped NW. (Adopted from Ref. 30.)

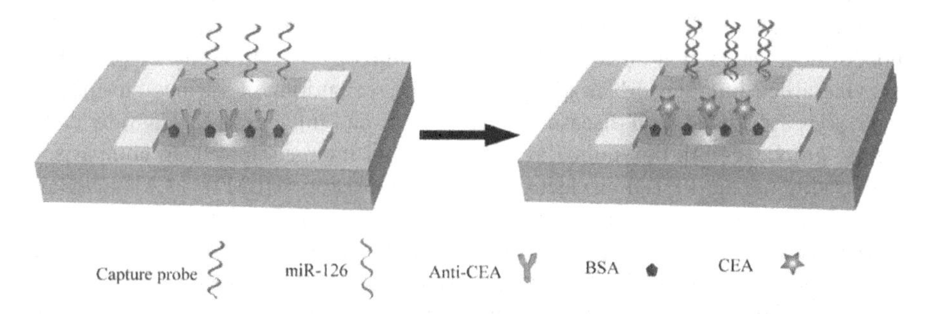

FIGURE 12.2 CEA and miRNA-126 label-free multiplexed electrical detection procedures. (Adopted from Ref. 37.)

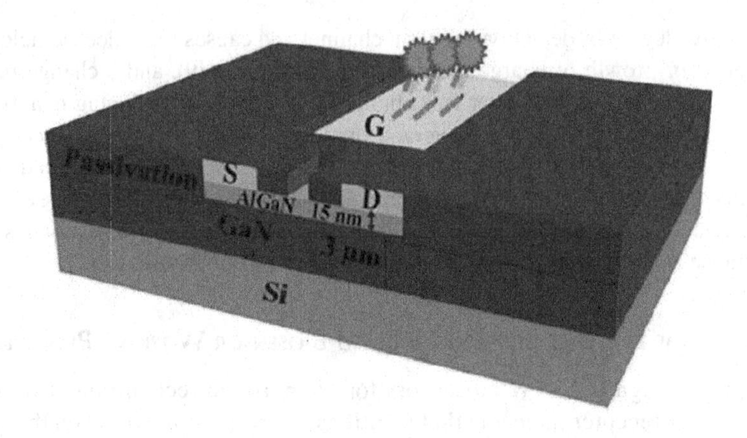

FIGURE 12.3 Active channel and high electron mobility transistor of the AlGaN/GaN NWFET-based biosensor. (Adopted from Ref. 19.)

to "a self-assembled monolayer with a terminal aldehyde group" (37). Separate DNA and anti-CEA biomolecules were etched on the surfaces of SiNW microchambers. The insertion of charged miRNA/CEA SiNW surfaces leads to the current shift in real time due to carrier depletion in the case of target biomarker insertion (37).

As illustrated in Figure 12.3, APTES and glutaraldehyde were used to modify the SiNW surface, resulting in a self-assembled monolayer with a terminal aldehyde group. Then, in different microchambers, imprisonment probe DNA and anti-CEA (an antibody against carcinoembryonic antigen) were covalently immobilized on SiNW surfaces. The n-type SiNW surface channel binds to gate dielectric due to insertion of negatively charged miRNA/CEA and causes carrier depletion leading to current change (38,47). The output I/V_{DS} curves of the SiNW biosensor devices were altered after various channel surface modifications. The interaction of different molecules creates a distinct electrostatic charge environment on the SiNW surface, which acts as a gate voltage. This result is similar to other semiconductor nanowire devices, and it is the semiconductor nanowire biosensor mechanism (37).

In addition to the NW FET biosensor, graphene FET-based biosensor is a new platform for sensing application due to high mechanical strength, extreme conductivity, configurable bandgap, customizable optical characteristics, and a huge specific surface area. Graphene FET-based biosensor is a crucial field-effect biosensor used to detect avidin and biotin biomolecules that have the most specific and powerful noncovalent interaction since avidin-biotin technology is commonly employed in enzyme-allied immunosorbent assay kits for the recognition of various bio-macromolecules associated with diseases such as cancer and influenza (32,47,48) by combining graphene with the unique avidin and biotin interaction (49). In comparison with traditional bio-FET, high electron mobility transistors (HEMTs) of AlGaN/GaN can not only sense protein in the physiological environment of human serum without diluting or washing but can also detect uncharged molecules due to its technological process (35); also, the streamlined fabrication process, high sensitivity and reproducibility, changeable signal magnitude, and a consistent baseline are additional advantages of the AlGaN/GaN transistor biosensor (19,32,45). Due to graphene's unique

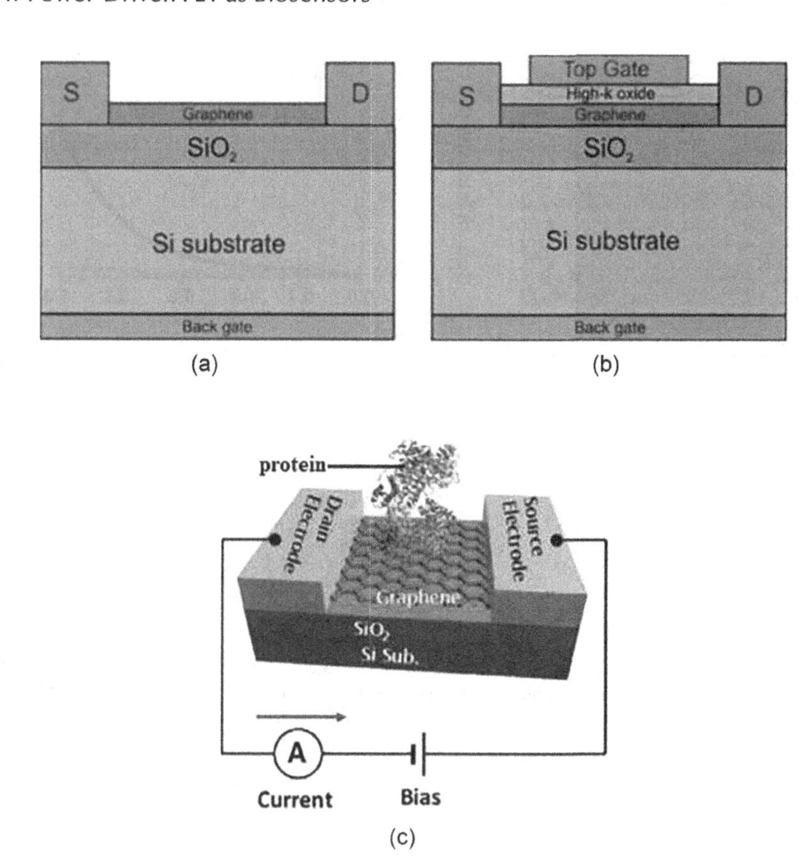

FIGURE 12.4 (a) Back-gated graphene FET structure, (b) top-gated graphene FET structure, and (c) a schematic design of graphene bio-FET. (Adopted from Ref. 49.)

electrical properties, bio-FET's performance and sensing capability have unquestionably increased to new heights (49). This is why recently, graphene FET-based biosensors have been successfully used to detect and analyze the virus quickly to meet the scientific community's interest. The detection principle of graphene-based bio-FET is illustrated in Figure 12.4. The applied voltage in the gate terminals of graphene FET devices alters the graphene channel's conductivity, which is interfaced between the source and the drain due to graphene's very high current conductivity. As a result of the very high conductivity of graphene channel FET devices, the response time is extremely short, which aids in rapidly identifying the virus (49). The virus is etched on the graphene surface in biosensor graphene FET devices. In contrast, the virus affects the conductivity of the graphene channel, which is quickly detected at the output, as illustrated in Figure 12.4b. Finally, graphene-bio-FETs are broadly applied for diagnosis of various viruses due to their rapid detection ability, commercial availability, downsizing capability, low cost with high yield, large-scale and high-quality production, scalability, early-stage detection, and reduced requirement of qualified personnel (47); in addition to the point-of-care virus detection, graphene-based biosensors are free of the risk of cross-contamination via the biosensor.

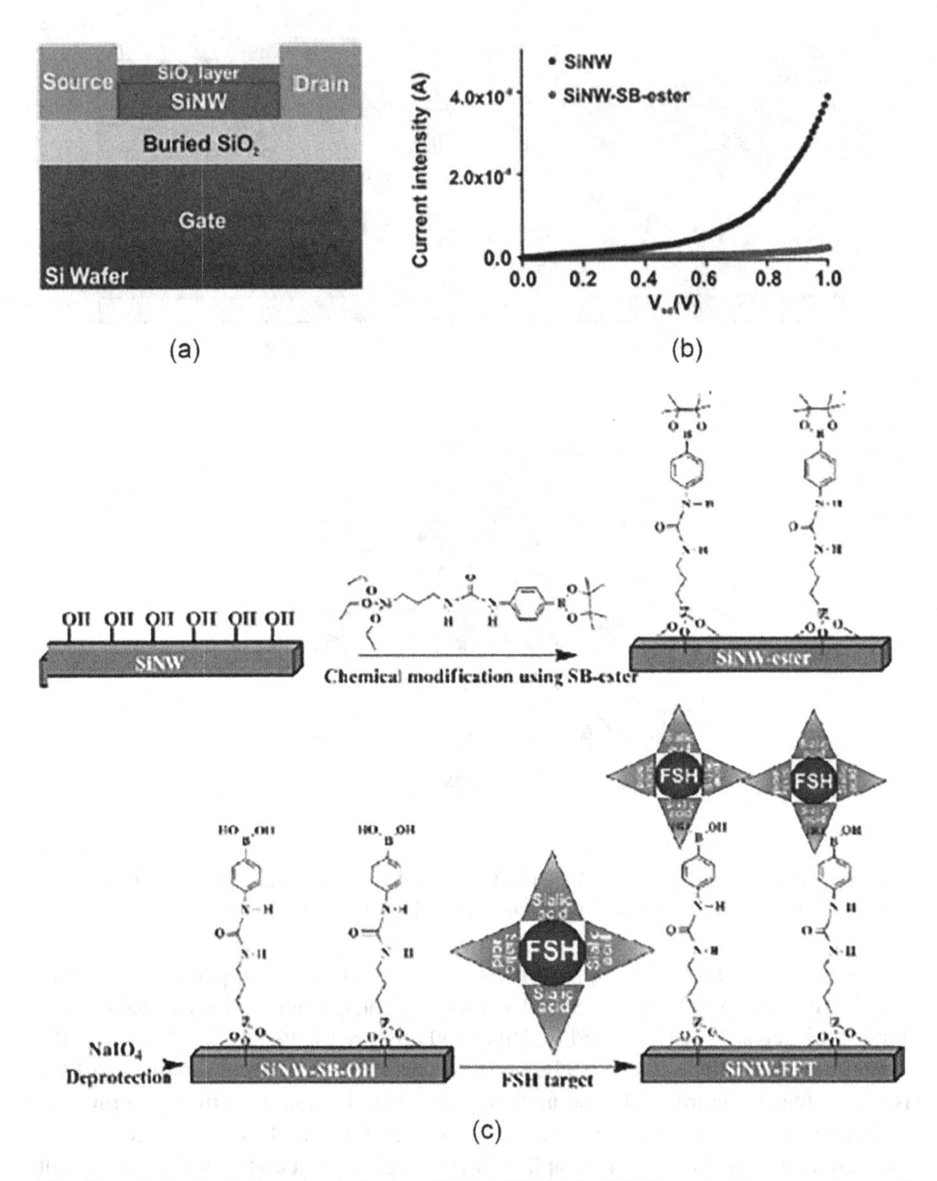

FIGURE 12.5 (a) Schematic design of SiNWFET, (b) SiNW-SB-ester current response of FET, and (c) SiNW-SB-ester chemical modification deprotection, and follicle-stimulating hormone (FSH) target for surface functionalization of SiNWFET devices. (Adopted from Ref. 50.)

Figure 12.5a shows a pictorial representation of SiNWFET-based biosensors with SiNW, source (S), drain (D), and electrodes. Ultrahigh-sensitivity integrated NWFET-based biosensor exhibits "detecting cytokeratin 19 fragment and prostate-specific antigen (PSA) in buffer solution" (50) and is highly selective in various cancer biomarker identification. The hydroxyl-terminated silicon dioxide of SiNW was chemically altered with SB-ester to generate SiNW SB-ester. The ethoxy group

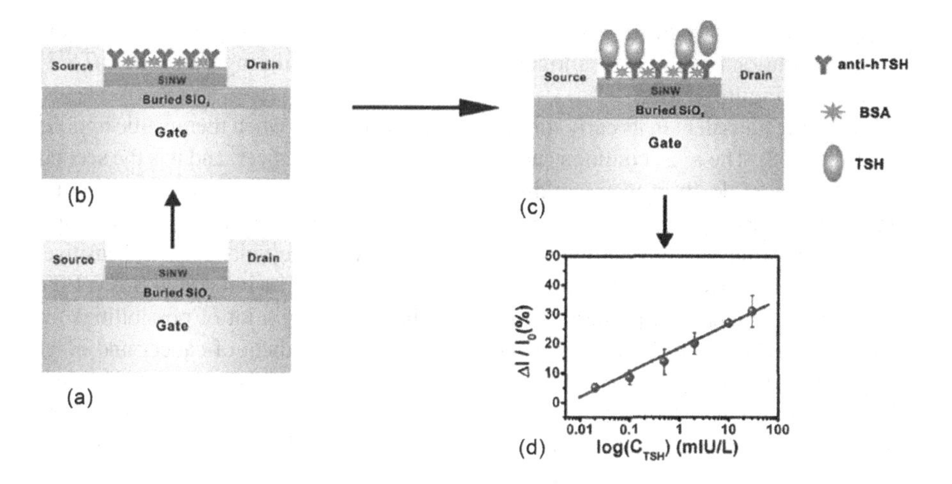

FIGURE 12.6 (a) Schematic of SiNWFET-based biosensor device. (b and c) Sensing platform for detecting hTSH on functionalization of SiNWFET. (d) Sensitivity versus the logarithm of hTSH concentration. (Adopted from Ref. 21.)

of SB-ester interacts with the hydroxyl group of SiNW (50). "Follicle-stimulating hormone (FSH) determination can benefit in the analysis of a variety of aspects, including reproductive system physiology, fertility maintenance, and the discovery or treatment of reproductive diseases" (50,51). For instance, sialic acids are negative-charged byproducts of neuraminic acids found at the end of sugar chains and connected to cell surfaces and glycoproteins (26,50). On the other hand, existing diagnostic methodologies face numerous challenges, like complicated handling techniques, expensive equipment, etc. Designing effective, low-cost, and rapid follicle-stimulating hormone (FSH) detecting biosensors is essential (50). A boronic acid-functionalized metal-oxide-semiconductor (MOS) SiNWFET-based biosensor device is a novel sensing, low-cost, and compact approach for accurate and efficient follicle-stimulating hormone (FSH) detection and menopause diagnosis (50).

The SiNWFET-based biosensor devices have been used in a wide range of applications, such as nucleic acid and protein, virus detection, analysis of small-molecule protein interactions, and cell detection, due to their effectiveness of sensing abnormal biomarker concentrations linked to their small size and the huge surface-to-volume ratio (SVR) (21). For instance, CMOS-compatible SiNWFET-based biosensor can detect easily and rapidly human thyroid-stimulating hormone (hTSH) in the absence of any nanoparticle pairings, as illustrated in Figure 12.6. Using amino acid APTES biomolecule, the coated oxide layer of SiNW surface should be functionalized initially. These functionalized SiNW chips have been employed for electrical toughness using a semiconducting analyzer parameter. The concentrations of the associated antigen human thyroid-stimulating hormone (hTSH) were detected employing anti-hTSH antibodies immobilized on the surface of the SiNWFET-based biosensor, as illustrated in Figure 12.6b and c. The interaction between negatively charged analytes and the appropriate antibody leads to charge carriers in the current channel to be depleted, leading to an increase in current in the

SiNWFET-based biosensor device, and this current variation's magnitude reveals the concentration of the target antigen human thyroid-stimulating hormone (hTSH) as shown in Figure 12.6d.

Cancer is often silent in its early stages and is identified later when therapeutic options are limited (38). The most common cancer in men is prostate cancer, and it is the second greatest cause of death in men worldwide (38). Due to existing methods for detecting cancer, biomarkers are insensitive and need biochemical labeling or nanoparticle conjugations. Early cancer diagnosis and prognosis require a direct, efficient, highly sensitive, and accurate biosensor for detecting cancer biomarkers (52). The integrated SiNWFET device, with its enhanced properties and scalability, opens up a lot of possibilities for point-of-care tests (POCTs) for fast screening and early treatment of cancer and other complicated diseases. The integrated SiNWFET-based biosensors showed not only ultrahigh sensitivity of cytokeratin (CYFRA21-1) and prostate-specific antigen (PSA) with detection in buffer solution but also high selectivity of discrimination from other similar cancer biomarkers. "Prostate-specific antigen(PSA) is a glycoprotein released by the prostate gland's epithelial cells, and it has been recognized as the most extensively utilized tumor marker for screening early prostate cancer and detecting recurrence after therapy" (38) due to its absolute tissue specificity as stated clearly in Figure 12.7.

Peptide nucleic acid (PNA)_DNA hybridization detection using reduced graphene oxide (R-GO) FET-based biosensor is illustrated in Figure 12.8. Due to its large detection area, ambipolar field effect, high carrier (electron) mobility, simplicity of functionalization, and low intrinsic electrical noises, graphene has the potential to outperform CNTs and SiNWs (49). In order to detect biomolecules, highly sensitive and selective reduced graphene oxide bio-FET have recently been developed. The reduced graphene oxide (R-GO) FET biosensor was first functionalized using peptide nucleic acid to enhance DNA detection. Figure 12.8 shows the basic operation of the R-GO FET biosensor for DNA detection based on peptide nucleic acid (PNA)-DNA hybridization. Drop-casting R-GO onto the sensing channel as the conducting material produces the R-GO FET biosensor. The PNA probe is then immobilized using covalent bonding between the amino group on PNA and an amide bond on the other end of PASE, followed by the use of an ethanolamine solution to minimize nonspecific binding (53). The target DNAs are then placed on the device for hybridization with probe peptide nucleic acid (PNA), and a silver wire is applied as the gate to create a liquid-gated FET for electrical measurements. The left shift of the Ids-Vg curves of the devices can be used to measure PNA-DNA hybridization. When noncomplementary DNA is added to PNA-functionalized devices, it does not produce a responsive signal, as clearly stated in Figure 12.8 (Table 12.1).

12.5 SENSITIVITY ENHANCEMENT OF NWFET-BASED BIOSENSOR

The electrical response of the NWFET remained essentially unaltered, while the buffer solution was flowing through the sensor surface in the case of the FET biosensor device. The reduction in electrical current is consistent with earlier findings, implying that negatively charged biomolecules binding to the gate dielectric of n-type NWFETs result in carrier depletion and thereby current reduction. Here are lists of various mechanisms that lead to enhancing NWFET-based biosensor sensitivity (59).

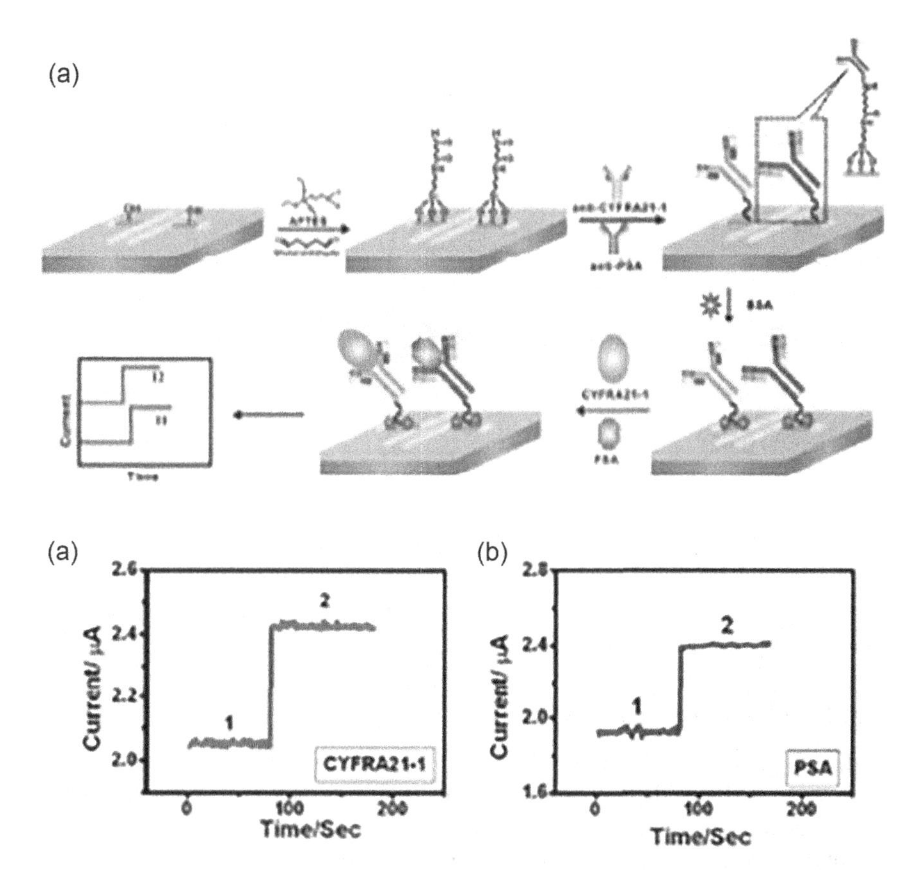

FIGURE 12.7 (a) Schematic diagram of microfluidic integrated SiNW biosensor CYFRA21-1 and PSA detection. (b) Plot of current versus time of anti-CYFRA21-1 functionalized SiNW, with regions one and two representing pure buffer solution and CYFRA21-1 introduction, respectively. (c) Plot of current versus time of an anti-PSA-functionalized SiNW, with regions one and two signifying buffer solution flow and PSA insertion, respectively. (Adopted from Ref. 38.)

12.5.1 Impact of Size on Sensitivity Capability

As we have various research articles, the sensitivity of NWFET devices is higher than that of microwire-FET devices (32,35). The SVR increases significantly as the wire diameter decreases, and the influence of the external electric field can reach the whole cross-section of the wire; this causes a considerable increase in the induced conductance change inside nanowire FETs compared to microwire-FETs. However, large internal wire sections are unaffected, leading to a less significant change in conductance. As a result, whenever charged particles encounter thicker wire, the area influenced by the electric field generated by the charged analyte is constrained to the wire surface.

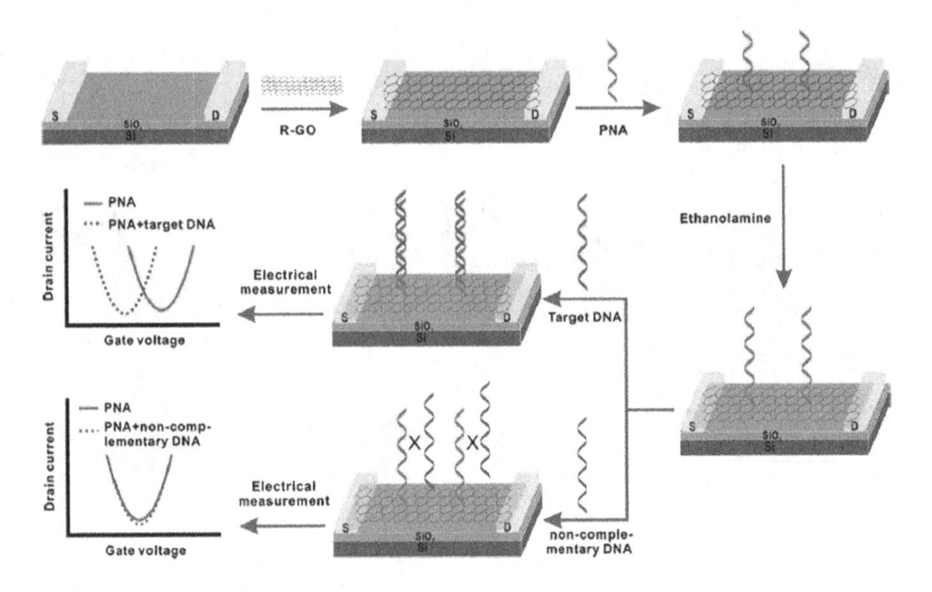

FIGURE 12.8 Reduced graphene oxide (R-GO) FET biosensor, which uses PNA-DNA hybridization to detect DNA. (Adopted from Ref. 53.)

12.5.2 Surface Probe Alignments

Randomly polarized and oriented linker biomolecules cause a problem due to unaligned monolayer; these melodramatically diminish the communal charge opposition carried about by disparaging interference, thereby less sensitive detection. As a result of high-quality surface modification on NW FET sensors, sensing capability and reliability have been improved (30).

12.5.3 SiNW Structure

The top-down strategy for NW production remains challenging due to lithography process constraints in shrinking wire size to 20 nm or below. Metal-assisted chemical etching (mac-etch) (60) is another approach to build nanostructure from the original nanowire leading to a rise in the SVR of nanowires and generation of nanoforest only in the channel region, with no misalignment due to the self-aligned Joule-heating process. Au or ZnO nanorod particles are synthesized precisely on NW (46).

12.6 NANO-FET BIOSENSOR DEVICE CHALLENGES

The preliminary step in developing a biosensor is ultrasmall scale device fabrication in recent technology. Nowadays, mass produced, highly sensitive, real-time measurement capability, label-free electrical sensors and cost-effective, scalable biosensors with low-power-driven FETs are perfect candidates for various medical and basic scientific research projects (54). However, individual nanostructures cannot be consistently placed at specified places on a substrate using current large-

TABLE 12.1

Applications of FET-based biosensors for early detection of various diseases.

Application	RE	Linker	Surface (Channel)	Sensor	Sample	Ref.
Lung cancer diagnosis	Probe DNA	APTES	SiNW	CMOS-FET	Buffer solution	(37)
Cancer diagnosis	Antibody	APTES	SiNW	CMOS -FET	Human serums	(38)
Nucleic acid analysis	Probe DNA	APTES	SiO_2/Si_3N_4/SWNT	SiNWFET	Buffer solution	(26)
DNA detection	PNA	APTES	Si/SiO_2/graphene	Graphene FET	Buffer solution	(54)
Recognition of biomolecules	Biotin	APTES	SiNW	SiNWFET	Buffer solution	(55)
Direct protein sensing	Biotin	Magnetic beads	SiNW	SiNWFET	Buffer solution	(56)
Sensing crossbreeding DNA	DNA probe	APTES	SiO2/SiNW	SiNWFET	Buffer solution	(57)
Thyroid diagnosis	Anti-hTSH	APTES	SiO2/SiNW	SiNWFET	Buffer solution	(21)
Diagnosis of Lyme disease	Borrelia antibody	Diazonium salts	Si/ SiO_{2-}Nanotube	SWNT-FET	Buffer solution	(58)
Follicle-stimulating hormone (FSH) detection	Boronic acid	SB-ester $NaIO_4$	SiNW	FET	Serum solution	(50)

scale fabrication technologies. As a result, current production procedures, which are highly vulnerable to processing conditions and probabilistic nanostructure positioning, produce nano-FET devices with significant unpredictability and device-to-device inconsistency. Some of the important challenges faced in the development of nano-bio-FET are detection time and specificity (describes the sensor's ability to distinguish between biological entities that are targeted and those that aren't in a sample), introducing significant challenges when designing nano-FET biosensor systems, such as finding a suitable technology while maintaining the highest sensitivity and specificity. As a result, in order to commercialize this technology on an ultrasmall scale, more study in this field is required. For instance, lab-on-chip devices and sensor specificity and sensitivity need to be improved in order to handle the whole blood or serum samples of humans. Due to these reasons, further research/efforts in nano-FET biosensor integration with microfluidic and lab-on-chip devices would be necessary to facilitate point-of-care diagnostic applications and real-time closed-loop drug therapy systems. By overwhelming these limitations, nano-FET biosensor devices can be used for various applications, including combinatorial chemical and biological research, high-throughput screening systems for drug development, and point-of-care diagnostics in upcoming technology.

12.7 CONCLUSIONS

This review addressed the most recent different biologically sensitive low-power-driven FET sensors for early diseases analysis and cell-based tests for drug screening investigations. When scalability of the design is an issue, the study on the structural evolution of FET-based biosensors indicates that focusing on CMOS-based FET sensors facilitates further standardization with well-developed CMOS circuitry and mass production in addition to adequate and rapid detection of biological analytes.

The study focus on recently published articles of low-power-driven FET-based biosensor discoveries in biomedical applications. As we have seen in recent articles, NWFET-based biosensors are highly selective and sensitive for detecting biological species in buffer solutions in real-time without labels. For instance, protein–protein interaction, viral detection, neuronal electrical response, human thyroid-stimulating hormone (hTSH), DNA hybridization, and clinical diagnosis are all detected with NWFET-based biosensors via direct electrical measurement, enabling label-free detection of biomolecules, which eradicates time and power consumption labeling experiments and decreases the cost of expensive diagnostic screening examinations.

However, there are still certain problems to overcome the sensitivity downside of ultrasmall device size. For instance, the high and nonspecificity of entire blood samples provide a strong background on the channel surface of NW FET-based biosensors. Even though some obstacles need to be overcome, the SiNWFET biosensor will play a key role in the future biological investigation and cellular inquiries.

Finally, we recommend that the NWFETs, which may be employed as low-cost, disposable, and label-free transportable chips showed promise as fast screening tests and point-of-care diagnostics of various diseases in biomedical applications with low-power consumption.

REFERENCES

1. Omura Y, Mallik A, Matsuo N. *MOS Devices for Low-Voltage and Low-Energy Applications*. Hoboken, NJ: Wiley; 2017.
2. Technologies E, Committee DS, Power E, Soldier D, Isbn C, Pdf T, et al. Low power electronics and design. In: *Energy-Efficient Technologies for the Dismounted Soldier*. 1997. Washington, DC: The National Academies Press. https://doi.org/10.17226/5905.
3. Gopalakrishnan K, Woo R, Jungemann C, Griffin PB, Plummer JD. Impact ionization MOS (I-MOS) - Part II : Experimental results. *IEEE Trans Electron Devices*. 2005;52(1):77–84.
4. Rony MW, Mondol P, Myler HR. Sensitivity of a 10nm dual-gate GAA Si nanowire nMOSFET to process variation. In: *19th Int Conf Comput Inf Technol ICCIT 2016*. 2017;557–62.
5. Veera Boopathy E, Raghul G, Karthick K. Low power and high performance MOSFET. In: *2015 International Conference on VLSI Systems, Architecture, Technology and Applications, VLSI-SATA 2015*. 2015.
6. Kumar A, Tripathi MM, Chaujar R. Comprehensive analysis of sub-20 nm black phosphorus-based junctionless-recessed channel MOSFET for analog/RF applications. *Superlattices Microstruct* [Internet]. 2018;116:171–80. Available from: https://doi.org/10.1016/j.spmi.2018.02.018
7. Chhabra A, Kumar A, Chaujar R. Sub-20 nm GaAs junctionless FinFET for biosensing application. *Vacuum* [Internet]. 2018;160:467–71. Available from: https://doi.org/10.1016/j.vacuum.2018.12.007
8. Kumar A, Tripathi MM, Chaujar R. Reliability issues of In_2O_5Sn gate electrode recessed channel MOSFET: Impact of interface trap charges and temperature. *IEEE Trans Electron Devices*. 2018;65(3):860–6.
9. Adnan MMR, Hafiz MS Bin, Tasneem N, Khosru QDM. Quantum ballistic transport in ultra-small silicon channel cylindrical gate-all-around junctionless nanowire transistor using NEGF formalism. In: *2016 IEEE Int Conf Electron Devices Solid-State Circuits, EDSSC 2016*. 2016;62–5.
10. Kumar A, Kaur D, Tripathi MM, Chaujar R. Reliability of high-k gate stack on transparent gate recessed channel (TGRC) MOSFET. In: *2017 Int Conf Microelectron Devices, Circuits Syst ICMDCS 2017*. 2017;1–4.
11. Houl T, Low T, Xul B, Ljl M, Samudral G, Kwong DL, et al. Impact of metal gate work function on nano CMOS device performance. In: *Proceedings. 7th International Conference on Solid-State and Integrated Circuits Technology*. 2004;57–60.
12. Lee BH, Mocuta A, Bedell S, Chen H, Sadana D, Rim K, et al. Performance enhancement on sub-70 nm strained silicon SOI MOSFETS on ultra-thin thermally mixed strained silicon/SiGe on insulator(TM-SGOI) substrate with raised S/D. In: *Digest. International Electron Devices Meeting*. 2002;946–8.
13. Garg N, Pratap Y, Gupta M, Kabra S. Analysis of interface trap charges of double gate junctionless nanowire transistor (DG-JNT) for digital circuit applications. In: *Proc Int Conf 2018 IEEE Electron Device Kolkata Conf EDKCON 2018*. 2018;563–7.
14. Goel A, Rewari S, Verma S, Gupta RS. Dielectric modulated triple metal gate all around MOSFET (TMGAA) for DNA Bio-molecule detection. In: *2018 IEEE Electron Devices Kolkata Conf*. 2019;1:337–40.
15. S.-H. Kim, J. G. Fossum, and VPT. Bulk inversion in FinFETs and the implied insignificance of the effective gate width. In: *IEEE International SO1 Conference*. 2004;145–7.
16. Auth C. 22-nm fully-depleted tri-gate CMOS transistors. In: *Proc Cust Integr Circuits Conf*. 2012.
17. Gupta N, Chaujar R. Optimization of high-k and gate metal work function for improved analog and intermodulation performance of Gate Stack (GS)-GEWE-SiNW MOSFET. *Superlattices Microstruct* [Internet]. 2016;97:630–41. Available from: http://dx.doi.org/10.1016/j.spmi.2016.07.021

18. Tasneem N, Adnan MR, Hafiz SB, Khosru QDM. Self-consistent determination of quantum- mechanical threshold voltage of gate-all-around junctionless nanowire transistor. In: *Proc 9th Int Conf Electr Comput Eng ICECE 2016*. 2017;135–8.
19. Syu Y-C, Hsu W-E, Lin C-T. Review—Field-effect transistor biosensing: Devices and clinical applications. *ECS J Solid State Sci Technol*. 2018; 7(7):Q3196.
20. Buitrago E, Badia MFB, Georgiev YM, Yu R, Lotty O, Holmes JD, et al. Electrical characterization of high performance, liquid gated vertically stacked SiNW-based 3D FET biosensors. *Sensors Actuators, B Chem*. 2014;199:291–300.
21. Lu N, Dai P, Gao A, Valiaho J, Kallio P, Wang Y, et al. Label-free and rapid electrical detection of hTSH with CMOS-compatible silicon nanowire transistor arrays. *ACS Appl Mater Interfaces*. 2014;6(22):20378–84.
22. Windbacher T, Sverdlov V, Selberherr S. Biotin-streptavidin sensitive BioFETs and their properties. *Commun Comput Inf Sci*. 2010;52:85–95.
23. Sadighbayan D, Hasanzadeh M, Ghafar-zadeh E. Biosensing based on field-effect transistors (FET): Recent progress and challenges. *TrAC Trends in Analytical Chemistry*. 2020;133:116067.
24. Goswami R, Bhowmick B. Dielectric-modulated TFETs as label-free biosensors. In: *Design, Simulation and Construction of Field Effect Transistors*, 2018;1:17–35.
25. Wang Y, Li G. Simulation of a silicon nanowire FET biosensor for detecting biotin/streptavidin binding. In: *2010 10th IEEE Conf Nanotechnology, NANO 2010*. London: IntechOpen; 2010;1036–9.
26. Lu N, Gao A, Dai P, Li T, Wang Y, Gao X, et al. Ultra-sensitive nucleic acids detection with electrical nanosensors based on CMOS-compatible silicon nanowire field-effect transistors. *Methods*. 2013;63(3):212–8.
27. Kim K, Park C, Rim T, Meyyappan M, Lee JS. Electrical and pH sensing characteristics of Si nanowire-based suspended FET biosensors. In: *14th IEEE Int Conf Nanotechnology, IEEE-NANO 2014*. 2014;3(1):768–71.
28. Cao A, Sudhölter EJR, de Smet LCPM. Silicon nanowire-based devices for gas-phase sensing. *Sensors (Switzerland)*. 2013;14(1):245–71.
29. Peesa RB, Panda DK. Rapid detection of biomolecules in a junction less tunnel field-effect transistor (JL-TFET) biosensor. *Silicon*. 2022;14:1705–1711.
30. Zhang GJ, Ning Y. Silicon nanowire biosensor and its applications in disease diagnostics: A review. *Anal Chim Acta* [Internet]. 2012;749:1–15. Available from: http://dx.doi.org/10.1016/j.aca.2012.08.035
31. Getnet Yirak M, Chaujar R. TCAD analysis and modelling of gate-stack gate all around junctionless silicon NWFET based bio-sensor for biomedical application. In: *Proceedings of 2nd International Conference on VLSI Device, Circuit and System, VLSI DCS 2020*. 2020:18–9.
32. Béraud A, Sauvage M, Bazán CM, Tie M, Bencherif A, Bouilly D. Graphene field-effect transistors as bioanalytical sensors: Design, operation and performance. *Analyst*. 2021;146(2):403–28.
33. Jokilaakso N. A biotechnology perspective on silicon nanowire FETs for biosensor applications (Doctoral dissertation, KTH Royal Institute of Technology); 2013.
34. Singh S, Raj B, Vishvakarma SK. Analytical modeling of split-gate junction-less transistor for a biosensor application. *Sens Bio-Sensing Res* [Internet]. 2018;18:31–6. Available from: https://doi.org/10.1016/j.sbsr.2018.02.001
35. Vu C, Chen WY. Field-effect transistor biosensors for biomedical applications: Recent advances and future prospects. *Sensors*. 2019; 19(19):4214.
36. Omar MN, Salleh AB, Lim HN, Ahmad Tajudin A. Electrochemical detection of uric acid via uricase-immobilized graphene oxide. *Anal Biochem* [Internet]. 2016;509:135–41. Available from: http://dx.doi.org/10.1016/j.ab.2016.06.030

37. Gao A, Yang X, Tong J, Zhou L, Wang Y, Zhao J, et al. Multiplexed detection of lung cancer biomarkers in patients serum with CMOS-compatible silicon nanowire arrays. *Biosens Bioelectron* [Internet]. 2017;91:482–8. Available from: http://dx.doi.org/10.1016/j.bios.2016.12.072
38. Lu N, Gao A, Dai P, Mao H, Zuo X, Fan C, et al. Ultrasensitive detection of dual cancer biomarkers with integrated CMOS-compatible nanowire arrays. *Anal Chem.* 2015;87(22):11203–8.
39. Piermarini S, Migliorelli D, Volpe G, Massoud R, Pierantozzi A, Cortese C, et al. Uricase biosensor based on a screen-printed electrode modified with Prussian blue for detection of uric acid in human blood serum. *Sensors Actuators, B Chem* [Internet]. 2013;179:170–4. Available from: http://dx.doi.org/10.1016/j.snb.2012.10.090
40. Chakraborty A, Sarkar A. Analytical modeling and sensitivity analysis of dielectric-modulated junctionless gate stack surrounding gate MOSFET (JLGSSRG) for application as biosensor. *J Comput Electron.* 2017;16(3):556–67.
41. Madan J, Pandey R, Chaujar R. Conducting polymer based gas sensor using PNIN- gate all around - tunnel FET. *Silicon.* 2020; 12(12):2947–2955.
42. Rahman E, Shadman A, Khosru QDM. Effect of biomolecule position and fi ll in factor on sensitivity of a dielectric modulated double gate junctionless MOSFET biosensor. *Sens Bio-Sensing Res* [Internet]. 2017;13:49–54. Available from: https://doi.org/10.1016/j.sbsr.2017.02.002
43. Yirak MG, Chaujar R. Interface trap charge analysis of junctionless triple metal gate high-k gate all around nanowire FET-based biotin biosensor for detection of cardiovascular diseases. In: *Microelectronics, Circuits, and systems.* Singapore: Springer; 2021;47–57.
44. Kaisti M. Detection principles of biological and chemical FET sensors. *Biosens Bioelectron.* 2017;98:437–448. Available from: http://dx.doi.org/10.1016/j.bios.2017.07.010
45. Im H, Huang XJ, Gu B, Choi YK. A dielectric-modulated field-effect transistor for biosensing. *Nat Nanotechnol.* 2007;2(7):430–4.
46. Chen K, Li B, Chen Y. Silicon nanowire field-effect transistor-based biosensors for biomedical diagnosis and cellular. *Nano Today* [Internet]. 2011;6(2):131–54. Available from: http://dx.doi.org/10.1016/j.nantod.2011.02.001
47. Szunerits S, Boukherroub R. Graphene-based biosensors. *Interface Focus.* 2018;8(3):20160132.
48. Wang S, Hossain Z, Shinozuka K, Shimizu N, Kitada S. Graphene field-effect transistor biosensor for detection of biotin with ultrahigh sensitivity and specificity. *Biosens Bioelectron J.* 2020;165:112363.
49. Sengupta J, Hussain CM. Graphene-based field-effect transistor biosensors for the rapid detection and analysis of viruses: A perspective in view of COVID-19. *Carbon Trends* [Internet]. 2021;2:100011. Available from: https://doi.org/10.1016/j.cartre.2020.100011
50. Lee M, Palanisamy S, Zhou BH, Wang LY, Chen CY, Lee CY, et al. Ultrasensitive electrical detection of follicle-stimulating hormone using a functionalized silicon nanowire transistor chemosensor. *ACS Appl Mater Interfaces.* 2018;10(42):36120–7.
51. Sarangadharan I, Pulikkathodi AK, Chu C-H, Chen Y-W, Regmi A, Chen P-C, et al. Review—high field modulated FET biosensors for biomedical applications. *ECS J Solid State Sci Technol.* 2018;7(7):Q3032–42.
52. Doucey MA, Carrara S. Nanowire sensors in cancer. *Trends Biotechnol* [Internet]. 2019;37(1):86–99. Available from: https://doi.org/10.1016/j.tibtech.2018.07.014
53. Cai B, Wang S, Huang L, Ning Y, Zhang Z, Zhang GJ. Ultrasensitive label-free detection of PNA-DNA hybridization by reduced graphene oxide field-effect transistor biosensor. *ACS Nano.* 2014;8(3):2632–8.

54. Zheng C, Huang L, Zhang H, Sun Z, Zhang Z, Zhang GJ. Fabrication of ultrasensitive field-effect transistor DNA biosensors by a directional transfer technique based on CVD-grown graphene. *ACS Appl Mater Interfaces*. 2015;7(31):16953–9.
55. Li BR, Chen CW, Yang WL, Lin TY, Pan CY, Chen YT. Biomolecular recognition with a sensitivity-enhanced nanowire transistor biosensor. *Biosens Bioelectron* [Internet]. 2013;45(1):252–9. Available from: http://dx.doi.org/10.1016/j.bios.2013.02.009
56. Generalov VM, Naumova O V., Fomin BI, P'yankov SA, Khlistun IV., Safatov AS, et al. Detection of ebola virus VP40 protein using a nanowire SOI biosensor. *Optoelectron Instrum Data Process*. 2019;55(6):618–22.
57. Wenga G, Jacques E, Salaün AC, Rogel R, Pichon L, Geneste F. Step-gate polysilicon nanowires field-effect transistor compatible with CMOS technology for label-free DNA biosensor. *Biosens Bioelectron*. 2013;40(1):141–6.
58. Lerner MB, Dailey J, Goldsmith BR, Brisson D, Charlie Johnson AT. Detecting Lyme disease using antibody-functionalized single-walled carbon nanotube transistors. *Biosens Bioelectron* [Internet]. 2013;45(1):163–7. Available from: http://dx.doi.org/10.1016/j.bios.2013.01.035
59. Chen Y, Ren R, Pu H, Guo X, Chang J, Zhou G, et al. Field-effect transistor biosensor for rapid detection of ebola antigen. *Sci Rep* [Internet]. 2017;7(1):4–11. Available from: http://dx.doi.org/10.1038/s41598-017-11387-7
60. Jayakumar G, Garidis K, Hellström PE, Östling M. Fabrication and characterization of silicon nanowires using STL for biosensing applications. In: *2014 15th International Conference on Ultimate Integration on Silicon (ULIS)*. 2014;109–12.

13 2D Materials for Spin Orbital Torque MRAM

A Path toward Neuromorphic Computing

Shashidhara M and Abhishek Acharya
SV National Institute of Technology

CONTENTS

13.1 INTRODUCTION

Smart electronic gadgets and the Internet of Things (IoT) generate massive amounts of data, which demands ultrafast scalable memory devices. The current availability of huge amounts of data and computer capacity, as well as neuromorphic computing architectures, has facilitated the resurgence of machine learning development of deep learning methodologies for practical uses [1,2]. Despite the fact that the solutions are mostly software-based, neural networks are nevertheless simulated in neuromorphic computing utilizing von Neumann architectures. Simulating a single synapse/neuron on a hardware level still necessitates a tremendous number of transistors, which occupy a large area and cause both integration and energy consumption issues [3–5]. Furthermore, due to the planar structure of complementary metal oxide semiconductor (CMOS) integrated circuit technology [6], hardware development of neural connection is a serious restriction. As a result, traditional computing is unable to meet the growing demand for hardware architectures for cognitive and recognition activities, which consume a significant amount of processing energy and cost [7]. Traditional von Neumann computing machines include data processing units (CPU)

DOI: 10.1201/9781003240778-13

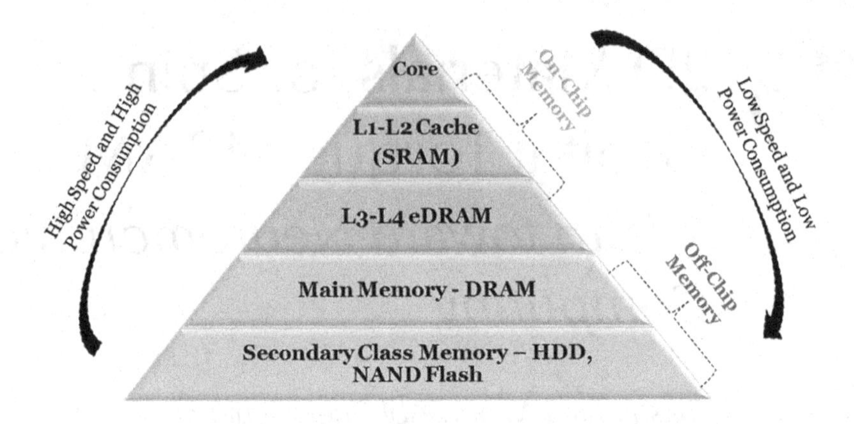

FIGURE 13.1 Hierarchy of computational memory.

and memory units that are physically distinct, resulting in high latency and energy consumption. Modern computing architectures such as multicore processors enable parallelization and combine CPU and memory units for high-demand data processing applications [8]. The on-chip cache was introduced in the CPU to keep up with the data processing speed and to reduce transmission latency. On-chip caches allow for data to be kept close to the CPU cores improving overall chip performance. Figure 13.1 shows the hierarchy of computational memory.

Several nonvolatile memory architectures have been invented in recent years and are undergoing extensive development and research [9]. There is no unique data storage device that can today meet all of the requirements simultaneously. Combining several memory architectures in a single computational system is a time-consuming and difficult task. It also has a negative impact on the system's overall performance. As a result, there is great interest in finding versatile memory architecture that has high storage density and nonvolatile memory, low latency, and high endurance. Many alternative memory technologies have been offered as potential versatile memories in recent decades. Resistive random-access memory (RRAM), phase change memory, racetrack memory [10], and magneto resistance random-access memory (MRAM) are some of the most well-known alternate memories. In MRAM, the magnetic state of a nanometer-scale magnet is used to store information. The magnetic tunnel junction (MTJ) is a fundamental storage element in MRAM.

The mechanics of spin-transfer torque (STT) and spin orbit torque (SOT) are employed to write MTJ. When an unpolarized charge current is injected into the first ferromagnetic (FM) layer of the MTJ, electrons become spin-polarized, exerting a torque on the magnetic moment of the second FM layer in the MTJ. Due to the Rashba–Edelstein effect (REE) and/or the spin Hall effect (SHE), unpolarized flow through a nonmagnetic (NM) material with a high spin orbit coupling (SOC) generates a strong spin current in the SOT mechanism [11–13]. As a result of the accumulation of spin-polarized electrons at the NM/FM heterojunction, current passing through NM material imposes a torque on the surrounding FM layer. To generate spin-polarized current in traditional SOT devices, a three-dimensional (3D) heavy metal with a high SOC, such as Pt, W, or Ta, is employed [14,15]. However, the main disadvantage of HMs is that the efficiency of damping-like torque (T_{DL}), which plays a major role in magnetization switching, ranges from 0.1 to 0.3.

Two-dimensional (2D) materials with strong SOC are prospective candidates to replace HMs, according to recent progress in the field of research in identifying materials that result in T_{DL} with high-efficiency and strong SOC [16,17]. As a result, 2D SOT-MTJ has piqued the curiosity of nonvolatile memory and logic designers. Two types of 2D materials have shown to be excellent prospects for SOT applications. One is transition metal dichalcogenides (TMDs), for example, 2D van der Waals (vdW) materials with controllable conductivity and high SOC [18]. TMDs have been demonstrated to generate out-of-plane T_{DL}, which could allow for efficient and reliable magnetic switching materials with perpendicular magnetic anisotropy (PMA). Perpendicularly magnetized magnets are efficient compared to in-plane magnetized magnets in MTJ because they require a significantly lower switching current [19,20]. The topological insulator (TI) is another type of 2D material. Because TIs have insulating bulk characteristics and conducting surface states, their SOC is several orders of magnitude more powerful than those of HMs [21]. This is due to spin momentum locking. Engineering chemical and electrical properties, spin transport, ferromagnetism and antiferromagnetism, and conversion efficiency of charge to spin are only a few of the spintronic properties given by 2D materials (TIs and TMDs). The improvements in SOT made possible by TIs and TMDs have prompted extensive study on SOT devices based on 2D materials. SOTs in TIs and TMDs are the subject of much 2D/FM heterostructure research.

The format of the chapter is outlined as follows. First, the basic working concept of MTJ will be understood, with a focus on the different writing processes, as well as the underlying physics of STT and SOT writing mechanisms, and materials employed in conventional SOT devices. The present state of research into 2D TIs and TMDs as a material for SOT devices is then discussed. Finally, we discuss current difficulties and future potential in the subject.

13.2 MTJ: A NONVOLATILE MAGNETIC MEMORY

The MTJ is made up of a barrier that separates two FM layers. As shown in Figure 13.2, the two layers are labeled free layer (FL) and pinned layer (PL). The direction of magnetization of the FL can change, whereas the magnetization of the PL cannot. MgO serves as a tunnel barrier between PL and FL in modern MTJs, providing stronger tunneling magneto resistance (TMR) [22–24]. In parallel configuration, both PL and FL magnetization directions are the same, while in antiparallel configuration, both PL and FL magnetization directions are opposite. When

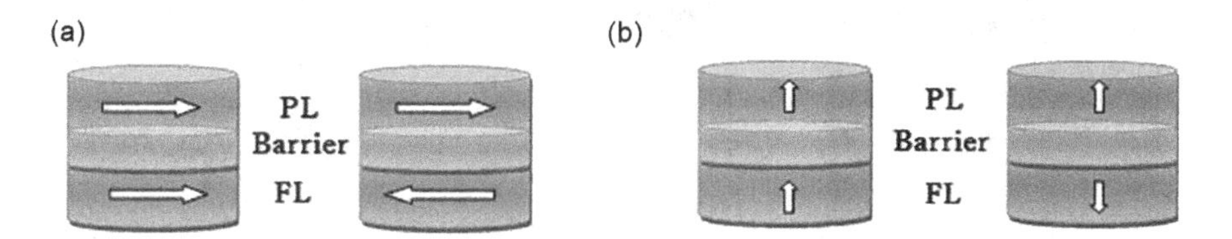

FIGURE 13.2 (a) i-MTJ with in-plane magnetization; magnetization direction is indicated by an arrow. (b) p-MTJ with perpendicular magnetization.

the direction of magnetization is on the plane, in-plane MTJ (i-MTJ) is used, while perpendicular plane MTJ (p-MTJ) is used when the direction of magnetization is perpendicular. As previously stated, p-MTJ is favored over i-MTJ because it requires less switching current [25]. TMR is defined as $\mathrm{TMR} = \frac{R_{AP} - R_P}{R_P}$, where R_{AP} and R_P denote MTJ resistance in antiparallel and parallel states, respectively. TMR is reported to be 604% with MgO as a tunnel barrier, compared to 70% with Al_2O_3. When it is operating in parallel mode, MTJ is said to be storing logic 1, and when it is operating in antiparallel mode, it is said to be storing logic 0. By 2032, MTJ devices are projected to reach a 1,000% TMR ratio, thanks to reasonable investigation of novel materials and combinations of oxides for barrier [26,27].

Tunneling is a quantum phenomenon that occurs when the density of states (DOS) of FM1 and FM2 differ. The DOS diagram, as illustrated in Figure 13.3a, can be used to analyze the tunneling effect. In parallel state, since the number of spin-polarized (spin-up and spin-down) electron states in two FM layers is the same, all spin-orientated electrons in one FM layer's Fermi level can easily tunnel into another FM layer's Fermi level. As a result, MTJ has a low-resistance property. In the electric transit of the antiparallel state, however, only few electrons from FM1 can reach FM2 because of a mismatch in the density of states among the two FM layers. It has the potential to provide a stronger resistance than the P state. The electrons travel through and emerge on the other side of a thin insulating layer when a significant voltage is applied across it, and this is known as the tunneling effect due to the quantum effect.

13.2.1 Field-Driven MRAM

Various types of MRAMs based on writing techniques in MTJ have been developed as spintronic research has progressed, whereas the read method in MTJ is common in all MRAMs [28]. As indicated in Figure 13.3b, the MTJ fixed layer is connected to the MRAM bit-line. The function of the bit-line is to read the state of MTJ; on applying bias voltages of 0.1–0.2V across the MTJ, charge current flows along the

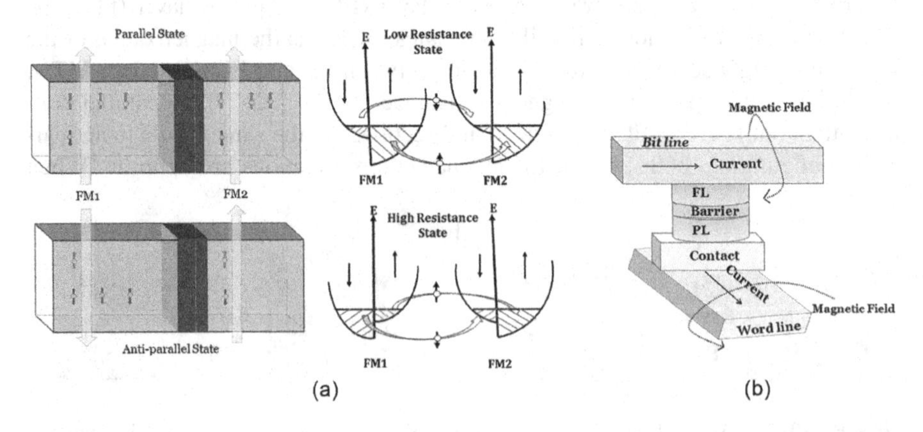

FIGURE 13.3 (a) Density of states in MTJ in parallel and antiparallel states. (b) Read and write mechanisms in MTJ using the field-driven method.

bit-line and tunnels through the MgO barrier into the FM layer (free). As stated earlier, the electrical resistance of the current route is determined by the direction of magnetization of PL and FL. At ambient temperature, the resistance between the two magnetic states can be increased by several orders of magnitude by choosing good magnetic materials and an excellent spin filter such as MgO as a barrier between the two FM layers.

To perform write operation, a field-driven method is used in early MRAM designs in which magnetic fields are used to change the magnetization direction of FL as shown in Figure 13.3b. The field-driven MRAM operation has a simple architecture, but it poses several challenges. The magnetic field generated cannot be restricted to a single bit-cell, and hence, it results in unwanted magnetization switching of the neighboring bit-cell. Also, magnetic field's generation is a power-hungry activity. The power required to switch magnetization of each bit grows dramatically when larger arrays are used. Switching the MTJ FL using spin-transfer torque is a more reliable and energy-efficient approach to write MRAM compared to the field-driven method. A more reliable and energy-efficient technique to write MRAM is to use spin-transfer torque to switch the magnetization of the MTJ FL.

13.2.2 Spin-Transfer Torque-Based MRAM (STT-MRAM)

J.C. Slonczewski. first proposed the theory of spin-transfer torque in 1989 by demonstrating the transfer spin angular momentum from one FM layer to another under a nonzero external bias [29]. Also, magnetic oscillations and even switching can be generated by transferring enough spin angular momentum. Much progress has been achieved in the development of STT-MRAM since then; Everspin introduced the first commercial STT-MRAM module in 2012, while Samsung and Avalanche Technology commercially introduced STT-MRAM in recent years.

In STT-MTJ, the magnetization of FL is switched via a STT mechanism. As shown in Figure 13.4a, orientation of electron spin in charge current is random. The magnetic moment of electrons and the magnetic moments of the FM layer impose torque on each other when charge current passes through a thick magnetic layer FM1. Inside a ferromagnet, all magnetic moments point in the same direction. Due

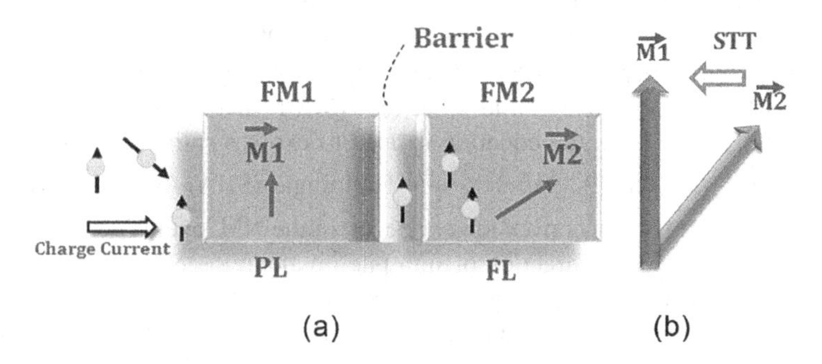

FIGURE 13.4 (a) Flow of random electrons from FM1 to FM2 with distinct magnetic moments. (b) Application of torque in the direction of FM1's magnetization due to mismatch in the magnetization orientation of FM2.

to random orientation of electron spin in charge current, the magnetization direction $\overrightarrow{M1}$ (of FM1) is unaffected by the STT. Rather, the majority of the electrons in charge current become spin-polarized in the direction of FM1's magnetization $\overrightarrow{M1}$. When spin-polarized electrons with magnetic moment $\overrightarrow{M1}$ pass through a second thin FM (FM2) layer with a distinct magnetic moment $\overrightarrow{M2}$, a torque will be applied to the magnetic moments of FM2 by electron spins that are mismatched with the magnetization of the FM2, and this torque is called as spin-transfer torque (STT). As shown in Figure 13.4b, the torque is proportional to -$\overrightarrow{M2} \times (\overrightarrow{M1} \times \overrightarrow{M2})$.

Higher efficiency in spin torque can be achieved using STT-MTJ. Since STT-MRAM devices share the same path for read and write operations, high writing current injected via the tunneling barrier may result in the breakdown of the barrier oxide. A voltage pulse of 0.45–0.65V is delivered across the MTJ to inject the desired amount of writing current for magnetization switching. As a result, the tunneling barrier is exposed to a strong electric field (~109 V/m), which can result in a dielectric breakdown. The separation of read and write routes in MTJ resolves the majority of STT-MTJ difficulties. STT-MRAMs are now limited to applications where memory stability is highly critical because of their less endurance, low density, and moderate power consumption. Several researchers have proposed a three-terminal MRAM cell based on SOT mechanism in order to have independent read and write paths in MTJ [30].

13.2.3 Spin Orbit Torque-Based Three-Terminal MRAM (SOT-MRAM)

A three-terminal SOT MRAM device is a heterostructure of FM, and a NM material is shown in Figure 13.5a. When the charge current is injected into the NM layer, spin-polarized electrons pile up near the heterojunction FM/NM. The spin-polarized electrons impose a torque on the adjacent FM layer, causing switching of magnetization and the movement of the FM layer's domain wall. The magnetic state of the device can be obtained by passing a small read current through the device (Figure 13.5b). The SHE and REE effects generate SOT due to polarized spin charge accumulation at the NM/FM heterostructure [31]. The angular momentum due to spin and the mechanical angular momentum of the lattice can exchange angular momentum thanks to SOC.

As a result, electrical approaches could be used to adjust magnetic order in a more energy-efficient manner. Due to significant SOC, unpolarized current flowing through NM materials causes deflection of spin-polarized electrons, resulting in polarized spin electron build-up at the NM/FM heterostructure as shown in Figure 13.6a. As a result, the equation is used to express spin current due to SHE is $Js = \frac{\hbar}{2e}\theta sh\ (Jc \times \sigma)$ [32,33]. The spin Hall angle (SHA) (θsh) is a parametric measure of charge-to-spin conversion efficiency in the NM layer. The sign and magnitude of θsh indicate the direction of spin pile up along the NM/FM interface and the spin current density for the in-plane current Jc, respectively. The inverse process in which spin current is converted into charge current is also called the inverse spin Hall effect (ISHE).

The REE is mainly due to broken inversion symmetry at the NM/FM heterojunction as shown in Figure 13.6b. [34–36]. Conduction electrons move with velocity **V** near the

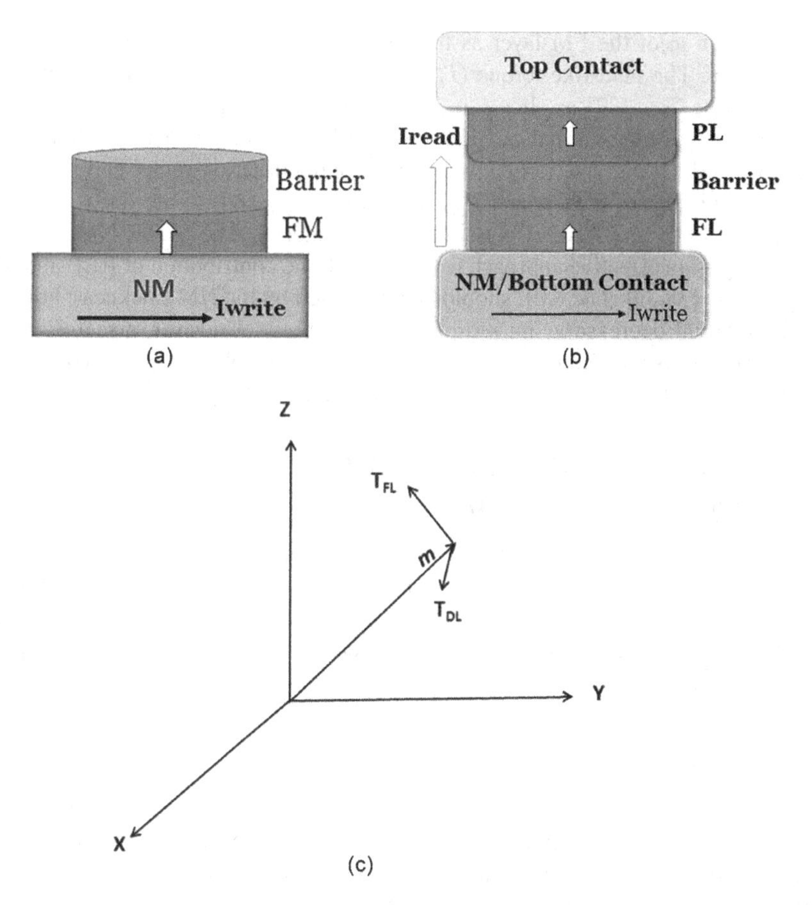

FIGURE 13.5 (a) NM/FM heterostructure. (b) Decoupled paths for read and write mechanisms in SOT-MRAM. (c) Decomposition of SOT torque components.

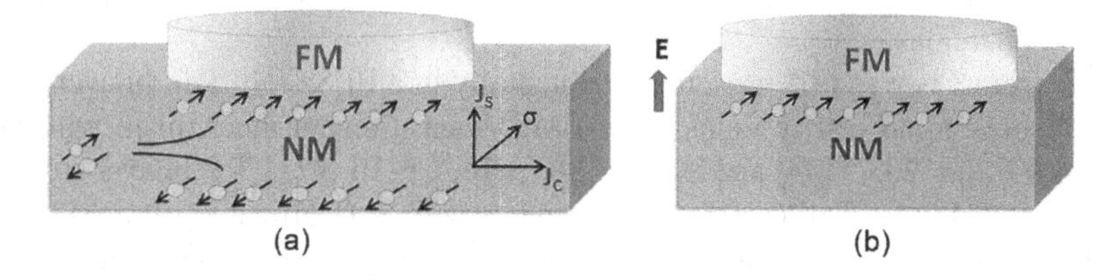

FIGURE 13.6 (a) Spin Hall effect and deflection of spin-polarized current. (b) Rashba–Edelstein effect and accumulation of spin electrons at the interface of the NM and FM layer.

NM/FM spin-polarized due to the magnetic field in the direction $\mathbf{E} \times \mathbf{V}$, where $\mathbf{E}$ is the electric field across the heterojunction due to the broken inversion symmetry structure as shown in Figure 13.6b. This magnetic field polarizes the conduction electrons' spin magnetic moments along $\mathbf{E} \times \mathbf{V}$ by coupling to their spin magnetic moments.

This interfacial SOC induces accumulation of spin electrons near the junction FM/NM and exerts a torque on the FM layer. The torque T_{SOT} is imposed on the magnetization m of the FM layer as the spin current is injected into the neighboring FM layer. The field-like torque (T_{FL}) and damping-like torque (T_{DL}) contributes to total torque T_{SOT} [37] as shown in Figure 13.5c. Damping-like torque $T_{DL}=\tau_{DL}$ $m\times(m\times\sigma)$ is the longitudinal or Slonczewski like torque, and field-like torque T_{FL} $=\tau_{FL}\, m\times\sigma$ is the transverse torque, where σ is the polarization vector and τ_{FL} and τ_{DL} are the magnitudes of the T_{FL} and T_{DL}, respectively. τ_{FL} and τ_{DL} are due to contribution of both SHE and REE effects in the NM material. The SOT's NM thickness dependency is an effective approach to distinguish the contribution of REE and SHE to τ_{FL} and τ_{DL} [38,39]. The REE should be unaffected by the NM thickness; however, the SHE should decrease as bulk dimensions decrease. However, research experimental data shows that the τ_{DL} in most systems is inextricably related to SHE and that REE mainly contributes in the development of the τ_{FL}.

13.3 SPIN ORBIT TORQUE APPLICATIONS BASED ON 2D MATERIALS

NM materials, such as TMDs and TIs, have shown substantial efficiency in charge-to-spin current conversion in recent advancements in discovering novel materials for SOT device use [40]. As a result of their excellent conversion efficiency of charge to spin, TIs and TMDs are being intensively pursued for SOT-based spintronic applications [39,41,42].

13.3.1 SOT FROM TOPOLOGICAL INSULATORS (TIs)

TIs are quantum materials with an insulator-like bulk and topologically shielded conducting edge states. Topological surface states (TSSs) are conducting states in which electrons moving across a surface lock their spin momentum due to spin polarization frozen in the opposite direction of their motion [43]. To create spin current, many investigations have looked into new high-efficiency TI materials. The SHA estimated in TIs is 3–4 orders of magnitude larger than those of typical HMs such as W, Pt, and Ta. In this light, TI-based memory devices are ideal candidates for low-power applications.

SOT from TIs was first proven experimentally in a $Bi_2Se_3/Ni_{80}Fe_{20}$ bilayer 2D TI using a spin-torque FM resonance (ST-FMR) measurement mechanism with SOT efficiency θ_{SOT}= 203.5 [44] and in a $(Bi_xSb_{1-x})_2Te_3$/Cr-$(Bi_xSb_{1-x)2}$Te3 bilayer structure by second-harmonic Hall (SHH) measurements with SOT efficiency θ_{SOT}=425 [45], which is several orders greater than the SOT efficiency in HM. The magnetization switching is achieved with a current density of 8.9104 A/cm^2, which is several orders lesser than those in conventional HMs. ST-FMR to investigate current-induced SOT in Bi_2Se_3/Py heterostructures is shown in Figure 13.7a. As seen in Figure 13.7b, the signal from the ST-FMR measurement displayed a massive symmetric peak, indicating a strong conventional damping-like SOT [46].

The SOTs T_{FL} and T_{DL} in Bi_2Se_3/Py bilayer TI is due to the interfacial character of the produced spin electrons from TSS and bulk SHE, respectively. However, as the temperature drops, the observed T_{DL} increases significantly, Figure 13.7c, which is not expected from the bulk SHE processes. As a result, T_{DL} was thought to be caused

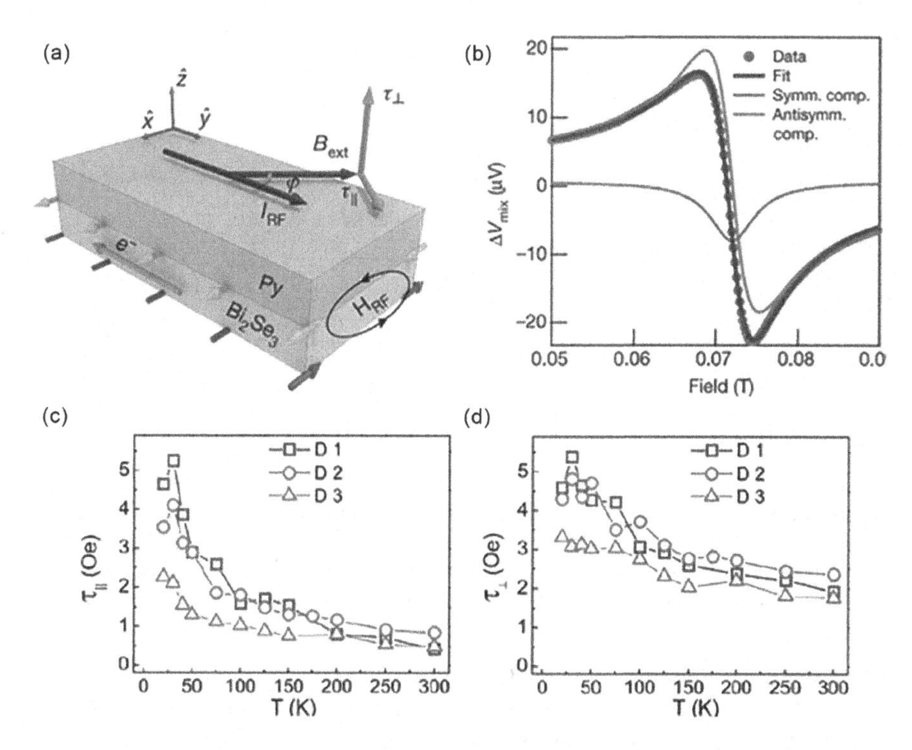

FIGURE 13.7 (a) Schematic of the TI Bi_2Se_3/Py heterostructure and (b) ST-FMRR signal for TI Bi_2Se_3/Py heterostructure at room temperature. Dependence on temperature of (c) SOT (in-pane) and (d) SOT (out of plane) measured using ST-FMR in Bi_2Se_3/CoFeB. (Reproduced with permission from A. R. Mellnik et al. [44] and Yi. Wang et al. [46].)

by the spin transfer of TSS generated spins. Figure 13.7c and d demonstrates the temperature dependency of the T_{DL} and T_{FL} components of SOTs, respectively. The TI $(Bi_xSb_{1-x})_2Te_3$ has lower switching current density (J_{sw}) than Bi_2Se_3 and larger SOT as it is more insulating in the bulk. Increase in SOT efficiency and reducing J_{sw} for higher efficiency are potential future research paths for adopting SOT for MRAM, which can be achieved by enhancing the interfacial spin transparency and improving the film quality. As already stated, in a TI-based heterostructure, the carrier density of TSS is critical in determining the SOT efficiency. Also, it has been demonstrated that altering the proportion of Bi and Sb in $(Bi_{1-x}Sb_x)_2Te_3$ can tune the carrier density of TSSs as illustrated in Figure 13.8a [47].

The highest effective field in SOT is likewise observed with the Fermi level very near to the Dirac point, and the TSSs dominate the transport characteristics. In a TI device with Cr-doped top gate voltage, SOT efficiency may be adjusted by a factor of 4 [45]. When the density of carriers at the top surface is the lowest and the total density of carriers of the surface state reaches its peak, as shown in Figure 13.8b, the TI device has the best spin conversion efficiency. In order to achieve PMA in the TI/FM heterostructure, a HM layer is introduced between TI/FM. However, a HM layer between the TI/FM heterostructure has the disadvantage of preventing the flow of spin current into the FM layer. It may be desirable to use a light metal layer with a low

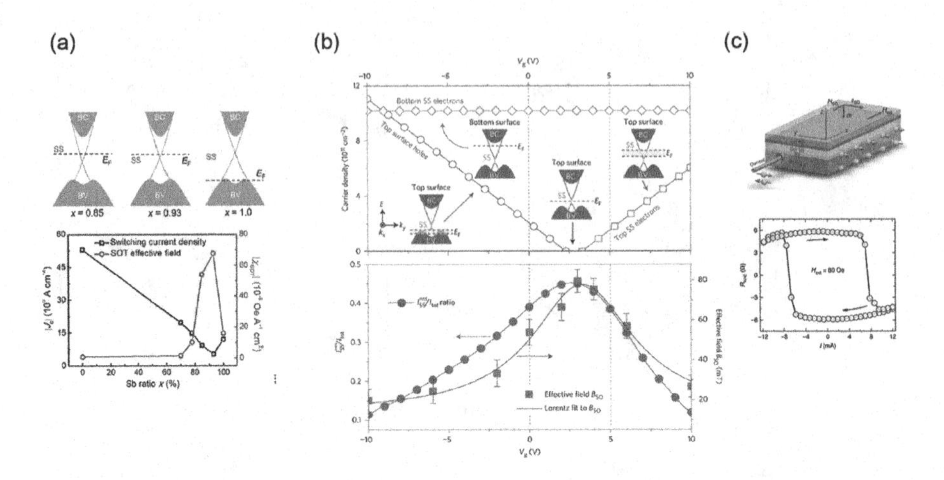

FIGURE 13.8 (a) Switching current density varies as a function of Sb percentage, along with the Fermi level shown. (b) Surface carrier density as a function of Vg. (c) Schematic of 3D $Bi_xSe_{(1-x)}$/CoFeB TI heterostructure and SOT-induced switching of the magnetization with bias. (Reproduced with permission from Wu et al. [47], Fan et al. [45], and Mahendra et al. [48].)

SOC strength to maintain a high SOT efficiency. PMA has been realized in Bi_xSe_{1-x} and CoFeB stack using a thin metal layer Ta inserted between the stack as shown in Figure 13.8c [48]. To alter the magnetization in this device, 4.3×10^5 A/cm^2 current density is required. Magnetization switching due to current induced in Bi_2Te_3/Mo/CoFeB, (BiSb)Te_3/Mo/CoFeB, and $BiSTe_3$/Mo/CoFeB structures has been demonstrated using PMA [39]. The $(BiSb)_2Te_3$/Mo/CoFeB device has a switching current as low as 3×10^5 A/cm^2 and a SOT efficiency of roughly 2.60.

13.3.2 Spin Orbit Torque from TMDs

Several 2D materials can be found in mass as layers closely packed with minimal interlayer attraction, enabling exfoliation into individual, molecularly thin sheets. Currently, graphene has been attracting most attention. TMDs, in particular, exhibit a diverse set of optoelectronic, photonic, physical, chemical, and thermal properties. TMDs have a hybrid structure with the typical molecular formula MX_2, where M is the transition metal from group IV (Hf, Zr, Ti, etc.), group V (Ta, Nb, V, etc.), or group VI (W, Mo, etc.), and X is the chalcogen (Te or Se, S). The chalcogen atoms are arranged in two hexagonal surfaces separated by a layer of transition metal atoms in these materials, forming a layered structure with the arrangement X-M-X. The adjacent layers, like graphene, are bound together by weak van der Waals interactions, allowing for physical or chemical abrasion into single or few-layer structures.

As per the stacking sequence and coordination of transition metal atoms, the bulk crystal can have a range of polytypes. TMDs have rhombohedral or hexagonal global symmetry, and the metal atoms are coordinated in octahedral (1T) or trigonal prismatic (2H) ways. Monolayer and few-layer TMDs display a variety of fascinating features not found in their conventional materials due to quantum confinement and surface impact at lower dimensions. Indirect band gaps exist in bulk semiconducting trigonal prismatic TMDs, although direct band gaps exist in monolayer forms. Logic circuits, increased photoluminescence, and field effect transistor applications can benefit from this alteration.

With ST-FMR measurement, current-induced SOT has been demonstrated in a 2D vdW heterostructure MoS2/Py [49]. In this technique, magnetization precession is driven by Oersted field torque T_h and SOT induced by in-plane current [50]. A DC voltage is created by combining microwave current and oscillating magnetoresistance, and it is made up of antisymmetric and symmetric Lorentzian functions. The antisymmetric and symmetric Lorentzian peaks are caused by T_{DL}, T_{FL}, and T_h. The symmetric peak's computed amplitude was around four times that of the antisymmetric peak. These findings show that in the MoS_2/Py 2D heterostructure, T_{DL} is dominant due to the interfacial effect, and the torque ratio is found to be $T_{FL}/T_{DL}=0.19 \pm 0.015$.

Current-induced SOT has been demonstrated in 2D heterostructures MOS_2/CoFeB and WSe_2/CoFeB using the SHH technique [51]. The torque T_{DL} makes a significant contribution to SOT, whereas T_{FL} contribution is negligible. With spin current conductivities of 2.9×103 and 5.5×10^3 $(\hbar/2e)/(\Omega m)$, respectively, the SOT efficiency is recognized in MOS_2 and WSe_2. In addition, as illustrated in Figure 13.9a and b, the torque ratios of T_{FL} and T_{DL} in WS_2/Py heterostructures are proportional to the change in the back-gate voltage [52]. The torque ratio can be improved by a factor of 4 by varying the gate voltage from −60 to 60 V, as shown in Figure 13.9c. The ability to modulate SOTs with electrical voltage effectively offers up new possibilities for building and using SOT devices in logic applications, data processing, and storage applications.

Energy-efficient SOT devices can be realized using 2D semimetals with high conductivity. There is significant power consumption in 2D vdW semiconductors due to the majority of the current flowing through the FM layer. 2D semimetals such as WTe_2 demonstrated a strong substantial spin orbit coupling as well as a spin momentum locking effect [53,54]. SOTs T_{DL} and T_{FL} are 0.029 and 0.033, respectively, according to ST-FMR measurements. SHH measurements are also used to verify the SOTs. Magnetization switching in WTe_2/Py heterostructures has been demonstrated with T_{DL} larger than 0.51 and a current density of 105 A/cm^2 [55] (Figure 13.10).

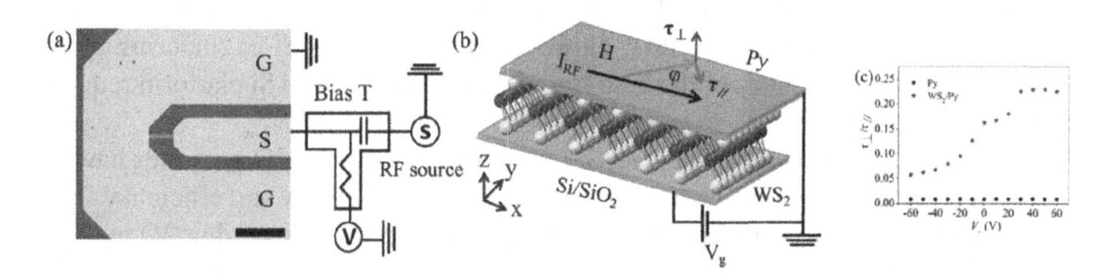

FIGURE 13.9 (a) ST-FMR measurement with WS_2/Pygeometry. (b) Schematic of the WS2/Py heterostructure device geometry. A gate voltage Vg is applied across SiO_2 and Py. (c) Ratio of T_{DL}/T_{FL} controlled by Vg. (Reproduced with permission from Lv et al., [52] Copyright 2018 American Chemical Society.)

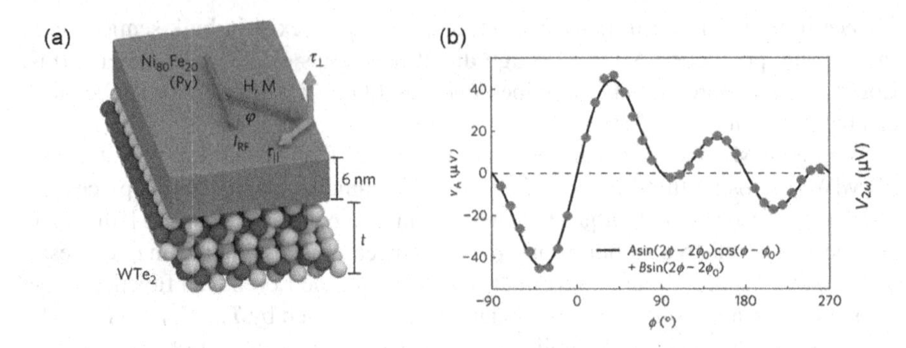

FIGURE 13.10 (a) Schematic of the WTe_2/Py bilayer. (b) ST-FMR shows antisymmetric components detected at a magnetic field with various orientations. (Reproduced with permission from MacNeill et al. [53].)

2D semimetals such as 1T-$MoTe_2$ maintains bulk inversion symmetry while having only one mirror plane for surface symmetry [56]. The conductivity of the spin current with the out-of-plane T_{DL} caused by inversion symmetry breaking at the interface of $MoTe_2$ is 1.0×10^3 (ħ/2e) $(\Omega m)^{-1}$, when the switching current is oblique to the plane of single mirror, which is about 34% that of WTe_2. Also, $MoTe_2$ has a large SHA of 0.32 and a long spin-diffusion length of 2.2 μm, indicating that $MoTe_2$ is an ideal option for simultaneous spin creation, detection, and transport [57]. It has been demonstrated that switching of magnetization in $MoTe_2$/Py, with a critical current density of 105 A/cm^2, is the same as that of heterostructureWTe_2/Py [58].

13.4 CONCLUSIONS

Nonvolatile memory MRAM plays a vital role in neuromorphic computing for high-demand data processing applications. SOT MRAM offers more versatility and reduces the risk of oxide breakdown as opposed to STT-MRAMs. In the conventional SOT design, however, a large current through the NM increases the risk of electromigration and SOT efficiency is also not significant in conventional SOT MRAMs with heavy metals. Techniques such as heterostructure engineering are being used in research to enhance the efficiency of SOT and hence minimize switching current. Novel 2D materials, such as TMDs and TIs, are also being investigated to further minimize the write current in SOT. However, unusual material engineering challenges such as compatibility of the material with the conventional Si platform require further investigation.

The materials available for creating SOTs are vast and varied. 2D materials having a high SOC, such as TIs and TMDs, have a high spin conductivity and efficiency. The field-like torque, along with damping-like torque, has been discovered in 2D materials to play a significant role in determining the switching dynamics in SOT, and it may also result in backward switching. The successful modulation of SOT performance by electrical voltage suggests that the concept of modulated SOT performance is feasible. Despite their benefits, most reported TIs and TMDs have low electrical conductivities, and finding TIs and TMDs with high conductivities and strong SOCs is difficult. We should expect the SOC phenomenon to continue to develop spintronic applications for new, long-lasting technologies if previous explosive research studies are any indicator of future scientific and engineering breakthroughs.

REFERENCES

[1] Zahedinejad, M., et al., Two-dimensional mutually synchronized spin Hall nano-oscillator arrays for neuromorphic computing. *Nature Nanotechnology*, 2020. **15**(1): p. 47–52.

[2] Chen, Y., et al., Neuromorphic computing's yesterday, today, and tomorrow–an evolutional view. *Integration*, 2018. **61**: p. 49–61.

[3] Boybat, I., et al., Neuromorphic computing with multi-memristive synapses. *Nature Communications*, 2018. **9**(1): p. 1–12.

[4] Grollier, J., D. Querlioz, and M.D. Stiles, Spintronic nanodevices for bioinspired computing. *Proceedings of the IEEE*, 2016. **104**(10): p. 2024–2039.

[5] Kurenkov, A., et al., Artificial neuron and synapse realized in an antiferromagnet/ferromagnet heterostructure using dynamics of spin–orbit torque switching. *Advanced Materials*, 2019. **31**(23): p. 1900636.

[6] Huff, H., et al., High-k gate stacks for planar, scaled CMOS integrated circuits. *Microelectronic Engineering*, 2003. **69**(2–4): p. 152–167.

[7] Keller, R., D. Kramer, and J.-P. Weiss, *Facing the Multicore-Challenge II: Aspects of New Paradigms and Technologies in Parallel Computing*. Vol. 7174. 2012: Springer.

[8] Null, L. and J. Lobur, *Essentials of Computer Organization and Architecture*. 2014: Jones & Bartlett Publishers.

[9] Endoh, T., et al., An overview of nonvolatile emerging memories—Spintronics for working memories. *IEEE Journal on Emerging and Selected Topics in Circuits and Systems*, 2016. **6**(2): p. 109–119.

[10] Bläsing, R., et al., Magnetic racetrack memory: From physics to the cusp of applications within a decade. *Proceedings of the IEEE*, 2020. **108**(8): p. 1303–1321.

[11] Rashba, E. and V. Sheka, Combinational resonance of zonal electrons in crystals having a zinc blende lattice. *Soviet Physics-Solid State*, 1961. **3**(6): p. 1257–1267.

[12] Fukami, S., et al., Magnetization switching by spin–orbit torque in an antiferromagnet–ferromagnet bilayer system. *Nature Materials*, 2016. **15**(5): p. 535–541.

[13] Han, X., et al., Spin-orbit torques: Materials, physics, and devices. *Applied Physics Letters*, 2021. **118**(12): p. 120502.

[14] Nguyen, M.-H., D. Ralph, and R. Buhrman, Spin torque study of the spin Hall conductivity and spin diffusion length in platinum thin films with varying resistivity. *Physical Review Letters*, 2016. **116**(12): p. 126601.

[15] Pai, C.-F., et al., Dependence of the efficiency of spin Hall torque on the transparency of Pt/ferromagnetic layer interfaces. *Physical Review B*, 2015. **92**(6): p. 064426.

[16] Gong, C. and X. Zhang, Two-dimensional magnetic crystals and emergent heterostructure devices. *Science*, 2019. **363**(6428).

[17] Cao, Y., et al., Prospect of spin-orbitronic devices and their applications. *IScience*, 2020. **23**(10): p. 101614.

[18] Hidding, J. and M.H. Guimarães, Spin-orbit torques in transition metal dichalcogenides/ferromagnet heterostructures. *Frontiers in Materials*, 2020. **7**: p. 383.

[19] Liu, L., et al., Current-induced switching of perpendicularly magnetized magnetic layers using spin torque from the spin Hall effect. *Physical Review Letters*, 2012. **109**(9): p. 096602.

[20] Miron, I.M., et al., Perpendicular switching of a single ferromagnetic layer induced by in-plane current injection. *Nature*, 2011. **476**(7359): p. 189–193.

[21] Manchon, A., Rashba spin–orbit coupling in two-dimensional systems, in *Spintronic 2D Materials*. 2020: Elsevier. p. 25–64.
[22] Baibich, M.N., et al., Giant magnetoresistance of (001) Fe/(001) Cr magnetic superlattices. *Physical Review Letters*, 1988. **61**(21): p. 2472.
[23] Wang, D., et al., 70% TMR at room temperature for SDT sandwich junctions with CoFeB as free and reference layers. *IEEE Transactions on Magnetics*, 2004. **40**(4): p. 2269–2271.
[24] Yuasa, S., et al., Giant room-temperature magnetoresistance in single-crystal Fe/MgO/Fe magnetic tunnel junctions. *Nature Materials*, 2004. **3**(12): p. 868–871.
[25] Johnson, M., et al., Magnetic anisotropy in metallic multilayers. *Reports on Progress in Physics*, 1996. **59**(11): p. 1409.
[26] Inomata, K., et al., Large tunneling magnetoresistance at room temperature using a Heusler alloy with the B2 structure. *Japanese Journal of Applied Physics*, 2003. **42**(4B): p. L419.
[27] Hirohata, A., et al., Roadmap for emerging materials for spintronic device applications. *IEEE Transactions on Magnetics*, 2015. **51**(10): p. 1–11.
[28] Engel, B., et al., A 4-Mb toggle MRAM based on a novel bit and switching method. *IEEE Transactions on Magnetics*, 2005. **41**(1): p. 132–136.
[29] Slonczewski, J.C., Current-driven excitation of magnetic multilayers. *Journal of Magnetism and Magnetic Materials*, 1996. **159**(1–2): p. L1–L7.
[30] Qiu, X., et al., Characterization and manipulation of spin orbit torque in magnetic heterostructures. *Advanced Materials*, 2018. **30**(17): p. 1705699.
[31] Koo, H.C., et al., Rashba effect in functional spintronic devices. *Advanced Materials*, 2020. **32**(51): p. 2002117.
[32] Hirsch, J., Spin hall effect. *Physical Review Letters*, 1999. **83**(9): p. 1834.
[33] Kato, Y.K., et al., Observation of the spin hall effect in semiconductors. *Science*, 2004. **306**(5703): p. 1910–1913.
[34] Dyakonov, M.I. and V. Perel, Current-induced spin orientation of electrons in semiconductors. *Physics Letters A*, 1971. **35**(6): p. 459–460.
[35] Edelstein, V.M., Spin polarization of conduction electrons induced by electric current in two-dimensional asymmetric electron systems. *Solid State Communications*, 1990. **73**(3): p. 233–235.
[36] Meijer, F., et al., Universal spin-induced time reversal symmetry breaking in two-dimensional electron gases with Rashba spin-orbit interaction. *Physical Review Letters*, 2005. **94**(18): p. 186805.
[37] Avci, C.O., et al., Fieldlike and antidamping spin-orbit torques in as-grown and annealed Ta/CoFeB/MgO layers. *Physical Review B*, 2014. **89**(21): p. 214419.
[38] Sánchez, J.R., et al., Spin-to-charge conversion using Rashba coupling at the interface between non-magnetic materials. *Nature Communications*, 2013. **4**(1): p. 1–7.
[39] Shao, Q., et al., Room temperature highly efficient topological insulator/Mo/CoFeB spin-orbit torque memory with perpendicular magnetic anisotropy. in *2018 IEEE International Electron Devices Meeting (IEDM)*. 2018: IEEE.
[40] Husain, S., et al., Emergence of spin–orbit torques in 2D transition metal dichalcogenides: A status update. *Applied Physics Reviews*, 2020. **7**(4): p. 041312.
[41] Liu, Y. and Q. Shao, Two-dimensional materials for energy-efficient spin–orbit torque devices. *ACS Nano*, 2020. **14**(8): p. 9389–9407.
[42] Shin, I., et al., Spin-orbit Torque Switching in an All-Van der Waals Heterostructure. arXiv preprint arXiv:2102.09300, 2021.
[43] Soumyanarayanan, A., et al., Emergent phenomena induced by spin–orbit coupling at surfaces and interfaces. *Nature*, 2016. **539**(7630): p. 509–517.
[44] Mellnik, A., et al., Spin-transfer torque generated by a topological insulator. *Nature*, 2014. **511**(7510): p. 449–451.

[45] Fan, Y., et al., Electric-field control of spin–orbit torque in a magnetically doped topological insulator. *Nature Nanotechnology*, 2016. **11**(4): p. 352–359.
[46] Wang, Y., et al., Topological surface states originated spin-orbit torques in Bi 2 Se 3. *Physical Review Letters*, 2015. **114**(25): p. 257202.
[47] Wu, H., et al., Room-temperature spin-orbit torque from topological surface states. *Physical Review Letters*, 2019. **123**(20): p. 207205.
[48] Mahendra, D., et al., Room-temperature high spin–orbit torque due to quantum confinement in sputtered Bi x Se (1–x) films. *Nature Materials*, 2018. **17**(9): p. 800–807.
[49] Zhang, W., et al., Research Update: Spin transfer torques in permalloy on monolayer MoS2. *APL Materials*, 2016. **4**(3): p. 032302.
[50] Manchon, A., et al., Current-induced spin-orbit torques in ferromagnetic and antiferromagnetic systems. *Reviews of Modern Physics*, 2019. **91**(3): p. 035004.
[51] Shao, Q., et al., Strong Rashba-Edelstein effect-induced spin–orbit torques in monolayer transition metal dichalcogenide/ferromagnet bilayers. *Nano Letters*, 2016. **16**(12): p. 7514–7520.
[52] Lv, W., et al., Electric-field control of spin–orbit torques in WS_2/permalloy bilayers. *ACS Applied Materials & Interfaces*, 2018. **10**(3): p. 2843–2849.
[53] MacNeill, D., et al., Control of spin–orbit torques through crystal symmetry in WTe_2/ferromagnet bilayers. *Nature Physics*, 2017. **13**(3): p. 300–305.
[54] MacNeill, D., et al., Thickness dependence of spin-orbit torques generated by WTe_2. *Physical Review B*, 2017. **96**(5): p. 054450.
[55] Shi, S., et al., All-electric magnetization switching and Dzyaloshinskii–Moriya interaction in WTe_2/ferromagnet heterostructures. *Nature Nanotechnology*, 2019. **14**(10): p. 945–949.
[56] Stiehl, G.M., et al., Layer-dependent spin-orbit torques generated by the centrosymmetric transition metal dichalcogenide β– $MoTe_2$. *Physical Review B*, 2019. **100**(18): p. 184402.
[57] Song, P., et al., Coexistence of large conventional and planar spin Hall effect with long spin diffusion length in a low-symmetry semimetal at room temperature. *Nature Materials*, 2020. **19**(3): p. 292–298.
[58] Liang, S., et al., Spin-orbit torque magnetization switching in $MoTe_2$/permalloy heterostructures. *Advanced Materials*, 2020. **32**(37): p. 2002799.

Index

Printed in the United States
by Baker & Taylor Publisher Services